MATHEMATICS
A Human Endeavor

"This is the part I always hate."

SECOND EDITION

MATHEMATICS
A Human Endeavor

A Book for Those Who Think They Don't Like the Subject

HAROLD R. JACOBS

W. H. FREEMAN AND COMPANY
San Francisco

The cover drawing is after a woodcut by Maurits Escher.
Reproduced by permission of the Escher Foundation,
The Hague, Netherlands.

Designer: Robert Ishi
Illustrators: Kelly Solis-Navarro, Dale Johnson, and John Johnson
Illustration coordinator: Cheryl Nufer
Permissions editor: Barbara Ferenstein
Production coordinator: Linda Jupiter
Compositor: York Graphic Services, Inc.
Printer and binder: R. R. Donnelley & Sons Company

Library of Congress Cataloging in Publication Data

Jacobs, Harold R.
    Mathematics, a human endeavor.

    Includes index.
    1. Mathematics—Popular works.   I. Title.
QA93.J33 1982   510   81-17499
ISBN 0-7167-1326-8   AACR2

1 2 3 4 5 6 7 8 9 0   DO   0 8 9 8 7 6 5 4 3 2

Contents

1

MATHEMATICAL WAYS OF THINKING   5

2

NUMBER SEQUENCES   57

6

MATHEMATICAL CURVES   311

7

METHODS OF COUNTING   377

8

THE MATHEMATICS OF CHANCE   423

9

AN INTRODUCTION TO STATISTICS   499

10

TOPICS IN TOPOLOGY   575

APPENDIX: BASIC IDEAS AND OPERATIONS   623

Martin Gardner

Foreword

"Well, three or four months run along, and it was well into the win-
ter, now. I had been to school most all the time, and could spell, and
read, and write just a little, and could say the multiplication table up to
six times seven is thirty-five, and I don't reckon I could ever get any farther than that if I was to live forever. I don't take no stock in mathematics, anyway."

—*Huckleberry Finn*

The surest way to quench the enthusiasm of students for mathematics is to
bombard them with dull, difficult problems unrelated to their interests. In
Sketches Old and New, Mark Twain reprints a number of letters from readers,
along with his replies. Here is an inquiry from a man in Nevada who signed
it "Arithmeticus":

If it would take a cannon-ball $3\frac{1}{8}$ seconds to travel four miles, and $3\frac{3}{8}$
seconds to travel the next four, and $3\frac{5}{8}$ to travel the next four, and if its
rate of progress continued to diminish in the same ratio, how long
would it take it to go fifteen hundred million miles?

Twain's answer was, "I don't know."

One suspects that Arithmeticus was a pen name for the author of some
dreary, long-forgotten textbook. Even today, thousands of people are being
permanently turned away from mathematics by texts with exercises so bor-
ing and meaningless that their readers, like Mark Twain and Huck, not only
don't know the answers, but don't even care about them.

Harold Jacobs is one of those rare teachers who knows all the best ways to open the eyes of students to the excitement, beauty, and mystery of mathematics. The First Edition of his *Mathematics: A Human Endeavor,* with its abundant use of jokes, games, puzzles, paradoxes, magic tricks, and other forms of recreational mathematics, quickly became one of the most successful of all introductory textbooks. It is hard to imagine that it could be improved. Nevertheless, Jacobs has done it, and he has done it throughout.

The First Edition provoked a flood of letters and comments from students, teachers, and others, and the new edition draws on this invaluable source of guidance. The difficulties noted by users have been smoothed away. New recreational material has been added, and recent developments in serious mathematics have been introduced where appropriate. For example, the computer proof of the notorious four-color map theorem is mentioned in the first chapter. New and funnier cartoons have replaced old ones. New illustrations, both drawings and photographs, have been added, and page layouts are strikingly improved.

Of course, the book's great virtue is that Jacobs, amidst all the fun, never loses sight of his main objective: to teach students what mathematics is all about. No games, puzzles, or paradoxes are here merely to entertain or challenge; they are here to draw the reader almost unwittingly into the fundamental ideas of mathematics. You cannot play the card game Eleusis without learning what inductive reasoning really is. You cannot understand a mind-reading trick, done using only a folder of matches, without learning some elementary number theory. You cannot fathom the mystery of vanishing-area paradoxes without learning some elementary geometry.

But why learn any mathematics at all? Don't millions of educated people get along as well as Huckleberry Finn, thank you, without knowing any mathematics beyond the simplest arithmetic? Yes, but the day is long gone when such persons deserve to be called educated. You cannot understand science without knowing some mathematics, and if you know nothing about science you are as much a Philistine as a person who never read a line of Shakespeare. Our time is often called the age of science. The sad truth is otherwise. The widespread belief in astrology alone tells us that science has only begun to penetrate the minds of the "educated."

Science obviously is now the chief cause of social change for both good and evil. It is certain that the fate of humanity depends not so much on the progress of science itself as on progress toward public understanding of science. We are in a race, as H.G. Wells liked to say, between education and catastrophe. Unfortunately, science speaks in a curious language, the language of mathematics, a language that remains to most people as incomprehensible as Latin or Esperanto.

Adults can often be heard to say, almost with pride, how much they hated math in school. Sometimes they follow with a remark that makes all mathematicians squirm: "Why, I can't even balance my checkbook," as though this had much to do with mathematics. I cannot believe that this distaste for mathematics is inborn. I believe it to be the result of deficiencies in our educational system—deficiencies that can be remedied.

So what to do? The answer may be found in enthusiastic and dedicated teachers like Harold Jacobs and in textbooks as stimulating as the one you are now holding.

Martin Gardner
Author
Originator of "Mathematical
Games," SCIENTIFIC AMERICAN

Photograph by Roy Bishop

A Letter to the Student

In 1623, the great Italian scientist Galileo wrote: "That vast book which stands forever open before our eyes, the universe, cannot be read until we have learned the language and become familiar with the characters in which it is written. It is written in mathematical language, without which means it is humanly impossible to comprehend a single word."

Mathematics has made possible the great advances in science and technology that have occurred since Galileo wrote these words. Indeed, it has become so important to so many fields, including those as diverse as psychology, economics, medicine, linguistics, and even history, that mathematics is now an integral part of most courses of study at universities.

Unfortunately, even though mathematics is a very broad subject, many people leave school with a rather narrow view of it. Someone who has studied only arithmetic may identify mathematics with computation. A student who has taken algebra or geometry may think of it as being limited to solving equations and proving theorems. Even people who have taken advanced courses in mathematics may not be aware of its many applications.

The goal of this book is to give you a much broader view of mathematics by introducing you to areas that you may never have thought about before. Fortunately, you can understand and enjoy mathematics without having to find square roots, or memorize the quadratic formula, or prove geometric facts.

Some of the topics in this book may seem to you to be of little practical use, but the significance of mathematics does not rest only on its practical value. It is hard to believe that someone flying over the Grand Canyon for the first time could remark, "What good is it?" Some people say the very same thing about mathematics. A great mathematician of our century, G. H. Hardy, said, "A mathematician, like a painter or a poet, is a maker of patterns." Some of these patterns have immediate and obvious applications; others may never be of any use at all. But, like the Grand Canyon, mathematics has its own kind of beauty and appeal to those who are willing to look.

Harold Jacobs

From *Medical Radiography and Photography*, published by Radiography Markets Division, Eastman Kodak Company

MATHEMATICS
A Human Endeavor

The radio telescope at Arecibo, Puerto Rico

Introduction:
Mathematics—A Universal Language

One of the largest radio telescopes in the world is located at Arecibo, Puerto Rico. It can communicate with an antenna of equal size anywhere in our galaxy and has been aimed at several stars in the hope of receiving signals from another civilization.

Is our civilization the only one in the universe? If life existed on a planet of another star and the beings of that far-off world tried to communicate with us, what kind of message would they send? If we sent a message to them, what should we say and how should we say it?

We could hardly expect that something like Morse code would work or that any earthly language would make sense. How, then, would it be possible to begin a conversation with another world? Scientists agree that the kind of message most likely to be understood would be a mathematical one.

Here are diagrams of radio signals once suggested by a British physicist as a way of starting a conversation. Each line of pulses represents a mathematical statement. Can you figure out what the statements say?

1. 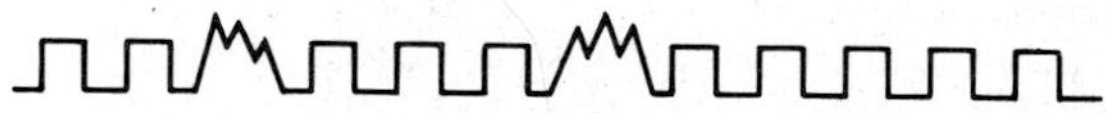 ⎍⎍⎍/\⎍⎍⎍⎍/\⎍⎍⎍⎍⎍⎍

Hint: This message seems to have three parts, separated by two zigzag patterns. What does each part mean?

2.  ⎍⎍⎍⎍/\⎍⎍⎍⎍⎍⎍/\⎍⎍⎍⎍⎍⎍⎍⎍⎍

3.  ⎍⎍⎍⎍⎍⎍⎍⎍/\⎍⎍⎍⎍⎍/\⎍⎍⎍

4.  ⎍⎍⎍⎍⎍⎍/\⎍⎍⎍⎍⎍⎍/_____

5.  ⎍⎍⎍/\⎍⎍⎍⎍⎍/\⎍⎍⎍⎍⎍⎍⎍

6.  ⎍⎍⎍⎍⎍/\⎍⎍/\⎍⎍⎍⎍⎍

7.  ⎍⎍⎍⎍⎍⎍⎍⎍/\⎍⎍⎍/\⎍⎍⎍⎍⎍⎍

8.  ⎍⎍⎍⎍⎍⎍/\⎍⎍⎍⎍/\⎍⎍⎍

A problem somewhat like that of understanding a message from outer space is that of making sense out of the messages left behind by an early civilization here on earth. The earliest known records of mathematics were made by the Babylonians about 4,000 years ago. They left behind thousands of clay tablets, some of which reveal their number system and their discoveries in algebra and geometry. The photograph at the top of the next page is of a Babylonian tablet of about 1800 B.C. The wedge-shaped writing, called cuneiform, was made by a stylus on wet clay.

A copy of the front and back of another tablet, dug up in the late-nineteenth century, is shown below. Can you translate the groups of wedge-shaped symbols into familiar symbols and explain what the tablet is about? *Hint:* Figure out what all the symbols in the left-hand columns mean before working on the right-hand columns.

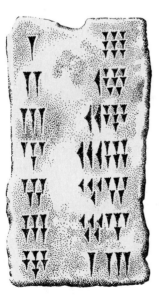

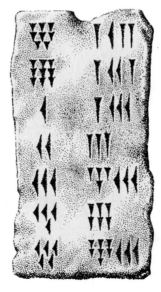

Interesting Reading

Cosmos, by Carl Sagan, Random House, 1980: Chapter 12, "Encyclopaedia Galactica."

Whispers From Space, by John W. Macvey, Macmillan, 1973: Chapter 11, "Toward a Cosmic Tongue."

Who Goes There?, by Edward Edelson, Doubleday, 1979.

The Search for Extraterrestrial Intelligence, prepared by the National Aeronautics and Space Administration and edited by Philip Morrison, John Billingham, and John Wolfe, Dover, 1979.

Chapter 1

MATHEMATICAL WAYS
OF THINKING

Lesson 1

The Path of a Billiard Ball

An expert billiard player's ability to control the path of a ball seems almost miraculous. Mathematicians like the game of billiards because the path of the ball can be precisely calculated by mathematical methods. The path is determined by how the ball is hit, by the shape of the table, and by the positions of the other balls.

An ordinary billiard table is twice as long as it is wide (10 feet by 5 feet) and, unlike a pool table, it does not have any pockets. Suppose that a ball is hit from one corner so that it travels at 45° angles with the sides of the table.* If it is the only ball on the table, where will it go?

The first figure at the top of the next page shows the direction that the ball takes as it is hit from the lower-left corner. The second figure shows that the ball hits the midpoint of one of the long sides of the table. In striking the cushion, the ball rebounds from it in a new direction but at the same angle. (The angles of hitting and rebounding have been marked with curved lines to show that they

*For a discussion of angles and their measurement, see pages 624-626.

6

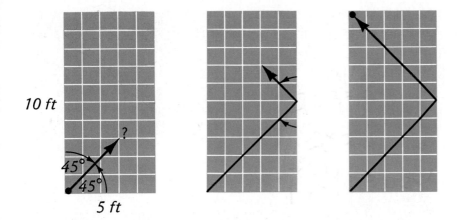

10 ft

5 ft

are equal.) The third diagram shows that the ball goes to the corner at the upper left, and we will assume that the ball stops when it comes to a corner.

What would the ball's path be if the table had a different shape? Suppose that the table was 10 feet by 6 feet and that the ball was again hit from the lower left corner at 45° angles with the sides as shown in the first figure below. This time, after the first rebound, it

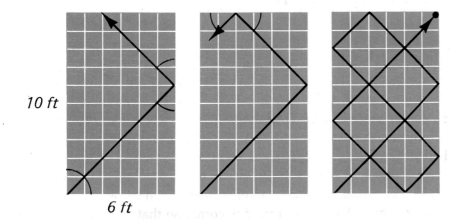

10 ft

6 ft

would miss the upper left corner and hit the top side as shown in the second figure. It would rebound from that side in a new direction, but the angles of hitting and rebounding would again be equal. The third figure shows that the ball would rebound several more times before finally ending up in the corner at the upper right.

These two tables suggest several questions about tables of other shapes. Would the ball always end up in a corner? Could it come back to the original corner? If it did end up in a corner, is it possible to predict which one without drawing a figure? Perhaps you can think of other questions as well. We are presented with quite a puzzle. Edward Kasner, in his book *Mathematics and the Imagination,* wrote:

> *Puzzles are made of the things that the mathematician, no less than the child, plays with, and dreams and wonders about, for they are made of the things and circumstances of the world he lives in.*

Exercises

Set I

On graph paper,* make a diagram of each of the numbered tables below. Be careful to use the same dimensions and write them along the sides of each table as shown.

---

* Four or five squares per inch or two squares per centimeter is convenient.

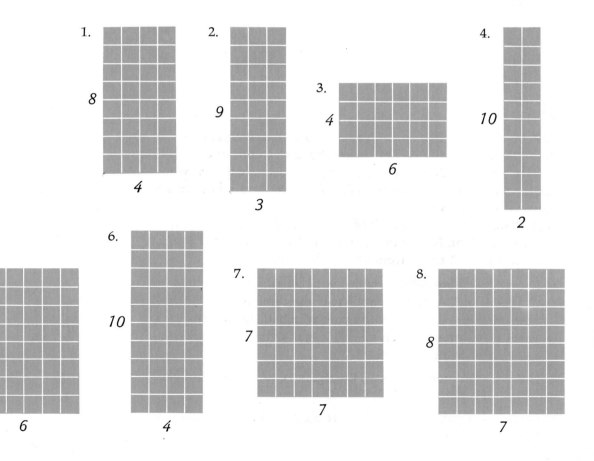

Chapter 1: MATHEMATICAL WAYS OF THINKING

Draw the path of a ball starting from the lower-left corner of each table, making 45° angles with the sides. (Because of this, the ball travels diagonally across each square of the grid through which it passes.) Continue each path as far as it can go. If the ball ends up in a corner, mark the corner with a large dot.

9. On which table does the ball have the simplest path? Why?

10. On which table does the ball have the most complicated path?

11. What do you notice about the paths on tables 3 and 5?

12. The paths on two other tables seem to be related in a similar way. Which are they?

13. Judging from the eight tables that you have drawn, do you think that a ball hit from the lower-left corner of a table at 45° angles with the sides will always end up in a corner?

14. Judging from the tables that you have drawn, do you think that a ball hit in this way can end up in any of the four corners? Explain.

Set II

It is hard, for the most part, to see any pattern in the paths on the tables you have drawn so far because the dimensions of the tables have varied at random. The following exercises will give you a chance to discover some patterns for tables whose dimensions are related in special ways.

A billiard table with a length of 12 units and a width of 1 unit is shown at the right. Notice that the ball follows a path that zigzags back and forth 12 times from beginning to end.

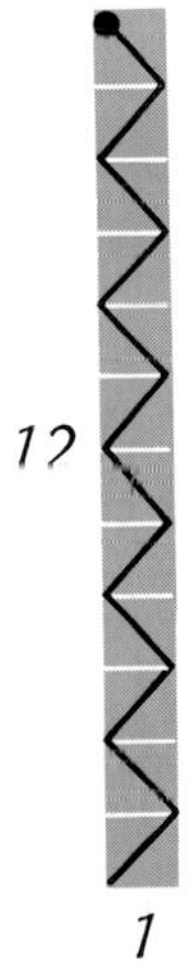

1. Draw four more billiard tables with lengths of 12 units and widths of 2, 3, 4, and 5 units. Write the dimensions along the sides of each table. Draw the path of a ball starting from the lower-left corner of each table and mark the corner where it ends up with a large dot.

2. How many times does the path zigzag back and forth on the tables with widths of 2, 3, and 4?

3. Do these numbers seem to be related to the dimensions of these tables? If so, how?

12

1

4. Why is the path of the ball on the table with the width of 5 so complicated?

5. Draw a table with a length of 12 units on which the path of the ball zigzags back and forth 2 times.

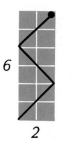

Two billiard tables on which the ball travels the same path are shown at the left.

6. Although the tables do not have the same dimensions, there *is* something about their dimensions that is the same. What is it?

7. Draw a billiard table with a length of 6 units and a width of 4 units and a billiard table with a length of 9 units and a width of 6 units. Draw the path of a ball starting from the lower-left corner of each table.

8. What do you notice about the paths on the two tables?

9. Draw a smaller billiard table on which the path of the ball is the same as on the tables you drew for exercise 7. What are its dimensions?

10. Draw a larger billiard table on which the path of the ball is the same. What are its dimensions?

On the set of tables below, the ball travels over every square.

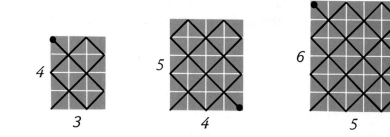

11. What do you think are the dimensions of the next table in this set for which the ball ends up in the upper-left corner?

12. Draw the table and show the path of the ball.

Chapter 1: MATHEMATICAL WAYS OF THINKING

In which corner do you think the ball would end up on a giant billiard table

13. with a width of 99 units and a length of 100 units?

14. with a width of 100 units and a length of 101 units?

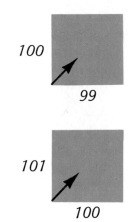

100

99

101

100

Set III

The first figure below shows the path of a ball on a square table formed from two 2-by-4 rectangles. The open circle shows the corner from which the ball begins and the solid circle shows the corner in which it ends.

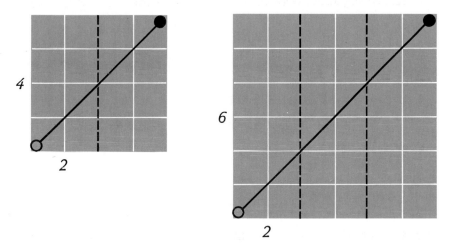

4

2

6

2

1. Put a piece of tracing paper over the figure and trace it. Then cut the tracing out and fold it along the dotted line. Tape the result on your paper.

2. What do you notice about the result?

Trace the second figure and cut it out. Fold the tracing along both dotted lines.

3. What does the result illustrate?

4. Without unfolding it, fold it a third time, this time so that the two shorter sides come together. Tape the result on your paper.

5. What does the result illustrate?

Lesson 2

More Billiard-Ball Mathematics

W. W. Sawyer, in his book *Mathematician's Delight*, wrote:

Everyone knows that it is easy to do a puzzle if someone has told you the answer. That is simply a test of memory. You can claim to be a mathematician only if you can solve puzzles that you have never studied before. That is the test of reasoning.

So far, in trying to solve the puzzle of the path of a ball on a billiard table, we have made several discoveries. From the examples that we have considered, it seems that, if the dimensions of the table are whole numbers and the ball is hit from the lower-left corner at 45° angles with the sides, it will end up in one of the other three corners. The path of the ball depends on the shape of the table, and the corner in which it ends up seems to be related to the dimensions of the table in some way.

The simplest possible path is clearly on a table in the shape of a square. On such tables, the ball travels diagonally from one corner to the opposite corner without rebounding from any of the sides. The fact that the path is so simple is related to the fact that the *ratio* of the length to the width of a square table is 1.

▶ The **ratio** of the numbers x to y is the number $\frac{x}{y}$.

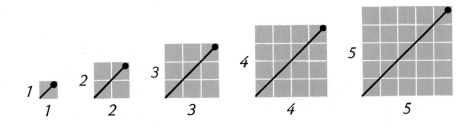

For the square tables shown in the figure above,

$$\frac{1}{1} = \frac{2}{2} = \frac{3}{3} = \frac{4}{4} = \frac{5}{5} = 1.$$

If the length of a billiard table is twice its width, a ball hit from the lower-left corner hits the midpoint of one of the longer sides and

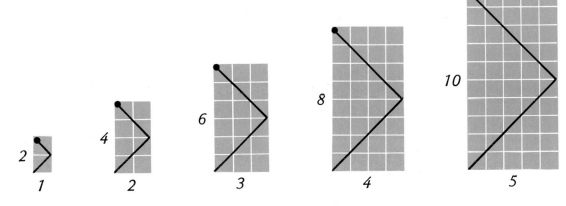

then goes to the upper-left corner. Again, the paths on such tables are the same because the ratio of the length to the width of each table is the same:

$$\frac{2}{1} = \frac{4}{2} = \frac{6}{3} = \frac{8}{4} = \frac{10}{5} = 2.$$

We have been representing the tops of the tables with rectangles. Rectangles for which the ratios of the lengths to the widths are the same have the same shape and are called *similar*. Because the paths of the billiard balls on tables that are similar are the same, the path on a table with large dimensions can be discovered by expressing the ratio of those dimensions in lowest terms.*

---

*The ratio $\frac{x}{y}$ is in lowest terms if there is no number larger than 1 that will divide evenly into both x and y.

Look, for example, at the two tables below. Because $\dfrac{12}{8} = \dfrac{3}{2}$,* the

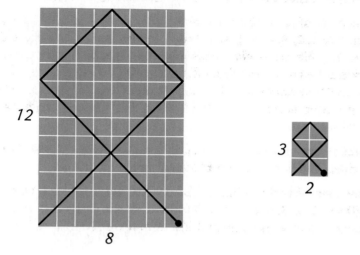

two tables have the same shape and hence the same path. The table with the smaller dimensions, however, is easier to draw. For this reason, we will proceed with the puzzle of predicting the corner in which the ball will end up by studying tables whose dimensions have been reduced to lowest terms.

*This is easy to check with a calculator. Each ratio is equal to 1.5.

Exercises

Set I

In the following exercises, we will draw the path of the ball on tables whose dimensions are in lowest terms.

1. Draw a set of seven billiard tables with widths of 1 unit and lengths of 1, 2, 3, 4, 5, 6, and 7 units. Show the path of the ball on each table, marking the corner in which the ball ends up with a large dot. The first two tables and paths are shown at the left.

Notice that, because the width is 1, the dimensions of all these tables are in lowest terms. The corner in which the ball ends up depends on whether the length is odd or even.

Chapter 1: MATHEMATICAL WAYS OF THINKING

2. In which corner does it end up if the length is odd?

3. In which corner does it end up if the length is even?

4. Draw a set of seven billiard tables with widths of 2 units and lengths of 1, 2, 3, 4, 5, 6, and 7 units. Show the path of the ball on each table *whose dimensions cannot be reduced to lower terms,* marking the corner in which the ball ends up with a large dot. The first two tables are shown at the right. (The path on the second table is not drawn because its dimensions are not in lowest terms.)

5. In which corner does the ball end up if the width is 2 units and the dimensions cannot be reduced to lower terms?

6. Draw a set of seven billiard tables with widths of 3 units and lengths of 1, 2, 3, 4, 5, 6, and 7 units. Draw the path of the ball on each table whose dimensions cannot be reduced to lower terms.

7. How is the corner in which the ball ends up on these tables related to the length?

8. Draw a set of seven billiard tables with widths of 4 units and lengths of 1, 2, 3, 4, 5, 6, and 7 units. Draw the path of the ball on each table whose dimensions cannot be reduced to lower terms.

9. In which corner does the ball end up on these tables?

Look again at the tables that you have drawn so far. Where does the ball end up if

10. the length of the table is odd and the width is even?

11. the length of the table is even and the width is odd?

12. both the length and the width of the table are odd?

Set II

The rules that you wrote for predicting the corner in which the ball will end up are based on the tables that are the easiest to draw: those whose dimensions cannot be reduced to lower terms. In the following exercises, we will consider tables whose dimensions *can* be reduced to lower terms.

Can you use your rules (Set I, exercises 10–12) to predict the correct corner for each of the following tables? Explain why or why not in each case.

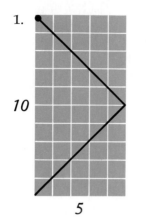

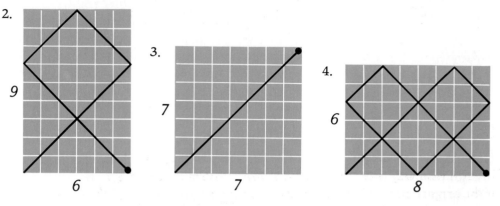

5. Why was no rule written for tables for which both length and width are even?

Here are three tables whose lengths and widths are even. How could the correct corner be predicted in each case if the paths were not drawn?

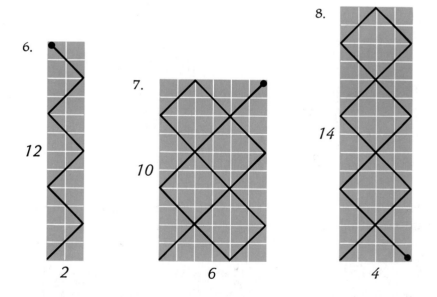

The figures at the top of the next page represent giant billiard tables and are not drawn to scale. In which corner of each table do you think the ball would end up? Explain in each case.

Chapter 1: MATHEMATICAL WAYS OF THINKING

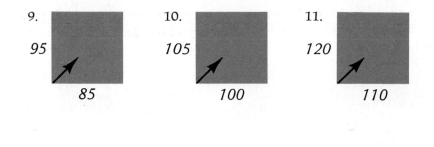

9.　95　85

10.　105　100

11.　120　110

Set III

How many times the ball hits the cushions during its trip on a billiard table depends on the dimensions of the table. For example, if the original and final corner positions are counted as "hits," then there are three hits on the first table below and five hits on the second.

1. Draw two tables having lengths of 6 units and widths of 5 and 6 units. Count the number of hits on each.

2. Referring to these four examples, write a rule for predicting the number of hits on a table on the basis of its dimensions.

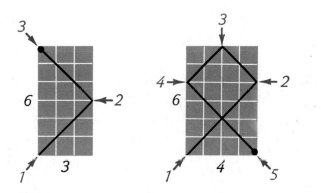

How many hits do you think there would be on each of these giant tables? (They are not drawn to scale.)

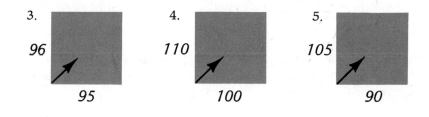

3.　96　95

4.　110　100

5.　105　90

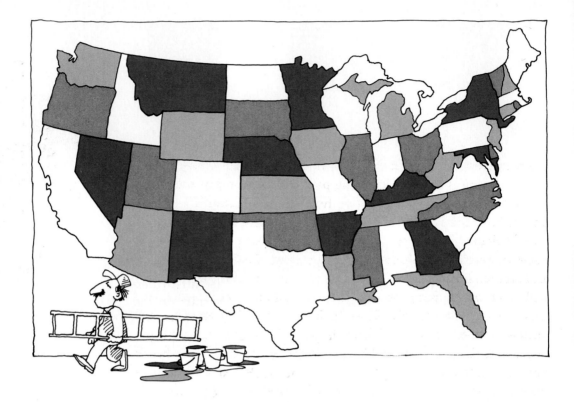

Lesson **3**

Inductive Reasoning: Finding and Extending Patterns

In our study of the path of a ball on a billiard table, we have collected evidence, noticed patterns, and drawn conclusions from these patterns. This method of reasoning is used by the scientist who makes observations, discovers regularities, and formulates general laws of nature. In science, this is called the experimental, or scientific, method. In mathematics, it is referred to as reasoning *inductively.*

▶ **Inductive reasoning** is a method of drawing general conclusions from a limited set of observations.

An interesting illustration of arriving at a conclusion by inductive reasoning is the map-coloring problem. How many colors are needed to color a map so that no two regions sharing a common border have the same color? Consider, for example, a map of the United States. It is easy to discover that at least four colors are necessary without coloring the entire map. California, Oregon, and Nevada have to be different colors because they share borders with each other. Suppose that they are colored white, gray, and dark brown, respectively. Then Idaho also can be colored white because it does not touch California, and Utah can be colored gray because it does not touch Oregon. Arizona, on the other hand, must be colored a fourth color because it touches states that have been colored the other three colors. Although four colors are necessary to color just these six states, the rest of the map can be colored without using any more.

Are there maps for which more than four colors are necessary? This problem first became well known in 1878 when the great English mathematician Arthur Cayley presented it to the London Mathematical Society. After that, so many people tried without success to draw maps for which more than four colors are needed that many mathematicians became convinced that every map can be colored with four colors or fewer. However, they could not be certain because, although inductive reasoning is of tremendous importance in *developing* mathematical ideas, it cannot *prove* that the ideas are correct. The possibility always exists that additional evidence will reveal that the conclusions are wrong.

In regard to the map-coloring problem, it was not until 1976 that two American mathematicians figured out how to use a computer to prove, using methods other than inductive reasoning, that maps requiring more than four colors do not exist.*

---

*Kenneth Appel and Wolfgang Haken, "The Solution of the Four-Color-Map Problem," *Scientific American*, October 1977.

Exercises

Set I

Number theory is the branch of mathematics dealing with properties of the whole numbers. An example of a pattern in number theory is shown at the left.

$$1 = 1 \times 1$$
$$1 + 3 = 2 \times 2$$
$$1 + 3 + 5 = 3 \times 3$$
$$1 + 3 + 5 + 7 = 4 \times 4$$

1. Write the next equation in this pattern.

2. Is the equation true?

3. Write what you think the seventh equation of the pattern should be.

4. Is it true?

5. Write what you think the tenth equation of the pattern should be.

6. Is it true?

7. Do you think that the pattern goes on indefinitely?

Do the indicated calculations to find the numbers that will make the following equations true.

8. $1 \times 8 + 1 = $ ▓

9. $12 \times 8 + 2 = $ ▓

10. $123 \times 8 + 3 = $ ▓

11. $1,234 \times 8 + 4 = $ ▓

From your results, guess the numbers that will make the following equations true.

12. $12,345 \times 8 + 5 = $ ▓

13. $123,456,789 \times 8 + 9 = $ ▓

14. Do the indicated calculation for the last equation to determine whether your guess is correct.

15. Do you think that the pattern goes on indefinitely?

Set II

The great Italian scientist Galileo used inductive reasoning to make several discoveries about the behavior of swinging weights—discoveries that led to the invention of the pendulum clock. One of these discoveries was of a relation between the *length* of the pendulum and the *time* of the swing.

This table lists the swing times of a series of pendulums having different lengths.

Length of pendulum	Time of swing
1 unit	1 second
4 units	2 seconds
9 units	3 seconds
16 units	4 seconds

Galileo Galilei

Courtesy of Columbia University Libraries.

1. From the pattern in the table, how does the length of the pendulum seem to be related to the time of the swing?

2. What do you think the length of a pendulum with a swing time of 5 seconds would be?

3. What do you think the length of a pendulum with a swing time of 10 seconds would be?

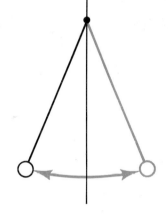

In 1661, the English chemist Robert Boyle did a series of experiments with the pressure of the air. One of his discoveries, by using inductive reasoning, was of a relation between the volume of a gas and the pressure exerted by it.

The table below lists some volumes and pressures for a gas at a given temperature.

Volume	Pressure
1 unit	60 units
2 units	30 units
3 units	20 units
4 units	15 units

From the pattern in the table, is it correct to say that, whenever the volume is doubled,

4. the pressure decreases 30 units?

5. the pressure becomes one half of what it was?

What happens to the pressure if the volume is

6. tripled?

7. multiplied by 4?

What would the pressure be when the volume is

8. 5 units?

9. 10 units?

Sun
4 • Mercury
7 • Venus
10 • Earth

15 • Mars

52 • Jupiter

96 • Saturn

In 1772, the German astronomer Johann Elert Bode used inductive reasoning to find a pattern in the distances of the planets from the sun. At that time, only six planets were known. The actual relative distances of the planets from the sun and his pattern are shown at the left.

Planet	Actual distance*	Bode's pattern
Mercury	4	0 + 4 = 4
Venus	7	3 + 4 = 7
Earth	10	6 + 4 = 10
Mars	15	12 + 4 = 16
		▦ + ▦ = ▦
Jupiter	52	48 + 4 = 52
Saturn	96	96 + 4 = 100
		▦ + ▦ = ▦

Notice that there is a pretty good match between the two sets of numbers.

10. What equation do you think belongs between Bode's equations for Mars and Jupiter?

11. What equation do you think belongs after Bode's equation for Saturn?

In 1781, William Herschel discovered Uranus, the next planet beyond Saturn. Because its distance of 192 units comes remarkably close to the number predicted by this equation, astronomers came to the conclusion that the equation between Bode's equations for Mars and Jupiter also must mean something.

---

* The distances are based on the distance from the earth to the sun being taken as 10 units.

Chapter 1: MATHEMATICAL WAYS OF THINKING

12. What do you suppose they thought it meant?

In 1801, the asteroid Ceres was discovered at a distance of 28 units from the sun.

Set III

In 1956, Robert Abbott invented a card game called Eleusis. According to Martin Gardner, writing in *Scientific American*, the game "is of special interest to mathematicians and scientists because it provides a model of induction, the process at the very heart of the scientific method."*

One player makes up a rule at the beginning of the game and the other players try to figure out what that rule is as the game is being played. An example of a rule too simple to be used in an actual game is illustrated in the figure below. The figure shows nine cards that have been played in succession and the rule is to play cards of alternating colors.

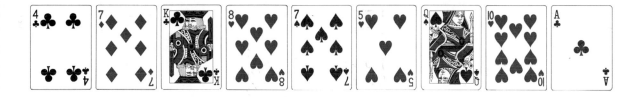

Other games played with different rules are shown below and on the next page. In these games, assume that the aces count as 1 and the jacks, queens, and kings as 11, 12, and 13, respectively.

Can you figure out rules that explain these sequences of cards? If so, tell what the rules are.

1.

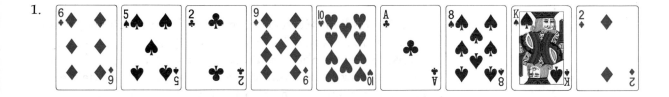

---

*Martin Gardner in his "Mathematical Games" column, *Scientific American*, October 1977.

2.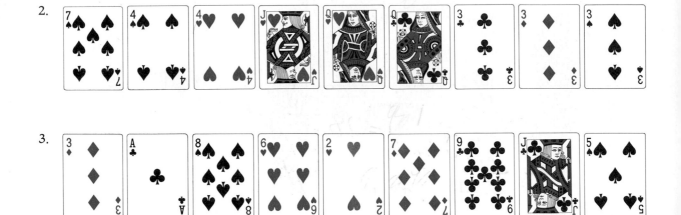

3.

4.

Chapter 1: MATHEMATICAL WAYS OF THINKING

Drawing by Ed Fisher; copyright 1966
Saturday Review, Inc.

"Water boils down to nothing . . . snow boils down to nothing . . . ice boils down to nothing . . . everything boils down to nothing."

Lesson 4

The Limitations of Inductive Reasoning

After discovering that water, snow, and ice boil down to nothing, the cave man in this cartoon has concluded that *everything* boils down to nothing. This illustrates the basic weakness of inductive reasoning. Because the conclusions arrived at by this method are drawn from a limited amount of evidence, the possibility always exists that more evidence may be discovered that will prove the conclusions to be incorrect.

An interesting example of this possibility from science concerns the "noble gases." In 1894, a new element was discovered. The element, a gas named argon, had never been found in any compounds because its atoms would not combine with those of any other element. Within the next six years, five more elements of the same type were discovered. Named helium, krypton, neon, xenon, and radon, these elements became known as the "noble gases" because they could not be made to form compounds with other elements. The conclusion that "noble gases cannot form compounds" became generally accepted by chemists and appeared in many books as if it were a statement of fact. In 1962, however, it was proved to be wrong. In that year, a compound of xenon was made for the first time and since then other compounds formed by these elements that "could not form compounds" have been made.

Conclusions arrived at by inductive reasoning in mathematics are no more certain. Consider, for example, our study of the behavior of a ball on a billiard table. If we had started by drawing a set of tables with lengths of 8 as shown below, we might have noticed that in every diagram the ball ends up in the same place, even after taking widely varying paths. A reasonable conclusion

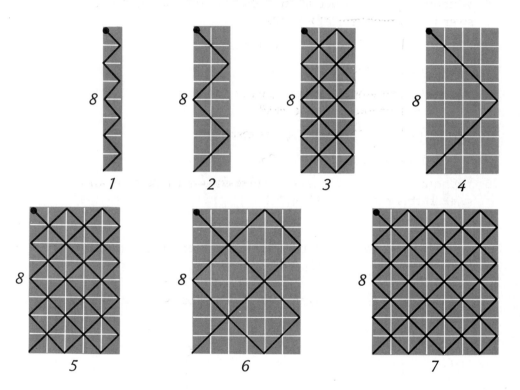

seems to be: "If the length of the table is 8, the ball will always end up in the upper-left corner." Although this is true for these seven cases, it is revealed to be false by the next case. If the width of the table is also 8, the table is square and the ball ends up in the upper-*right* corner.

Inductive reasoning, then, is of great importance in *suggesting* conclusions in mathematics. To be certain that the conclusions are reliable, however, mathematicians use another method called *deductive reasoning*. Deductive reasoning will be considered in the lessons following this one.

Chapter 1: MATHEMATICAL WAYS OF THINKING

Exercises

Set I

Here is an interesting pattern that you may never have noticed before. Do each of the following problems without a calculator.

1. 112 × 124   2. 211 × 421

3. What do you notice about these problems and their answers?

4. Now do this problem: 312 × 221.

5. Without doing any calculations, what do you think is the answer to this problem: 213 × 122?

6. Do the problem. Was your guess correct?

7. Now do this problem: 411 × 102.

8. Without doing any calculations, what do you think is the answer to this problem: 114 × 201?

9. Do the problem. Was your guess correct?

10. Now do this problem: 113 × 223.

11. Without doing any calculations, what do you think is the answer to this problem: 311 × 322?

12. Do the problem. Was your guess correct?

13. What do exercises 1 through 12 illustrate about the results obtained by inductive reasoning?

The upside down T's in the figure below are the basis of a popular optical illusion.

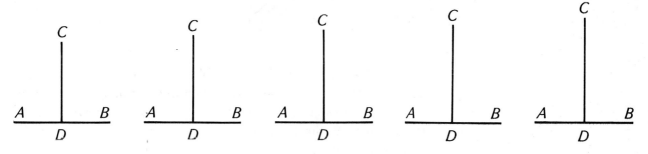

14. What seems to be true about the lengths marked AB?

15. What seems to be true about the lengths marked CD?

16. In which figure do the lengths AB and CD *appear* to be equal?

17. In which figure are the lengths AB and CD *actually* equal?

```
      .01234...
81 ) 1.00000...
      81
     ───
      190
      162
      ───
       280
       243
       ───
        370
        324
        ───
         46
```

The result of dividing 81 into 1 is an interesting pattern. The beginning of the pattern can be seen here.

18. What do you think the next five digits of the answer are?

19. Continue the division to see if you are correct.

Set II

Make a neat drawing of the figure below on one-half of a sheet of graph paper.

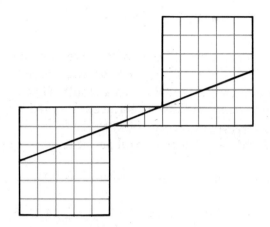

1. Explain how it is possible to tell, without having to count all of the small squares, that the area of the figure is 63 square units.

Cut the figure out and into its four separate pieces.

2. Rearrange the four pieces to form a square. Make a drawing of the arrangement on the other half of the sheet of graph paper.

3. What does the area of the square seem to be?

4. Rearrange the pieces again to form a long rectangle with a width of 5 units. Make a drawing of the arrangement. (The outline of the rectangle should look like this.)

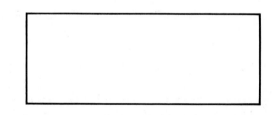

5. What does the area of this rectangle seem to be?

6. What seems to happen to the total area of the four pieces each time they are rearranged?

7. Do you have any ideas of why this happens? If you do, explain.

Here is a set of three circles. Two points have been chosen on the first circle and a straight line segment drawn between them. The circle is separated into two regions as a result. Three points were chosen on the second circle and connected with three line segments to form four regions. Four points were chosen on the third circle and, after being connected in all possible ways, eight regions resulted.

Here is a table that includes the results so far.

Number of points connected	2	3	4	5	6
Number of regions formed	2	4	8	‖‖‖‖	‖‖‖‖

Two more cases have been added.

8. Guess from the pattern in the second line of numbers what the missing numbers are.

9. Draw a pair of large circles and choose five points on one circle and six points on the other. Join the points of each *in every possible way.*

10. How many regions are formed in each?

11. Do both results agree with your guesses?

Set III

Mel Stover has created several clever puzzles based on geometrical principles. One of these puzzles, consisting of black and white pencils, is shown on the next page.

Use a sheet of tracing paper to trace the figure and then glue the paper to a stiff card. Check to see that you have shaded six of the pencils black as shown. Cut the figure into three pieces as indicated by the brown lines. Reverse the two pieces labeled A and B and compare the resulting figure to the original on this page. What is strange about the result? Can you explain how this odd result is achieved?

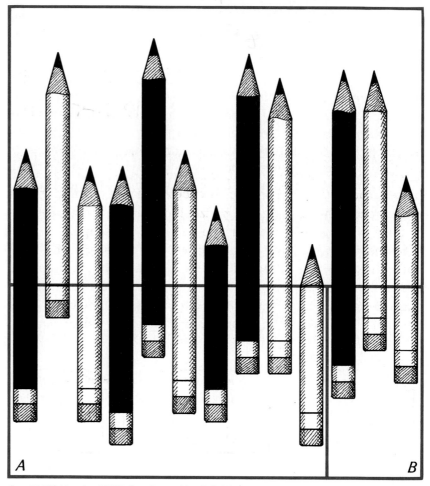

Interesting Reading

"The Disappearing Man and Other Vanishing Paradoxes" by Mel Stover, *Games* magazine, November/December 1980, pages 14–18.

Deductive Reasoning: Mathematical Proof

The Soma cube is a popular puzzle made from 27 small cubes. Created by Piet Hein, a noted Danish inventor, it consists of seven pieces made by gluing the small cubes together in different arrangements. The object of the puzzle is to put the pieces together to form either a large cube or another specified shape.

Suppose that we have a block of wood in the shape of a cube with which to make a Soma cube. First, we have to cut it up into the 27 small cubes. One way to do this is to make a series of six cuts through the cube while keeping it together in one block. The cuts are indicated by the arrows in the diagrams below.

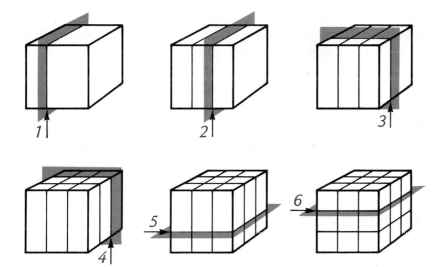

Now suppose that, instead of keeping the cube together in one block, the pieces are rearranged between each cut. Could the cube be cut into the 27 smaller cubes with less than six cuts? The two diagrams below show one way in which the first two pieces might be rearranged. Notice that the second cut will now cut through

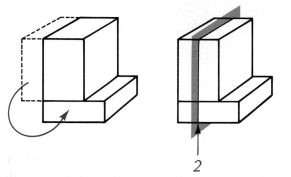

more wood than it would have if the piece at the left had not been moved to the bottom.

The number of pieces increases with each cut and there are many ways of rearranging them; so the number of different ways of making the cuts is very large. Suppose that we had a thousand blocks of wood and tried a thousand different ways to cut them up in less than six cuts without success. We still would not know for certain that it could not be done because this conclusion is based only on inductive reasoning.

Instead of reasoning inductively by testing many different cases, we can reason *deductively.*

▶ **Deductive reasoning** is a method of drawing conclusions from facts that we accept as true by using logic.

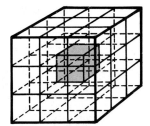

By reasoning deductively, we can prove that the cube cannot be cut into the 27 smaller cubes in less than six cuts. The key to the proof is the cube in the center, shown in brown in the figure at the left. Every one of its faces must be formed by a cut. Because it has *six* faces, *six* cuts are required to form it regardless of the ways in which the other pieces may be rearranged.

In contrast with inductive reasoning, which helps us find what *may* be true, deductive reasoning tells us what *must* be true. Because of this, mathematicians usually use deductive reasoning when they want to prove that something is true. Mathematical

statements that are proved true by deductive reasoning are called
theorems.

In 300 B.C., Euclid, in his book the *Elements* (the most widely
known textbook ever written), used deductive reasoning exclusively in proving statements about geometric figures. His success
in doing so helped to establish deductive reasoning as a fundamental tool of mathematics.

Exercises

Set I

Here is a game with a winning strategy that can be discovered by
using inductive reasoning. Deductive reasoning proves that this
strategy will always win. The game is played on a small checker-
board containing sixteen squares as shown at the right.

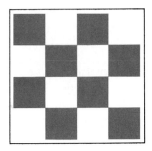

One player, A, begins by placing pennies on any two of the
squares. The other player, B, then places paper clips on the board
so that each clip lies on two squares that share a common side. The
clips may not overlap each other. To win, player B has to place
seven paper clips in this way so that they lie on the fourteen
squares not occupied by the pennies. If he cannot do this, player A
wins. The figures below show the outcomes of eight games.

B wins

B wins

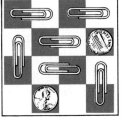

A wins

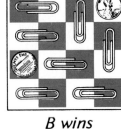

B wins

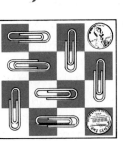

A wins

B wins

A wins

A wins

1. Look at the squares on which the pennies were placed on each board. How do their colors seem to be related to who wins?

2. Do these eight examples prove that your answer to the first question is correct for every possible game?

3. According to the rules, each paper clip has to be placed on two squares that share a common side. What can be concluded about the colors of two such squares?

4. If player B is able to place seven paper clips so that they cover fourteen squares, what must be true about the colors of the squares?

What are the colors of the fourteen squares if A begins by placing the pennies on

5. two white squares?

6. two brown squares?

7. a white square and a brown square?

8. Do your answers to questions 4 through 7 prove that your answer to the first question is correct for every possible game?

The following puzzle, which has appeared in a variety of forms in many different puzzle books, can be solved by reasoning deductively.

The figure at the left shows three match boxes. One contains two red marbles, one contains two white marbles, and one contains a red marble and a white marble. The labels telling the contents of the boxes have been switched, however, so that *the label on each box is wrong.*

You are permitted to choose one box and open it far enough to see just one marble. The puzzle is to explain how you can figure out from this what is in each box.

Would you know the color of the other marble in the box if you opened the box labeled "2 red" and saw

9. a red marble?

10. a white marble?

Would you know the color of the other marble in the box if you opened the box labeled "2 white" and saw

11. a red marble?

12. a white marble?

Would you know the color of the other marble in the box if you opened the box labeled "1 red, 1 white" and saw

13. a red marble?

14. a white marble?

15. Which box would be the best choice to look inside?

Suppose that you see a red marble when you open that box. What can you conclude about the color of

16. the other marble in the box?

17. the marbles in the box labeled "2 white"?

18. the marbles in the box labeled "2 red"?

Set II

The figures below illustrate the names of different parts of a cube.

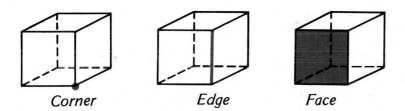

Corner   *Edge*   *Face*

1. How many corners does a cube have?

2. How many edges?

3. How many faces?

Suppose that a wooden cube is painted brown and then cut up into smaller cubes as shown in the first figure at the right. Use deductive reasoning to answer the following questions about the results.

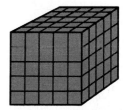

4. How many small cubes are produced?

Some of the small cubes have three faces painted brown. One of them is shown in color in the second figure.

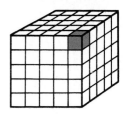

5. Where are these cubes located?

6. How many of these cubes are there altogether?

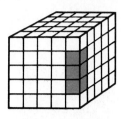

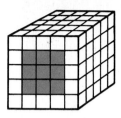

Some of the small cubes have two faces painted brown. Three of them are shown in color in the first figure at the left.

7. Where are these cubes located?

8. How many of these cubes are there altogether?

Some of the small cubes have exactly one face painted brown. Several are shown in color in the second figure.

9. Where are these cubes located?

10. How many of these cubes are there altogether?

Some of the small cubes have no faces painted brown.

11. Where are these cubes located?

12. How many of these cubes are there altogether?

Check to see if you have accounted for all of the cubes.

Suppose that a large wooden cube is painted brown and then cut up into smaller cubes as shown in this figure.

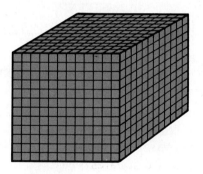

13. How many small cubes are produced?

How many of the small cubes have

14. three painted faces?

15. two painted faces?

16. one painted face?

17. no painted faces?

Check to see if you have accounted for all of the cubes.

Chapter 1: MATHEMATICAL WAYS OF THINKING

Set III

The best-known mathematical theorem is named for Pythagoras, a Greek mathematician of about 500 B.C. The Pythagorean Theorem states that, if squares are drawn on the three sides of a triangle that has a right (90°) angle, the square on the longest side of the triangle (called the hypotenuse) will be equal in area to the other two squares put together.

An example of this is shown on this postage stamp from Greece. Notice that the square on the longest side of the triangle contains 25 small squares, that the squares on the other two sides contain 16 squares and 9 squares, and that $25 = 16 + 9$.

The figure on the postage stamp does not prove that the Pythagorean Theorem is true for *all* right triangles because right triangles come in many different shapes. The figures below, however, can be used to make a general proof.

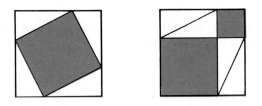

The first figure consists of four identical right triangles and a square.

1. On which side of the triangles is the square drawn?

The second figure consists of four triangles identical with those in the first figure and two squares.

2. On which sides of the triangles are these squares drawn?

3. Given that the two figures have the same area, explain why it follows that the large square must be equal in area to the other two squares put together.

Lesson 6

Number Tricks and
Deductive Reasoning

In the preface to *The Rhind Mathematical Papyrus*, a translation of the ancient Egyptian document named after A. Henry Rhind, who bought and gave it to the British Museum, Arnold Chace wrote:

I venture to suggest that if one were to ask for that single attribute of the human intellect which would most clearly indicate the degree of civilization of a race, the answer would be, the power of . . . reasoning, and that this power could best be determined in a general way by the mathematical skill which members of the race displayed. Judged by this standard the Egyptians of the nineteenth century B.C. had a high degree of civilization.

Two of the problems in the papyrus are number tricks. One of them is reproduced at the top of this page; it shows how it was written by a scribe many centuries ago. According to its directions, a number is chosen and then several things are done with it. At the end of the trick, the result is the same regardless of the number chosen.

Here is an example of a similar trick. Each step of the trick is described at the left and the results for four numbers chosen at random are shown in the columns at the right.

Choose a number.	4	7	12	35
Add five.	9	12	17	40
Double the result.	18	24	34	80
Subtract four.	14	20	30	76
Divide by two.	7	10	15	38
Subtract the number first thought of.	3	3	3	3
The result is three.				

The fact that the result is three in all four cases shown does not prove that the trick will work for every number. Although we can reason inductively that it may work, we need to use the deductive method to prove that it does.

We will go through the trick again, but this time with one slight change. Rather than choosing a specific number at the start, we will use a box to represent the number originally chosen. Throughout the trick, each box represents this same number. To represent numbers we know, we will use circles.

Choose a number.   □

Add five.   □ ○○○○○

Double the result.   □□ ○○○○○ / ○○○○○

Subtract four.   □□ ○○○ / ○○○

Divide by two.   □ ○○○
Subtract the number
    first thought of.   ○○○
The result is three.

Here we have a proof that the result is *always* three. Now it is easy to see why the choice of the original number makes no difference in the result.

Boxes and circles are somewhat clumsy and so mathematicians prefer instead to use a letter of the alphabet to represent the original number and ordinary numerals for the other numbers. Here is the same proof written in the symbols of algebra.

Choose a number. n
Add five. $n + 5$
Double the result.* $2(n + 5)$ or $2n + 10$
Subtract four. $2n + 6$
Divide by two. $n + 3$
Subtract the number
  first thought of. 3
The result is three.

---

*The principle being used here is called the *distributive rule*. Look on pages 626–628 if you are not familiar with this idea.

Exercises

Set I

If the original number in a number trick is represented by ☐, what would represent

1. the result of doubling the number?

2. the result of adding two to the number?

3. the next larger number?

If the number resulting from one of the steps of a number trick is represented by ☐ ☐ ☐ o o o , what would represent

4. the result of subtracting three from this number?

5. the result of dividing this number by three?

6. the result of adding the original number to this number?

In algebraic symbols, ☐ ☐ o o o o would be written as $2n + 4$. Write each of the following in algebraic symbols.

7. ☐ ☐ ☐ ☐ oo

8. ☐ ooooo

9. ooooo

10. ☐ ☐ ☐ ☐ ☐

The following number trick is illustrated with two different numbers. Proofs that it will always work are shown with boxes and circles and with algebraic symbols.

	Examples		Proofs	
Choose a number.	5	8	□	n
Multiply by three.	15	24	□ □ □	$3n$
Add six.	21	30	□ □ □ ∘∘∘/∘∘∘	$3n + 6$
Divide by three.	7	10	□ ∘∘	$n + 2$
Subtract the number first thought of.	2	2	∘ ∘	2

The result is two.

Copy each of the following number tricks. Using the same format as in the example above, illustrate each trick with two different numbers. Then write proofs with boxes and circles and with algebraic symbols that it will always work.

11. Choose a number.
    Add three.
    Multiply by two.
    Add four.
    Divide by two.
    Subtract the number
      first thought of.
    The result is five.

12. Choose a number.
    Double it.
    Add nine.
    Add the number
      first thought of.
    Divide by three.
    Add four.
    Subtract the number
      first thought of.
    The result is seven.

13. Choose a number.
    Add the next larger number.
    Add seven.
    Divide by two.
    Subtract the number
      first thought of.
    The result is four.

14. Choose a number.
    Triple it.
    Add the number one larger
      than the number first thought of.
    Add eleven.
    Divide by four.
    Subtract three.
    The result is the original number.

Set II

The following number tricks are of a different type from those in Set I.

Trick 1

Choose any three-digit number.
Multiply it by seven.
Multiply the result by eleven.
Multiply the result by thirteen.

An example of carrying out these steps is shown below.

$$
524 \quad
\begin{array}{r}
524 \\
\times \quad 7 \\
\hline
3{,}668
\end{array}
\qquad
\begin{array}{r}
3668 \\
\times \quad 11 \\
\hline
3668 \\
3668 \\
\hline
40{,}348
\end{array}
\qquad
\begin{array}{r}
40348 \\
\times \quad 13 \\
\hline
121044 \\
40348 \\
\hline
524{,}524
\end{array}
$$

1. Without using a calculator, do the trick again with a different three-digit number. Show all of the steps.

2. What relation does the result seem to have to the number that you have chosen?

3. Suppose that you did this trick with many different three-digit numbers and it always worked. Would that prove that it will work for every three-digit number?

4. Find $7 \times 11 \times 13$.

   The number 1,001 is the key to this number trick.

5. Without using a calculator, multiply the three-digit number that you chose for exercise 1 by 1,001. Show your work.

6. What do you notice about the result?

7. Does this result seem to suggest that the trick will work for every three-digit number?

Trick 2

Choose any two-digit number.
Multiply it by 13.
Multiply the result by 21.
Multiply the result by 37.

8. Do this trick with a two-digit number. Show all of the steps.

9. Do it again with a different two-digit number.

10. What do you notice about the results?

11. Find $13 \times 21 \times 37$.

12. Why does the trick work?

Set III

Here is a trick based on the number of matches in a matchbook.*

Take an unused matchbook that contains twenty matches and tear out any number of matches up to nine. Throw them away and count the matches that remain.

Add the two digits of this number and tear out as many additional matches from the book. Tear out two more matches.

1. Try this a couple of times, with or without a matchbook. What is the result?

2. Can you think of a way to prove that this trick will always work? If so, show your proof.

3. Will the trick still work if any of the directions are changed? If so, which ones?

---

*From "Mathematical Games" by Martin Gardner, *Scientific American*, August 1973.

Chapter 1 / Summary and Review

In this chapter we have been introduced to:

Inductive reasoning (*Lessons 1, 2, 3, and 4*)   Inductive reasoning is a method of drawing general conclusions from a limited set of observations. Although inductive reasoning is of great importance in developing mathematical ideas, it cannot prove that they are correct.

Deductive reasoning (*Lessons 5 and 6*)   Deductive reasoning is a method of drawing conclusions from ideas that we accept as true by using logic. It is used for mathematical proofs. Statements that are proved true by using deductive reasoning are theorems.

Exercises

Set I

Albrecht Dürer, a German artist of the sixteenth century, made a famous engraving titled *Melancholy*, which contains an interesting square of numbers. Two of the numbers in the square have been blacked out in this copy of that engraving.

1. What number do you think appeared in the corner?

2. What number do you think appeared inside?

3. If you assume from your observations that there is a pattern in the arrangement of the numbers in the square, what kind of reasoning are you using?

4. If you use the pattern observed to figure out what the missing numbers are, what kind of reasoning are you using?

Draw two billiard tables having the dimensions shown in the figures at the right.

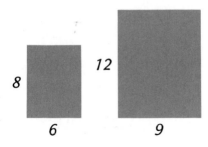

8

12

6

9

5. Show the path of a ball hit from the *upper-left* corner of each table at 45° angles with the sides.

6. What do you observe about the paths on the two tables?

7. Use the dimensions of the tables to explain why.

The following questions refer to this number trick.

> Choose a number from two to nine.
> Multiply it by 41.
> Multiply the result by 271.

8. Do this trick without using a calculator. Show the steps.

9. Do it again with a different number.

10. What do you notice about the results?

11. Find 41×271.

12. Why does the trick work?

Three golfers named Tom, Dick, and Harry are walking to the clubhouse. Tom, the best golfer of the three, always tells the truth. Dick sometimes tells the truth, while Harry, the worst golfer, never does.

13. To figure out who is who, it is best to first determine which one is Tom. Why?

14. How can you deduce which one is Tom from what each golfer says?

15. How can you determine which one is Harry?

16. Is Dick lying or telling the truth?

Set II

Here is an interesting number trick.

> Choose any three-digit number whose first and last digits differ by more than one.

Write down the same three digits in reverse order to form another three-digit number.

Subtract the smaller of your two numbers from the larger and circle your answer.

Reverse the digits in the number you circled to form another number and circle it.

Add the two circled numbers together.

An example of carrying out these steps is shown below.

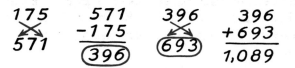

1. Without using a calculator, do the trick with a different number. Show all of the steps.

2. What do you notice about the results?

3. If you did this trick with many different numbers and it always worked, would that prove that it will work for every such number?

The first step in the trick was to choose a number in which the first and last digits differ by *more than* one. Suppose the first and last digits differed by *exactly* one. An example of what would happen is shown below.

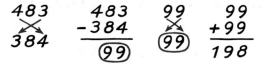

4. Do the trick with a different number whose first and last digits differ by exactly one. Show all of the steps.

5. What do you notice about the results?

6. What would happen if you started with a number in which the first and last digits were the *same?*

Look at the following equations.

$$99 \times 99 = 9{,}801$$
$$999 \times 999 = 998{,}001$$
$$9{,}999 \times 9{,}999 = 99{,}980{,}001$$
$$99{,}999 \times 99{,}999 = 9{,}999{,}800{,}001$$

Chapter 1: MATHEMATICAL WAYS OF THINKING

On the basis of these equations, guess the numbers that will complete the following equations.

7. $999,999 \times 999,999 =$ ▥

8. $9,999,999,999 \times 9,999,999,999 =$ ▥

9. On what kind of reasoning have you based your guesses?

The following puzzle is by Kobon Fujimura, Japan's leading inventor of puzzles.[*]

This figure shows eight square sheets of paper, all the same size, that have been placed on a table. The square in the center can be seen completely. The other seven squares overlap and can be only partly seen.

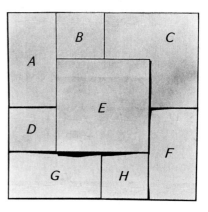

10. Give the letters of the squares in order from the top square to the bottom one.

11. On what kind of reasoning have you based your solution to this puzzle?

The following puzzle is by Pierre Berloquin, the author of several very successful puzzle books in France.[†]

Andre is a butcher and president of the street storekeepers' committee, which also includes a grocer, a baker, and a tobacconist. All of them sit around a table.

Andre sits on Charmeil's left.
Berton sits at the grocer's right.
Duclos, who faces Charmeil, is not the baker.

12. Make a drawing to show who sits where.

13. What kind of store does Berton have?

This puzzle appeared in an article by Martin Gardner titled "My Ten Favorite Brainteasers."[‡]

Miss Green, Miss Black, and Miss Blue are out for a stroll together. One is wearing a green dress, one a black dress, and one a blue dress.

"Isn't it odd," says Miss Blue, "that our dresses match our last names, but not one of us is wearing a dress that matches her own name?"

"So what?" said the lady in black.

14. What color is each lady's dress?

---

[*] *The Tokyo Puzzles* (Scribner's, 1978).
[†] *One Hundred Games of Logic* (Scribner's, 1977).
[‡] *Games* magazine, January/February 1978, page 16.

Set III

If a wooden cube is cut up into 8 small cubes, the cubes can be glued together in pairs to form 4 bricks that can be put back together to form the original cube.

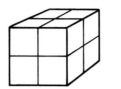 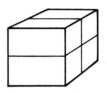

Suppose that a cube is cut up into 27 small cubes. If these cubes are glued together in pairs to form 13 bricks, there will be one cube left over. Suppose that the extra cube is thrown away. Can the 13 bricks be put back together to form the original cube with a hole in the center?

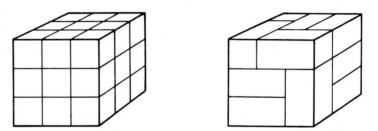

If it is possible to do this, then the 26 cubes that make up the bricks can be painted alternately brown and white as shown here.

1. How many white cubes would there be altogether?

2. How many brown cubes would there be altogether? (Remember that there is no cube in the center.)

Notice that each brick is made of two cubes that are different colors.

3. How many cubes of each color would be necessary to form the 13 bricks?

4. How many cubes of each color did you say were in the large cube with the hole in the middle?

5. Is it possible to solve the puzzle if you are very patient and keep trying out different arrangements of putting the bricks together?

"For a minute I thought we had him stymied."

Drawing by Tom Henderson

Chapter 1 / Problems for Further Exploration

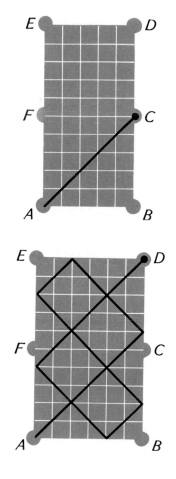

Lesson 1

1. A pool table has six pockets: one at each corner and two at the midpoints of the longer sides. The first figure at the right shows a pool table of standard dimensions whose pockets have been labeled for reference. If a ball is hit from the corner labeled A at 45° angles with the sides, it will end up in the pocket labeled C. The second figure shows that, on a table one unit wider, a ball hit from corner A would end up in pocket D instead.

   a) Draw pool tables with other shapes and show the path of a ball from corner A at 45° angles from the sides of each table. Try to find examples in which the ball ends up in each of the other pockets.

   b) If the ball is hit from corner A at 45° angles from the sides, do you think there are any pockets on the table in which it cannot end up? If you do, explain.

2. In 1961, Dr. Andrés Zavrotsky of the University of the Andes in Venezuela patented a device that can be used to find the greatest common divisor of a pair of numbers.* It consists of four

---

* The greatest common divisor of a pair of numbers is the largest number that can be divided into both of them evenly. For example, the greatest common divisor of 10 and 25 is 5.

Dr. Zavrotsky's device is described in *Martin Gardner's Sixth Book of Mathematical Games from Scientific American* (W. H. Freeman and Company, 1971).

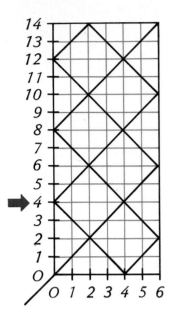

A

B

C

D

adjustable mirrors that can be used to form a rectangle with length and width equal to the two numbers. The figure at the left shows an overhead view of the mirrors adjusted to form a rectangle measuring 6 by 14.

A beam of light is sent through a crack at the lower left corner at 45° angles with the sides of the rectangle. It is reflected from mirror to mirror until it lights up one of the other three corners.

The greatest common divisor of the numbers under investigation is related to the position of the lighted point on the longer side closest to the corner where the light begins. This point is marked with an arrow in the figure.

In the illustration, the greatest common divisor of 6 and 14 is 2 and the closest lighted point on the longer side is 4 units from the corner.

Show the path of the light when the mirrors are adjusted to form rectangles of the following dimensions.

a) 3 × 6
b) 4 × 6
c) 5 × 6
d) 6 × 6

With one exception, the greatest common divisor of each pair of numbers can be determined from the lighted point on the longer side closest to the corner where the light begins.

e) How?
f) What is the exception?

Lesson 2

So far in our study of the path of a billiard ball, we have not encountered any situations in which the ball came back to the point from which it was hit. If the ball is hit from a point other than a corner, it is possible for this to happen, in which case the ball travels in an endless loop.

Look, for example, at the four drawings at the left. If the ball is hit from corner A, it travels diagonally across the table, ending up in the opposite corner. If it is hit from points B, C, or D at a 45° angle from the lower side, however, it could theoretically continue rebounding from the walls forever.

Is such a loop possible on a table that is not square? If it is, can the existence of such a path be predicted from the dimensions of the table?

Make as many drawings as you need to arrive at what you feel are reasonable conclusions. For simplicity, draw only paths that make 45° angles with the sides of the tables.

Chapter 1: MATHEMATICAL WAYS OF THINKING

Lesson 3

1. The equations below, even though all the numbers are missing in some, illustrate an interesting pattern. After you have solved a few equations, you will know what the pattern is and can then write those in which all the numbers are missing.

a) $12 \times 42 = $ ▦
b) $21 \times 24 = $ ▦
c) $26 \times 93 = $ ▦
d) $62 \times 39 = $ ▦
e) $14 \times 82 = $ ▦

f) $41 \times 28 = $ ▦
g) $46 \times 96 = $ ▦
h) ▦ $\times$ ▦ $= $ ▦
i) $24 \times 63 = $ ▦
j) ▦ $\times$ ▦ $= $ ▦

k) $12 \times 84 = $ ▦
l) ▦ $\times$ ▦ $= $ ▦
m) $23 \times 96 = $ ▦
n) ▦ $\times$ ▦ $= $ ▦
o) ▦ $\times$ ▦ $= $ ▦

p) $21 \times 36 = $ ▦
q) ▦ $\times$ ▦ $= $ ▦
r) $63 \times 48 = $ ▦
s) $13 \times 62 = $ ▦
t) ▦ $\times$ ▦ $= $ ▦

Unfortunately, this pattern does not hold for all equations of this sort. If you can figure out a way to rearrange the equations above in a logical order, however, you may be able to discover eight more equations for which the pattern does work.

u) Show a logical rearrangement of the equations.

v) What are the other eight equations for which the pattern is true?

2. The figure at the right represents three intersections along a city street. All three intersections are controlled by traffic lights and twelve consecutive times that each signal was observed to change are shown in the table below.

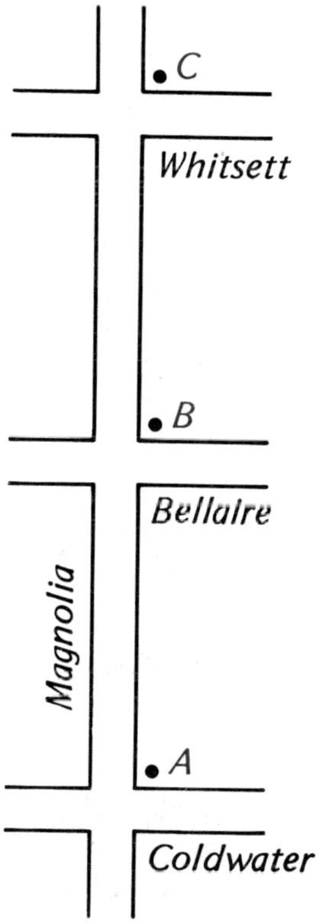

Color	Signal A	Signal B	Signal C
Green	9:01:10	9:01:20	9:01:10
Amber	9:01:34	9:05:09	9:01:39
Red	9:01:37	9:05:12	9:01:42
Green	9:02:10	9:05:20	9:02:10
Amber	9:02:34	9:08:09	9:02:39
Red	9:02:37	9:08:12	9:02:42
Green	9:03:10	9:08:24	9:03:10
Amber	9:03:34	9:14:09	9:03:39
Red	9:03:37	9:14:12	9:03:42
Green	9:04:10	9:14:22	9:04:10
Amber	9:04:34	9:16:09	9:04:39
Red	9:04:37	9:16:12	9:04:42

a) What conclusions can you draw about the traffic on these streets?

From this information, does it seem reasonable to predict the color of each signal at each of the following times? If you think so, tell the color. If you do not, explain why not.

b) 9:30:00   c) 9:30:15   d) 9:30:35

Lesson 4

1. A popular puzzle is about a dresser drawer full of socks, some of which are white and the rest black. Someone takes some socks out of the drawer in the dark, and the colors of the socks cannot be seen.

   a) How many socks have to be taken to be sure of getting a matching pair? Explain.

   b) How many socks have to be taken to be sure of getting two matching pairs? Explain.

   c) How many socks have to be taken to be sure of getting any given number of matching pairs? Explain.

2. The prime numbers are numbers that cannot be divided evenly by any whole numbers other than themselves and 1. A list of the first hundred prime numbers is shown below.

2	73	179	283	419
3	79	181	293	421
5	83	191	307	431
7	89	193	311	433
11	97	197	313	439
13	101	199	317	443
17	103	211	331	449
19	107	223	337	457
23	109	227	347	461
29	113	229	349	463
31	127	233	353	467
37	131	239	359	479
41	137	241	367	487
43	139	251	373	491
47	149	257	379	499
53	151	263	383	503
59	157	269	389	509
61	163	271	397	521
67	167	277	401	523
71	173	281	409	541

The following procedure can be used to produce some of the prime numbers in this list.

Choose any counting number.
Multiply by the next larger number.
Add 17.

Chapter 1: MATHEMATICAL WAYS OF THINKING

For example, if 6 is chosen, the next larger number is 7, $6 \times 7 = 42$, and $42 + 17 = 59$, a prime number.

a) Carry out this procedure for each counting number from 1 through 15. List the results in a table.

b) What conclusion seems reasonable on the basis of these results?

c) Do you think every counting number would yield the same result?

Lesson 5

1. Many logic puzzles are about people who do or do not tell the truth. Here is an especially clever puzzle of this type.*

   Heracles lies on Monday, Tuesday, and Wednesday. Theseus lies on Thursday, Friday, and Saturday. At all other times Heracles and Theseus tell the truth.

   "Yesterday was one of my lying days," says Heracles. "Yesterday was one of my lying days too," says Theseus.

   a) What day of the week is it? Explain.

   b) What if each of them claimed that yesterday was one of his truth-telling days? Can any conclusion be drawn from this?

2. The desk calendar shown in this photograph contains four cubes.† The two black cubes on the ends show the month and the day of the week and the two white cubes in the middle show the date.

   Each face of each white cube contains a single digit and the two white cubes can be arranged so that their front faces indicate every date from 01 to 31.

   Can you figure out what digits are on each cube, given that the one on the right contains the digits 3, 4, and 5? If so, tell what they are.

---

*This puzzle, by Raymond M. Smullyan, is from his book titled *What is the Name of This Book?* (Prentice-Hall, 1978).

†This puzzle is by Martin Gardner and appears in his book titled *Mathematical Circus* (Knopf, 1979).

Lesson 6

1. This number trick once appeared in the *Reader's Digest*.
   a) Try it out to find out what happens.
   b) How does the trick work?

A teaser—for men only

What's Your Wife's Name?

Condensed from THE IRISH DIGEST

IF YOU'RE a bachelor, and you answer these questions honestly, you'll be told the name of your future wife. If you're married, you'll be told the name of your present wife—and let's hope it's the right one.

1. Write down the number corresponding to the month of your birth from Table 1.

2. Add the number corresponding to your favorite dish from Table 2.

3. Multiply the answer by ten. Then add three if you want to know the name of your future wife, two for your present wife, one for your last wife.

4. Reverse the order of the figures, and subtract the result from the number you had before reversing. (Thus, if the number is 521, it becomes 125 on reversing, which is subtracted from 521 —leaving 396.)

5. Reverse this answer and add the result to the number it was before reversing.

6. Add 52,205,197 if you are a Brit-ish Lord; otherwise, add 423,571.

7. Look up the number corresponding to the first letter of your surname in Table 3, and place it on the right-hand side of the previous answer. If, for example, your name is Smith, and the previous answer was 123,456, you would place the two figures corresponding to S, which are 60, on the right, giving you 12,345,660.

8. Repeat this with the next letter of your surname, and continue for all the letters, in order, in your surname.

9. Halve the answer.

10. Divide the answer into groups of two figures. Each group represents a letter of your wife's name, when referred to Table 4. Thus, if the answer is 21-10-23-43, her name will be Mary.

TABLE 1		
January 90	July 70	
February 80	August 80	
March 70	September 90	
April 60	October 80	
May 50	November 70	
June 60	December 60	

TABLE 3			
A — 20	G — 64	M — 42	T — 44
B — 40	H — 68	N — 66	U — 28
C — 48	I — 24	O — 26	V — 84
D — 62	J — 80	P — 82	W — 38
E — 22	K — 54	Q — 90	X — 58
F — 50	L — 52	R — 46	Y — 86
		S — 60	Z — 56

TABLE 2	
Steak and Onions 8	
Hamburger and French Fries 7	
Ham and Eggs 6	
Chicken and Rice 5	
Franks and Beans 4	
Tuna Fish Casserole 3	
Roast Beef and Potatoes 9	

TABLE 4			
		27 — K	34 — H
10 — A	21 — M	28 — Z	40 — J
11 — E	22 — T	29 — X	41 — P
12 — I	23 — R	30 — S	42 — V
13 — O	24 — C	31 — D	43 — Y
14 — U	25 — F	32 — G	44 — W
20 — B	26 — L	33 — N	45 — Q

Reprinted with permission from the March 1969 *Reader's Digest*

2. Here is a clever card trick invented by Mark Wilson, the first magician to be invited to take his magic act to China since 1949.*

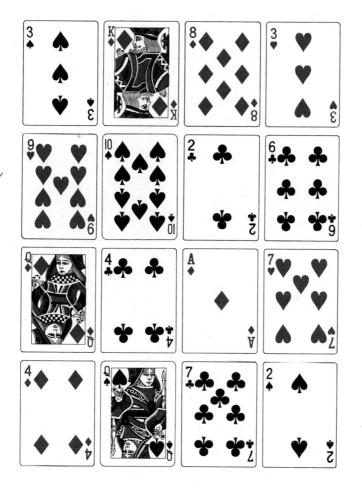

First, pick any brown card shown in this figure and place a coin on it. Now, take the coin and move it left or right to the nearest black card. Next, move the coin vertically up or down to the nearest brown card. Next, move diagonally to the nearest black card. Finally, move down or to the right to the nearest brown card.

a) Try the trick several times. What happens?

b) Can you figure out how the trick works? If so, explain.

---

*The trick appeared in the November/December 1980 issue of *Games* magazine.

Chapter 2

NUMBER SEQUENCES

Lesson 1

Arithmetic Sequences: Growth at a Constant Rate

A child first becomes aware of numbers through counting. Arranged in order, the counting numbers form a *number sequence:*

$$1 \quad 2 \quad 3 \quad 4 \quad 5 \quad 6 \quad 7 \quad 8 \quad 9 \quad 10 \quad 11 \quad \ldots$$

▶ A **number sequence** is an arrangement of numbers in which each successive number follows the last according to a uniform rule.

The numbers are the *terms* of the sequence. For the sequence of counting numbers, the rule is "add one to each term to get the next term." In symbols, if n represents any term of the sequence, the next term is $n + 1$. The fact that the sequence of counting numbers can be continued indefinitely is indicated by the three dots following the last term shown. A mathematician once said: "A number sequence is like a bus; nobody ever doubts that there is always room for one more."

When a parachutist jumps from an airplane, the distances in feet traveled during the first few seconds are

$$16 \quad 48 \quad 80 \quad 112 \quad 144 \quad \ldots$$

This sequence is similar to the sequence of counting numbers in that each successive term can be found by *adding the same number* to the preceding term.

$$16 \quad 48 \quad 80 \quad 112 \quad 144 \quad \ldots$$
$$+32 \quad +32 \quad +32 \quad +32 \quad +32$$

▶ A number sequence in which each successive term may be found by *adding the same number* is an **arithmetic sequence.**

58

By permission of Johnny Hart and Field Enterprises, Inc.

If the number added is positive, the sequence grows at a constant rate. The left-hand figure below illustrates an arithmetic sequence formed by adding 3. If the number added is negative,* the sequence shrinks at a constant rate. The right-hand figure illustrates an arithmetic sequence formed by adding –2.

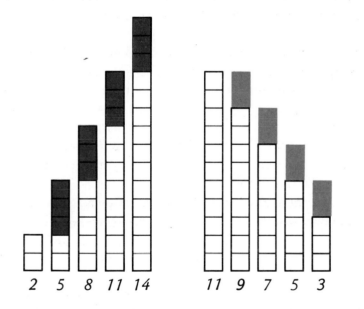

The number added to each term of an arithmetic sequence to form the next term is also the *difference* between each pair of successive terms. It is called the *common difference* of the sequence. The common difference of the first sequence illustrated above is 3, and the common difference of the second sequence is –2.

---

*Negative numbers are discussed in the appendix, beginning on page 628.

Exercises

Set I

In the *Chóu-peï,* an ancient Chinese book on numbers written about 1100 B.C., the following number sequences appear:

$$1 \quad 3 \quad 5 \quad 7 \quad 9 \quad 11 \quad \ldots$$

and

$$2 \quad 4 \quad 6 \quad 8 \quad 10 \quad 12 \quad \ldots$$

1. Are these sequences arithmetic? Explain why or why not.

2. What is the name of the set of numbers that make up the first sequence?

3. What is the name of the set of numbers that make up the second sequence?

The figure at the left illustrates the first three terms of an arithmetic sequence.

4. What are the terms?

5. What is their common difference?

6. What are the next three terms of the sequence?

January 1, 2000, will fall on a Saturday.

7. List the other days of that month that also fall on Saturday.

8. If *n* represents the date of a given Saturday of that month, what represents the date of the following Saturday of the month?

Copy the following arithmetic sequences, writing in the missing terms.

9. 2 7 12 17 ▓▓▓ ▓▓▓

10. 5 13 21 ▓▓▓ ▓▓▓

11. 11 15 ▓▓▓ 23 ▓▓▓

12. ▓▓▓ ▓▓▓ 20 29 38

13. 4 ▓▓▓ 18 ▓▓▓ 32

14. ▓▓▓ 33 ▓▓▓ 65 ▓▓▓

Chapter 2: NUMBER SEQUENCES

Find the missing terms in each of these arithmetic sequences.

15. 10 ▦ 70   18. 10 ▦ ▦ ▦ ▦ 70

16. 10 ▦ ▦ 70   19. 10 ▦ ▦ ▦ ▦ ▦ 70

17. 10 ▦ ▦ ▦ 70

Tell whether or not each of the following number sequences is an arithmetic sequence. For those that are arithmetic, give the common difference.

20. 4   24   44   64   84   . . .

21. 3   6   12   24   48   . . .

22. 2   15   28   41   54   . . .

23. 5   5   5   5   5   . . .

24. 10   100   1,000   10,000   100,000   . . .

25. 20   21   23   26   30   . . .

26. 15   14   13   12   11   . . .

If each term of the arithmetic sequence

$$1 \quad 5 \quad 9 \quad 13 \quad 17 \quad . . .$$

is increased by 6, the resulting sequence is

$$7 \quad 11 \quad 15 \quad 19 \quad 23 \quad . . .$$

27. Is this sequence also an arithmetic sequence?

28. If each term of an arithmetic sequence is increased by the same number, do you think the resulting sequence is always arithmetic?

If each term of the arithmetic sequence

$$1 \quad 5 \quad 9 \quad 13 \quad 17 \quad . . .$$

is multiplied by 2, the resulting sequence is

$$2 \quad 10 \quad 18 \quad 26 \quad 34 \quad . . .$$

29. Is this sequence also an arithmetic sequence?

30. If each term of an arithmetic sequence is multiplied by the same number, do you think the resulting sequence is always arithmetic?

The 100th term of the arithmetic sequence

$$1 \quad 2 \quad 3 \quad 4 \quad 5 \quad \ldots$$

is obvious: it is 100. The 100th term of the arithmetic sequence

$$2 \quad 5 \quad 8 \quad 11 \quad 14 \quad \ldots$$

however, is not obvious at all. One way to find out what it is would be to continue writing the sequence until we arrive at it. There is an easier way, however. Look at the diagram at the left and the accompanying pattern below.

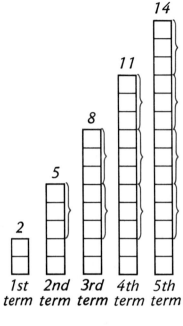

1st term	$2 + 0 \cdot 3 = 2$
2nd term	$2 + 1 \cdot 3 = 5$
3rd term	$2 + 2 \cdot 3 = 8$
4th term	$2 + 3 \cdot 3 = 11$
5th term	$2 + 4 \cdot 3 = 14$

The 100th term must be

$$2 + 99 \cdot 3 = 2 + 297 = 299$$

1. Find the 10th term of the arithmetic sequence

$$3 \quad 10 \quad 17 \quad 24 \quad \ldots$$

by writing down the next six terms.

2. Find the 10th term of the same sequence by using the shortcut suggested by the pattern:

1st term	$3 + 0 \cdot 7 = 3$
2nd term	$3 + 1 \cdot 7 = 10$
3rd term	$3 + 2 \cdot 7 = 17$
4th term	$3 + 3 \cdot 7 = 24$

3. Find the 20th term of the arithmetic sequence

$$11 \quad 15 \quad 19 \quad 23 \quad \ldots$$

by writing down the next sixteen terms.

4. Find the 20th term of the same sequence by using the shortcut suggested by the pattern:

$$
\begin{array}{ll}
\text{1st term} & 11 + 0 \cdot 4 = 11 \\
\text{2nd term} & 11 + 1 \cdot 4 = 15 \\
\text{3rd term} & 11 + 2 \cdot 4 = 19 \\
\text{4th term} & 11 + 3 \cdot 4 = 23
\end{array}
$$

Find the indicated terms of the following arithmetic sequences.

5. The 11th term of   5   11   17   23   . . .

6. The 7th term of   5   15   25   35   . . .

7. The 25th term of   7   9   11   13   . . .

8. The 50th term of the same sequence.

9. The 15th term of   100   97   94   91   . . .

10. The 34th term of the same sequence.

11. The 10th term of   24   35   46   57   . . .

12. The 111th term of the same sequence.

The diagram below represents an arithmetic sequence whose first term is a and whose common difference is d. Expressions for the

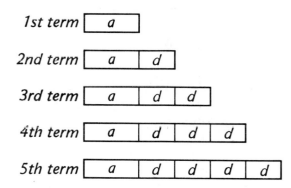

first three terms are a, $a + d$, and $a + 2d$, respectively. Write expressions for

13. the fourth and fifth terms.

14. the tenth term.

15. the nth term.

Set III

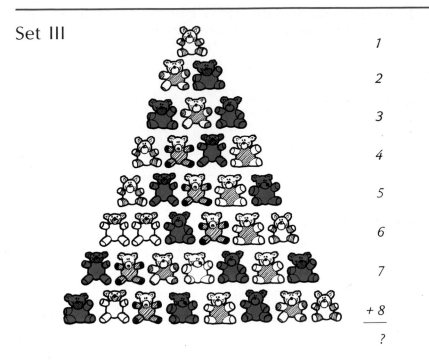

1

2

3

4

5

6

7

+ 8
―――
?

How many teddy bears were needed to build the pyramid shown above? One way to find out would be to add the numbers in the rows in order as indicated at the right of the picture. Another way is suggested by the figure below.

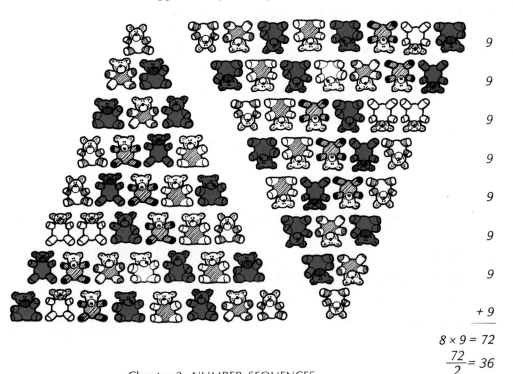

9

9

9

9

9

9

9

+ 9
―――

$8 \times 9 = 72$

$\dfrac{72}{2} = 36$

Chapter 2: NUMBER SEQUENCES

1. Find the sum of the seven terms shown of this arithmetic sequence by adding them in order from left to right.

$$3 \quad 7 \quad 11 \quad 15 \quad 19 \quad 23 \quad 27$$

2. Find the same sum by using the shortcut suggested by this pattern:

3	7	11	15	19	23	27
+27	+23	+19	+15	+11	+ 7	+ 3
30	30	30	30	30	30	30

3. Find the sum of the twelve terms shown of this arithmetic sequence by adding them in order from left to right.

$$1 \quad 3 \quad 5 \quad 7 \quad 9 \quad 11 \quad 13 \quad 15 \quad 17 \quad 19 \quad 21 \quad 23$$

4. Find the same sum by using the shortcut suggested by this pattern:

1	3	5	. . .
+23	+21	+19	. . .
24	24	24	. . .

As every bowler knows, the pins are set up in four rows with one pin in the first row, two pins in the second row, three pins in the third, and four in the last. Suppose that a gigantic set of pins was set up in 20 rows, the first row having one pin and each successive row having one more, so that the last row had 20 pins.

5. How many pins would you have to knock over to make a strike?

6. If pins were set up in the same fashion in 100 rows, how many would you have to knock over to make a strike?

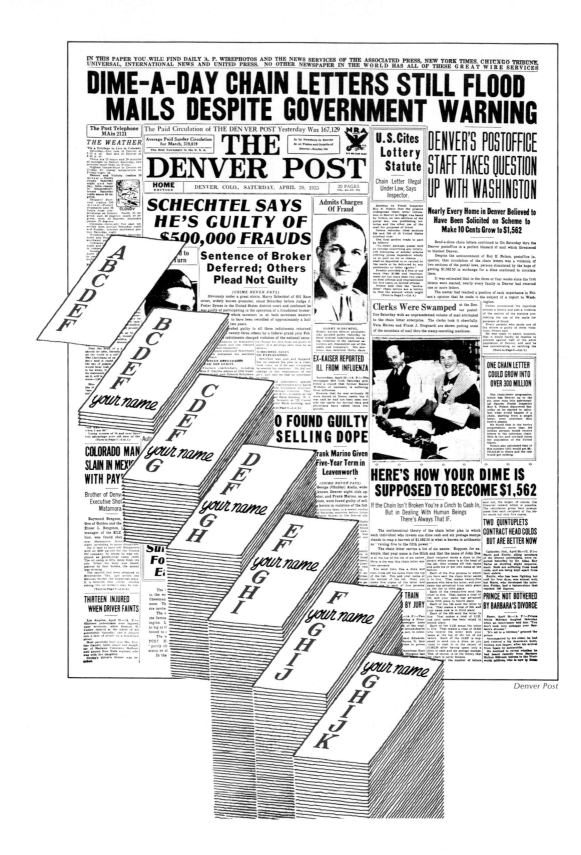

Denver Post

Geometric Sequences: Growth at an Increasing Rate

In 1935, a chain letter craze started in Denver and swept across the country. It worked like this. You receive a letter with instructions and a list of six names. You send a dime to the person named at the top, cross the name out, and add your own name at the bottom. Then you send a copy of the new list of names to each of five friends with instructions to do the same. When your five friends send 5 letters each, there will be 25 letters in all. If all twenty-five people getting these letters cooperate, 125 letters will be sent, and so on.

A list of these numbers in order, starting with one for the letter you receive and ending with the number of lists on which your name appears at the top, is shown here.

$$1 \quad 5 \quad 25 \quad 125 \quad 625 \quad 3{,}125 \quad 15{,}625$$

If each of the people receiving one of the 15,625 lists with your name at the top sends you a dime, you will receive $1,562.50. Not bad for an investment of only a dime and a few stamps. It is no wonder that this scheme, even though it was illegal, became popular when the United States was going through the Great Depression.

The terms in the list of chain letters sent form a number sequence because each successive number follows the preceding one according to a uniform rule. The rule, however, is to *multiply* each term by 5 to get the next term; so this is not an arithmetic sequence.

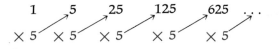

In symbols, if *n* represents any term of the sequence, the next term is 5*n*. The sequence is geometric and, as the stacks of letters shown on the page 66 indicate, it grows faster and faster.

▶ A number sequence in which each successive term may be found by *multiplying by the same number* is a **geometric sequence.**

If the number by which each term is multiplied is greater than 1, the sequence grows at an increasing rate. The left-hand figure below illustrates a geometric sequence formed by multiplying by 3. If each term is multiplied by a positive number that is less than 1, the sequence shrinks at a decreasing rate. The right-hand figure illustrates a geometric sequence formed by multiplying by $\frac{1}{2}$.

The ratios of successive terms in a geometric sequence are the same. Called the *common ratio* of the sequence, it is the number by which each term is multiplied to get the next. The common ratio of the left-hand sequence is

$$\frac{3}{1} = \frac{9}{3} = \frac{27}{9} = 3$$

and the common ratio of the right-hand sequence is

$$\frac{8}{16} = \frac{4}{8} = \frac{2}{4} = \frac{1}{2}.$$

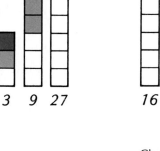

1   3   9   27   16   8   4   2

Chapter 2: NUMBER SEQUENCES

Exercises

Set I

The following number sequences appear on Babylonian cunei-form tablets of about 1800 B.C.

1   3   9   27   81   . . .

and

1   4   16   64   256   . . .

1. What is the rule for the first sequence?

2. What is the rule for the second sequence?

3. What kind of number sequences are they?

The figure at the right illustrates the first three terms of a geometric sequence.

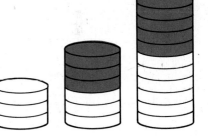

4. What are the terms?

5. What is their common ratio?

6. What are the next three terms of the sequence?

In a popular radio quiz program of the 1940s, the contestants were asked a series of as many as seven questions. Each question was worth twice as much as the preceding one, beginning with $1 for answering the first question correctly.

7. List the amounts that each of the other six questions were worth.

8. If n represents the value of a given question in the series, what represents the value of the next question?

Copy the following geometric sequences, writing in the missing terms.

9. 4   12   36   ▥   ▥

10. 11   121   ▥   ▥

11. 5   20   ▥   320   ▥

12. ▥   ▥   72   144   288

13. 0   ▥   ▥   ▥

14. 16   40   100   ▥   ▥

Find the missing terms in each of these geometric sequences.

15. 1 ||||||| 1,000,000

16. 1 ||||||| ||||||| 1,000,000

17. 1 ||||||| 64

18. 1 ||||||| ||||||| 64

19. 1 ||||||| 64,000,000

20. 1 ||||||| ||||||| 64,000,000

Tell whether or not each of the following number sequences is a geometric sequence. For those that are geometric, give the common ratio.

21. 5 10 15 20 25 . . .   24. 1 4 9 16 25 . . .

22. 2 8 32 128 512 . . .   25. 80 40 20 10 5 . . .

23. 7 7 7 7 7 . . .   26. 16 24 36 54 81 . . .

If each term of the geometric sequence

$$1 \quad 4 \quad 16 \quad 64 \quad . . .$$

is increased by 2, the resulting sequence is

$$3 \quad 6 \quad 18 \quad 66 \quad . . .$$

27. Is this sequence also a geometric sequence?

28. If each term of a geometric sequence is increased by the same number, do you think the resulting sequence is always geometric?

If each term of the geometric sequence

$$1 \quad 4 \quad 16 \quad 64 \quad . . .$$

is multiplied by 3, the resulting sequence is

$$3 \quad 12 \quad 48 \quad 192 \quad . . .$$

29. Is this sequence also a geometric sequence?

30. If each term of a geometric sequence is multiplied by the same number, do you think the resulting sequence is always geometric?

Chapter 2: NUMBER SEQUENCES

Set II

United States currency notes have been printed in the twelve denominations shown here.

Some of the denominations form the geometric sequence

$$1 \quad 10 \quad 100 \quad \ldots$$

1. What is the common ratio of this sequence?

2. How many terms of this sequence are denominations of currency?

3. Find another geometric sequence of currency denominations whose common ratio is 10.

Some of the denominations form the geometric sequence

$$2 \quad 100 \quad 5,000 \quad \ldots$$

4. What is the common ratio of this sequence?

5. How many terms of this sequence are denominations of currency?

Find a geometric sequence of currency denominations whose common ratio is

6. 2.

7. 5.

8. 100.

Compare the following patterns of the terms of an arithmetic sequence and a geometric sequence.

An arithmetic sequence: 2   5   8   11   14   . . .

1st term	2		$= 2 + 0 \cdot 3 =$	2
2nd term	$2 + 3$		$= 2 + 1 \cdot 3 =$	5
3rd term	$2 + 3 + 3$		$= 2 + 2 \cdot 3 =$	8
4th term	$2 + 3 + 3 + 3$		$= ▓▓▓▓▓ =$	11
5th term	$2 + 3 + 3 + 3 + 3$		$= ▓▓▓▓▓ =$	14

A geometric sequence: 2   6   18   54   162   . . .

1st term	2	$= ▓▓▓▓▓ =$	2
2nd term	$2 \cdot 3$	$= ▓▓▓▓▓ =$	6
3rd term	$2 \cdot 3 \cdot 3$	$= 2 \cdot 3^2 =$	18
4th term	$2 \cdot 3 \cdot 3 \cdot 3$	$= 2 \cdot 3^3 =$	54
5th term	$2 \cdot 3 \cdot 3 \cdot 3 \cdot 3$	$= 2 \cdot 3^4 =$	162

9. What expressions should be written in the two indicated spaces to complete the first pattern?

10. What expressions should be written in the two indicated spaces to complete the second pattern?*

11. Use the pattern for the arithmetic sequence to write an expression for its 10th term.

12. Use the pattern for the geometric sequence to write an expression for its 10th term.

The first term of another geometric sequence is 5 and the common ratio of the sequence is 2.

13. Write the first five terms of the sequence.

14. Using the pattern for the geometric sequence above as a guide, write the pattern for the first five terms of this sequence.

15. Use the pattern to write an expression for its 100th term.

---

* This pattern is based on *exponents*, which will be studied further in Chapter 4.

Write expressions for the indicated terms of the following geometric sequences.

16. The 11th term of   7   28   112   448   . . .

17. The 11th term of   4   28   196   1,372   . . .

18. The 20th term of   2   20   200   2,000   . . .

The first term of a geometric sequence is a and its common ratio is r. Expressions for its first three terms are a, $a \cdot r$, and $a \cdot r^2$, respectively. Write expressions for

19. the eighth term.

20. the nth term.

Set III

Although they are illegal, pyramid schemes for making money were a widespread phenomenon in 1980. The operation of a typical pyramid is illustrated in the diagram below.

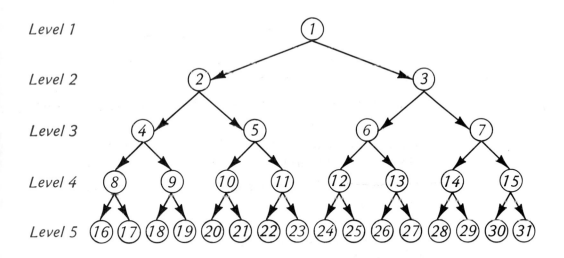

Thirty-one people are represented in the figure, each one on levels 1 through 4 having recruited the two people indicated by the arrows. Each person on level 5 gave $500 to the person on level one and $500 to the person that recruited him.

1. How much money did the person at the top get?

2. How much money did each person on level 4 get?

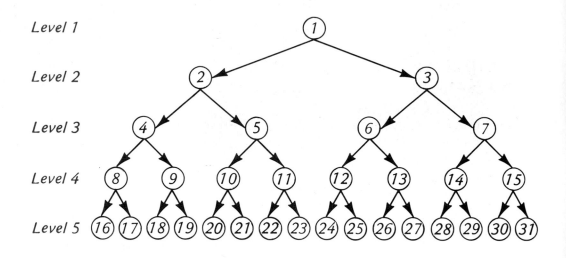

Level 1

Level 2

Level 3

Level 4

Level 5

Upon receipt of the money, the person at the top left the pyramid. In order for the two people on level 2 to make money, a sixth level of people had to join the pyramid.

3. How many people were needed for the sixth level?

In order for the people on level 5 to make money, three more levels of people had to join the pyramid.

4. How many people were needed for each of these levels?

5. After these people joined, how many people had joined the pyramid altogether?

6. How many people in the pyramid had made money at this point?

7. How much money had they made altogether?

8. Where did most of this money come from?

9. Through how many levels would the pyramid have to continue for everyone who joined it to make money?

Lesson 3

The Binary Sequence

According to legend, the game of chess was invented for a Persian king by one of his servants. The king was so pleased that he asked the servant what he would like as a reward. The man's request seemed very reasonable. He asked that one grain of wheat be placed on the first square of the chessboard, two grains on the second square, four grains on the third, and so on, each square having twice as many grains as the square before. The king was surprised, thinking that the servant had asked for very little. He was even more surprised when he found out how much wheat the man actually wanted.

The numbers of grains of wheat on the squares,

$$1 \quad 2 \quad 4 \quad 8 \quad 16 \quad 32 \quad 64 \quad 128 \quad . \ . \ ,$$

form a geometric sequence, which grows at an increasing rate. Because there are 64 squares on a chessboard, the inventor's request was for as many grains of wheat as the sum of the first 64 terms of this sequence. This number,

$$18{,}446{,}744{,}073{,}709{,}551{,}615{,}$$

is equivalent to approximately 175 billion tons of wheat, more than that which has been produced on the earth in recorded history.

75

▶ The sequence

$$1 \quad 2 \quad 4 \quad 8 \quad 16 \quad \ldots$$

is called the **binary sequence.**

Number	Binary sequence
1	1 2 4 8 ...
2	1 2 4 8 ...
3	1 2 4 8 ...
4	1 2 4 8 ...
5	1 2 4 8 ...
6	1 2 4 8 ...
7	1 2 4 8 ...
8	1 2 4 8 ...
9	1 2 4 8 ...
10	1 2 4 8 ...

The binary sequence has a remarkable property: every counting number can be expressed as the sum of one or more of its terms. This is illustrated for the numbers from one through ten in the table at the left.

If an electric circuit consisted of a sequence of switches corresponding to the terms of the binary sequence, the counting numbers could be represented by turning the switches on or off as illustrated in the table. For example, to represent the number 3, the switches for 1 and 2 would be turned on. To represent the number 10, the switches for 2 and 8 would be turned on, and so forth.

For this reason, the binary sequence is used in the representation of numbers in the circuits of electronic computers and calculators. The photograph at the left shows a small part of a microprocessor from a computer greatly enlarged so that the circuits representing the terms of the binary sequence can be seen.

To write a number in the form in which it is used in a computer, two digits are used: 1 to show that a switch is on and 0 to show that it is off. Because of this, such numbers are said to be written in *base 2,* or as *binary numerals.*

The binary numerals for the numbers from one to ten are given in the table below. Note that 0's to the left of the first 1 are customarily omitted.

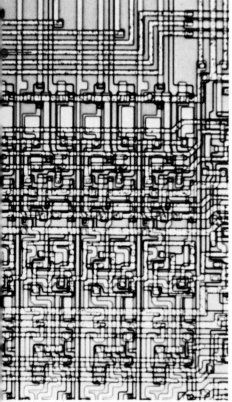

Courtesy of Texas Instruments, Inc.

Number	Binary numeral 8 4 2 1
1	1
2	1 0
3	1 1
4	1 0 0
5	1 0 1
6	1 1 0
7	1 1 1
8	1 0 0 0
9	1 0 0 1
10	1 0 1 0

Chapter 2: NUMBER SEQUENCES

Exercises

Set I

The photograph at the right shows the pattern in the lengths of the marks along one inch of a ruler. Along the edge, there are 16 small units, which have been numbered.

1. Look at the increasingly longer marks, shown in color. What do you notice about the numbers of these marks?

The table below shows the binary numerals of the numbers from zero to five, each written as four digits.

Number	Binary numeral
	8 4 2 1
0	0 0 0 0
1	0 0 0 1
2	0 0 1 0
3	0 0 1 1
4	0 1 0 0
5	0 1 0 1

2. Copy the table and continue it through the number 15.

Notice that the last digit of each binary numeral is alternately 0 and 1.

3. What other patterns do you notice in the digits of the binary numerals in this table?

To convert a binary numeral such as 11001 into decimal form, we first list the place value of each digit above it.

$$\begin{array}{ccccc} 16 & 8 & 4 & 2 & 1 \\ 1 & 1 & 0 & 0 & 1 \end{array}$$

Adding the place values of the digits that are 1's, we get

$$16 + 8 + 1 = 25.$$

Use the same procedure to convert the following binary numerals into decimal form.

4. 1111   5. 10000   6. 11111   7. 100000

"For example, Pop, if we write 210 in base 2 it looks like this—11010010."

8. Write the binary numeral for the number that is one more than the binary numeral 111111.

Convert the following binary numerals into decimal form.

9. 10100

10. 101000

11. 1101001

12. 11010010

Adding a zero to the end of a number in decimal form is equivalent to multiplying the number by ten. Compare, for example, 57 and 570.

13. What is adding a zero to the end of a binary numeral equivalent to? (Look at exercises 9 and 10 and your answers, as well as exercises 11 and 12.)

In 1605, the English statesman Francis Bacon devised a code for sending secret diplomatic messages. Each letter of the alphabet was represented by a five-letter group of *a*'s and *b*'s as shown in this table. (The letters *j* and *v* were not in the alphabet at that time.)

A	B	C	D	E	F
aaaaa	aaaab	aaaba.	aaabb.	aabaa.	aabab.
G	H	I	K	L	M
aabba	aabbb	abaaa.	abaab.	ababa.	ababb.
N	O	P	Q	R	S
abbaa.	abbab.	abbba.	abbbb.	baaaa.	baaab.
T	U	W	X	Y	Z
baaba.	baabb.	babaa.	babab.	babba.	babbb.

Courtesy of F.G. Heath

Chapter 2: NUMBER SEQUENCES

14. What connection do these five-letter groups have to binary numerals?

15. How many five-letter groups of a's and b's are possible in all?

Set II

The number of grains of wheat said to have been requested by the inventor of chess is the sum of the first 64 terms of the binary sequence. The following pattern suggests a way to find that sum without doing any addition.

$$1 \quad 2 \quad 4 \quad 8 \quad 16 \quad 32 \quad \ldots$$
$$1 + 2 = 3$$
$$1 + 2 + 4 = 7$$
$$1 + 2 + 4 + 8 = 15$$

1. Find the sum $1 + 2 + 4 + 8 + 16$ by adding the five numbers.

2. Show how to find the same sum without doing any addition.

3. Copy this list of the first six terms of the binary sequence and continue it by writing the next five terms.

$$1 \quad 2 \quad 4 \quad 8 \quad 16 \quad 32 \quad \ldots$$

4. Find the sum of the first seven terms of the binary sequence by adding them.

5. Show how to find the same sum without doing any addition.

6. What is the sum of the first ten terms of the binary sequence?

7. How could the sum of the first 64 terms of the binary sequence be found without doing any addition?

8. The 65th term of the binary sequence is

$$18,446,744,073,709,551,616.$$

What is the sum of the first 64 terms?

The pattern of sums of terms of the binary sequence is related to the binary numerals for those terms.

Decimal numerals	1	2	4	8	16	32	. . .
Binary numerals		1	10	100	1000	10000	100000 . . .

Translate each of the following equations from binary into decimal numerals.

9. $1 + 10 = 11$

10. $1 + 10 + 100 = 111$

11. $1 + 10 + 100 + 1000 = 1111$

12. $1 + 10 + 100 + 1000 + 10000 = 11111$

Translate each of the following equations from binary into decimal numerals.

13. $100 - 1 = 11$

14. $1000 - 1 = 111$

15. $10000 - 1 = 1111$

16. $100000 - 1 = 11111$

What would each of the following look like written as a binary numeral?

17. The sum of the first 64 terms of the binary sequence.

18. The 65th term of the binary sequence.

19. The 65th term of the binary sequence minus one.

Set III

After thinking that he had invented the binary numerals, the seventeenth-century German mathematician Gottfried Leibniz was astonished to find that an ancient Chinese book, the *I Ching*, contained a set of numbered figures, called hexagrams. Each hexagram consists of six lines, each of which is either solid or broken. The first eight, together with their names, are shown below.

Ch'ien	Kuai	Ta Yu	Ta Chuang	Hsiao Ch'u	Hsü	Ta Ch'u	Ta'i
0	1	2	3	4	5	6	7

Chapter 2: NUMBER SEQUENCES

1. The hexagrams are related to the binary sequence in a simple way. What is it?

2. How many such hexagrams are possible? Explain.

3. The hexagrams corresponding to the numbers 10, 20, 30, 40, and 50 have the names K'uei, Chia jên, I, Sung, and Lü, respectively. If you know how they are related to the binary sequence, you should be able to draw them.

Lesson 4

The Sequence of Squares

Among the many clay tablets made by the Babylonians about 4,000 years ago are some that contain number sequences. In addition to tablets containing arithmetic and geometric sequences, one tablet shows the first sixty terms of the sequence that begins

$$1 \quad 4 \quad 9 \quad 16 \quad 25 \quad 36 \quad 49 \quad \ldots$$

This sequence, formed by multiplying the consecutive counting numbers by themselves, is called the **sequence of squares.** It can also be written as

$$1^2 \quad 2^2 \quad 3^2 \quad 4^2 \quad 5^2 \quad 6^2 \quad 7^2 \quad \ldots$$

Much later (about 600 B.C.), the Greek mathematicians often represented numbers with dots arranged in geometric shapes. For example, the figures at the left show how they represented the sequence of square numbers. These figures reveal how the square numbers got their name. The square numbers are the numbers that can be represented by dots in square arrays. The number 25, for example, is a square number because it is the number of dots in a square array with a side of 5. In addition to calling 25 the *square of 5*, we call 5 a *square root* of 25. In symbols, $25 = 5^2$ and $5 = \sqrt{25}$.

The sequence of squares has an interesting relation to an arithmetic sequence—the sequence of odd numbers:

$$1 \quad 3 \quad 5 \quad 7 \quad 9 \quad 11 \quad \ldots$$

The sums of consecutive odd numbers starting with one are squares.

$$
\begin{aligned}
1 &= 1 = 1^2 \\
1 + 3 &= 4 = 2^2 \\
1 + 3 + 5 &= 9 = 3^2 \\
1 + 3 + 5 + 7 &= 16 = 4^2 \\
1 + 3 + 5 + 7 + 9 &= 25 = 5^2
\end{aligned}
$$

The basis for this relation is shown in the figures at the right.

Exercises

Set I

This figure represents a ball rolling down a ramp 25 meters long. It takes 5 seconds for the ball to roll from the top to the bottom and the positions of the ball after each second are shown.

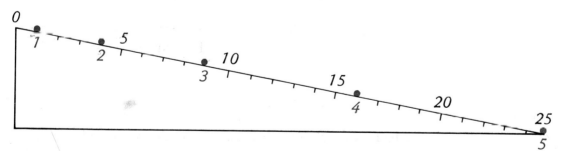

1. Write a number sequence representing the total distances traveled by the ball in the first second, the first two seconds, the first three seconds, the first four seconds, and the first five seconds.

2. What is the name of this sequence?

3. Write a number sequence representing the distances traveled by the ball in the first second, the second second, the third second, the fourth second, and the fifth second.

4. What is the name of this sequence?

5. The increase in the distance traveled in each successive second is called the *acceleration* of the ball. What is the acceleration of the ball?

The *perimeter* of a square is the number of linear units in its border. The figures below illustrate the perimeters of a series of squares having sides of lengths 1, 2, 3, 4, 5, and 6.

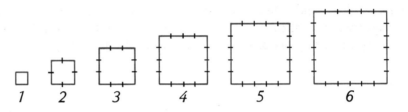

6. Copy and complete the following number sequence of the perimeters of these squares.

The *area* of a square is the number of square units inside. The figures below illustrate the areas of the same squares.

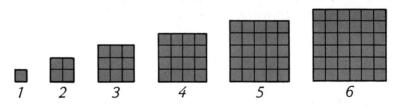

7. Copy and complete the following number sequence of the areas of these squares.

$$1 \quad 4 \quad \text{||||||} \quad \text{||||||} \quad \text{||||||} \quad \text{||||||}$$

Refer to the figures above and the sequences that you completed for exercises 6 and 7 to answer the following questions.

8. What happens to the *perimeter* of a square if the length of its side is doubled?

9. What happens to the *area* of a square if the length of its side is doubled?

If *x* represents the length of the side of a square, how would you express

10. its perimeter in terms of *x*?

11. its area in terms of *x*?

The numbers in the sequence

$$1 \quad 3 \quad 6 \quad 10 \quad 15 \quad \ldots$$

are called *triangular* because they are the numbers of dots in the triangular arrays shown here. The 10 pins in bowling are set up in a triangular array, as are the 15 numbered balls in pool.

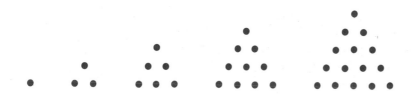

12. Copy the five terms (numbers) of the sequence listed above and continue the sequence to show the next five terms.

13. Now add each pair of consecutive terms of the sequence as shown below to make a new sequence.

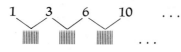

14. What do you notice about the resulting sequence?

15. How do the figures below illustrate your answer to exercise 13?

Set II

A table of the squares of the numbers from 1 through 40 is shown below.

No.	Square	No.	Square	No.	Square	No.	Square
1	1	11	121	21	441	31	961
2	4	12	144	22	484	32	1,024
3	9	13	169	23	529	33	1,089
4	16	14	196	24	576	34	1,156
5	25	15	225	25	625	35	1,225
6	36	16	256	26	676	36	1,296
7	49	17	289	27	729	37	1,369
8	64	18	324	28	784	38	1,444
9	81	19	361	29	841	39	1,521
10	100	20	400	30	900	40	1,600

Refer to this table to answer the following questions.

1. Can the square of an even number be odd?

2. Can the square of an odd number be even?

3. If the last digit of a number is 4, what is the last digit of its square?

4. For which digits is the last digit of a number the same as the last digit of its square?

5. Can a square number end in any digit? If not, in which digits can it not end?

6. Can a square number of two or more digits have all even digits? If so, give an example.

7. Do you think that a square number of two or more digits can have all odd digits? If so, give an example.

8. Make a table of the squares of the numbers from 41 through 50.

9. Do you notice a pattern in your table? If so, describe it.

The last digits of the squares of the numbers from 1 through 9 form a pattern that reads the same forward and backward:

$$1 \quad 4 \quad 9 \quad 6 \quad 5 \quad 6 \quad 9 \quad 4 \quad 1$$

Such a pattern is called a *palindrome*.

10. The last two digits of the squares of the numbers from 1 through a certain number also form a palindrome.* What is that number?

$$01 \quad 04 \quad 09 \quad 16 \quad 25 \quad 36 \quad 49 \quad \ldots$$

The **digital root** of a number is found by adding its digits, adding the digits of the resulting number, and so forth, until the result is a single digit. The digital roots of the squares of the numbers from 1 through 10 are shown in the table below.

Number	Square	Calculation	Digital root of square
1	1		1
2	4		4
3	9		9
4	16	$1 + 6 = 7$	7
5	25	$2 + 5 = 7$	7
6	36	$3 + 6 = 9$	9
7	49	$4 + 9 = 13; 1 + 3 = 4$	4
8	64	$6 + 4 = 10; 1 + 0 = 1$	1
9	81	$8 + 1 = 9$	9
10	100	$1 + 0 + 0 = 1$	1

11. Find the digital roots of the squares of the numbers from 11 through 20.

12. If a number is a square, what numbers can be its digital root?

The digital roots of the squares of the numbers from 1 through 8 form a palindrome:

$$1 \quad 4 \quad 9 \quad 7 \quad 7 \quad 9 \quad 4 \quad 1$$

13. The digital roots of the squares of the numbers from 1 through another number also form a palindrome. What is that number?

---

* We are considering the last two digits of 1, 4, 9 to be 01, 04, and 09, respectively.

Set III

Number sequences sometimes appear in science in mysterious ways. An interesting example is the relation of the elements in chemistry.

The elements can be listed in a table according to the arrangements of the electrons in their atoms so that elements with similar properties appear in columns. The first six rows of this table, called the periodic table, are shown below.

Subgroups of elements within the rows are named *s*, *p*, *d*, and *f*. Only the sixth row has elements in all four subgroups.

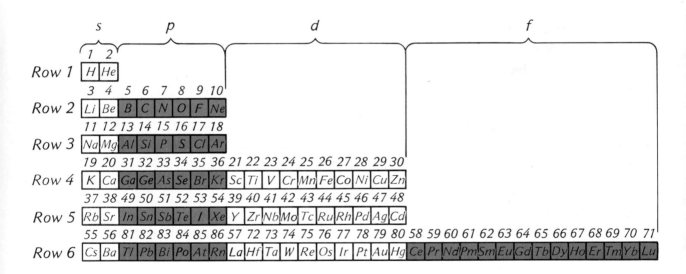

1. Copy and complete the following table showing the number of elements in each subgroup and in each row.

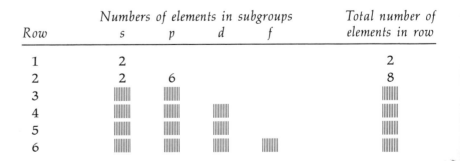

Row	Numbers of elements in subgroups				Total number of elements in row
	s	*p*	*d*	*f*	
1	2				2
2	2	6			8
3					
4					
5					
6					

All of the numbers in this table are even because the elements in each subgroup come in pairs.

2. Divide each number in half to make a new table as shown below.

Row	Number of pairs of elements in subgroups				Total number of pairs of elements in row
	s	p	d	f	
1	1				1
2	1	3			4
3	▥	▥			▥
4	▥	▥	▥		▥
5	▥	▥	▥		▥
6	▥	▥	▥	▥	▥

3. What do the numbers of *pairs* of elements in the subgroups have in common?

4. What do the numbers of *pairs* of elements in each row have in common?

At present, 105 elements are known, of which 7 have been created in the laboratory since 1950. If the number of new elements to be created were to increase indefinitely, how many elements would you predict for

5. the seventh row of the periodic table?

6. the eighth row of the periodic table?

Lesson **5**

The Sequence of Cubes

In addition to picturing the numbers called squares with square arrays of dots, the ancient Greeks pictured other numbers with cubic arrays of dots. The number 8, for example, was pictured as a cube having 2 dots on each edge. It can also be represented as $2 \cdot 2 \cdot 2$ or 2^3, which is read as "2 cubed." Cubes having 3 and 4 dots on each edge contain $3 \cdot 3 \cdot 3 = 3^3 = 27$ dots and $4 \cdot 4 \cdot 4 = 4^3 = 64$ dots, respectively. For this reason, 27 and 64 are called cube numbers.

▶The **sequence of cubes** is

$$1^3 \quad 2^3 \quad 3^3 \quad 4^3 \quad 5^3 \quad \ldots$$

or, equivalently,

$$1 \quad 8 \quad 27 \quad 64 \quad 125 \quad \ldots$$

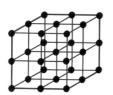

In the same way that square numbers are related to the *area* of a square (that is, the amount of *surface* that it occupies), cube numbers are related to the *volume* of a cube (the amount of *space* that it occupies). A square measuring 4 centimeters along each side has an area of $4^2 = 16$ square centimeters; a cube measuring 4 centimeters along each edge has a volume of $4^3 = 64$ cubic centimeters.

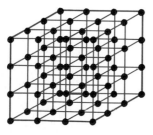

By permission of Johnny Hart and Field Enterprises, Inc.

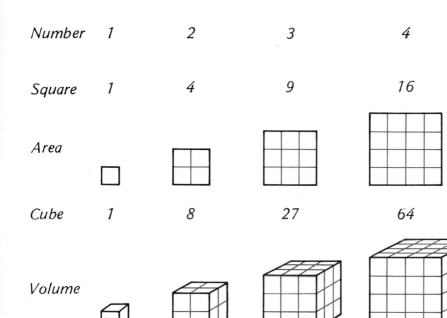

Number	1	2	3	4
Square	1	4	9	16
Area				
Cube	1	8	27	64
Volume				

Because the sequence of squares can be written as

$$1^2 \quad 2^2 \quad 3^2 \quad 4^2 \quad 5^2 \quad \ldots$$

and the sequence of cubes as

$$1^3 \quad 2^3 \quad 3^3 \quad 4^3 \quad 5^3 \quad \ldots,$$

they are also referred to as the sequences of second and third powers. We have seen that second powers can be pictured in two-dimensions using squares and third powers in three-dimensions

Lesson 5: The Sequence of Cubes

91

using cubes. Although we cannot picture sequences of powers higher than the third with arrays of dots in more than three dimensions, such sequences are both interesting and useful. The **sequence of fourth powers,**

$$1^4 \quad 2^4 \quad 3^4 \quad 4^4 \quad 5^4 \quad \ldots$$

or

$$1 \quad 16 \quad 81 \quad 256 \quad 625 \quad \ldots,$$

can be applied, for example, to relating the luminosity of a star to its temperature.

Sequences of even higher powers also have been found to have useful applications. The higher the power, the greater the rate at which the sequence grows:

Sequence	First five terms				
Fifth powers	1	32	243	1,024	3,125
Sixth powers	1	64	729	4,096	15,625
Seventh powers	1	128	2,187	16,384	78,125

Notice that we calculate powers by multiplying a number by itself an appropriate number of times. For example, the fifth power of 2 is $2^5 = 2 \times 2 \times 2 \times 2 \times 2 = 32$ and the sixth power of 3 is $3^6 = 3 \times 3 \times 3 \times 3 \times 3 \times 3 = 729$.

Exercises

Set I

The figures below illustrate cubes having edges of lengths 1, 2, 3, 4, and 5.

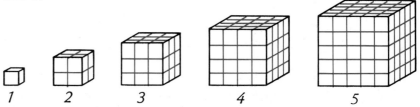

The surface area of a cube is the number of square units in its six faces. For example, the surface area of the second cube is 24 because each of its six faces contains four squares and $6 \times 4 = 24$.

1. Copy and complete the following number sequence of the surface areas of the cubes shown.

6   24   ▓▓▓   ▓▓▓   ▓▓▓

The volume of a cube is the number of cubic units that it contains.

2. Copy and complete the following number sequence of the volumes of the cubes shown.

1   8   ▓▓▓   ▓▓▓   ▓▓▓

Refer to the figures on page 92 and your completed sequences for exercises 1 and 2 to answer the following questions.

3. What happens to the *surface area* of a cube if the length of its edge is doubled?

4. What happens to the *volume* of a cube if the length of its edge is doubled?

If x represents the length of the edge of a cube, how would you express

5. its surface area in terms of x?

6. its volume in terms of x?

A table of the cubes of the numbers from 1 through 20 is shown below.

Number	Cube	Number	Cube
1	1	11	1,331
2	8	12	1,728
3	27	13	2,197
4	64	14	2,744
5	125	15	3,375
6	216	16	4,096
7	343	17	4,913
8	512	18	5,832
9	729	19	6,859
10	1,000	20	8,000

Refer to this table to answer the following questions.

7. If the last digit of a number is 2, what is the last digit of its cube?

8. If the last digit of a number is 8, what is the last digit of its cube?

9. If the last digit of a number is 3, what is the last digit of its cube?

10. If the last digit of a number is 7, what is the last digit of its cube?

11. For which digits is the last digit of a number the same as the last digit of its cube?

12. Can a cube number end in any digit? If not, in which digits can it not end?

The digital roots of the cubes of the numbers from 1 through 5 are shown in the table below.

Number	Cube	Calculation	Digital root of cube
1	1		1
2	8		8
3	27	$2 + 7 = 9$	9
4	64	$6 + 4 = 10; 1 + 0 = 1$	1
5	125	$1 + 2 + 5 = 8$	8

13. Find the digital roots of the cubes of the numbers from 6 through 12.

14. What pattern do you notice?

A table of the fourth and fifth powers of the numbers from 1 through 20 is shown below.

Number	4th power	5th power	Number	4th power	5th power
1	1	1	11	14,641	161,051
2	16	32	12	20,736	248,832
3	81	243	13	28,561	371,293
4	256	1,024	14	38,416	537,824
5	625	3,125	15	50,625	759,375
6	1,296	7,776	16	65,536	1,048,576
7	2,401	16,807	17	83,521	1,419,857
8	4,096	32,768	18	104,976	1,889,568
9	6,561	59,049	19	130,321	2,476,099
10	10,000	100,000	20	160,000	3,200,000

Chapter 2: NUMBER SEQUENCES

15. What do you notice about the last digits of fourth powers?

16. What do you notice about the last digits of fifth powers?

17. Each fourth power is also a square. Why?

18. Each power of 1 is also 1. Why?

Set II

The following exercises refer to this pattern.

$$
\begin{aligned}
1 &= 1 \\
3 + 5 &= |||||| \\
7 + 9 + 11 &= |||||| \\
13 + 15 + 17 + 19 &= ||||||
\end{aligned}
$$

1. Copy it, filling in the missing numbers.

2. What sequence do the numbers on the left sides of the equations form?

3. What sequence do the numbers on the right sides of the equations form?

4. Write the next two lines of the pattern.

5. Are they also true?

The following exercises refer to this pattern.

$$
\begin{aligned}
1 &= 1 \\
1 + 8 &= |||||| \\
1 + 8 + 27 &= |||||| \\
1 + 8 + 27 + 64 &= ||||||
\end{aligned}
$$

6. Copy it, filling in the missing numbers.

7. To what sequence do the numbers on the left sides of the equations belong?

8. To what sequence do the numbers on the right sides of the equations belong?

9. Rewrite the pattern in exercise 6, using exponents. (Write the first line as $1^3 = 1^2$.)

10. What is a shortcut for finding the last number on each line?

Lesson 5: The Sequence of Cubes

95

11. Write the next two lines of the pattern.

12. Are they also true?

The following exercises refer to this pattern.

$$3^2 + 4^2 \qquad = \text{|||||||}^2$$
$$3^3 + 4^3 + 5^3 = \text{|||||}^3$$

13. Copy it, filling in the missing numbers.

14. Write what you think is the next line of the pattern.

15. Is it also true?

Set III

Several science-fiction films and television programs have been based on the theme of human beings changing size.* If people could become larger or smaller, their physical characteristics would not change at the same rate. Their weight, for example, would vary with the *cube* of their height, whereas the strength of their bones would vary with its *square*. This means that someone who became 2 times his normal height would have bones $2^2 = 4$ times as strong and be $2^3 = 8$ times as heavy.

1. What would happen to the strength and weight of a man who became 12 times his normal height?

2. A person's thighbones can support as much as 10 times his weight. Why would the legs of a man who became 12 times his normal height break when he stood up?

3. What would happen to the strength and weight of a woman who became one-tenth her normal height?

4. To become stronger in proportion to his or her weight, which should a person do: grow or shrink? Explain.

---

* Among them are *The Incredible Shrinking Woman*, *The Amazing Colossal Man*, "Land of the Giants," and *Attack of the Puppet People*.

Chapter 2: NUMBER SEQUENCES

Leonardo of Pisa, who was called Fibonacci
Courtesy of Columbia University Libraries

Lesson 6

The Fibonacci Sequence

The great mathematician of the Middle Ages was Leonardo of Pisa, called Fibonacci. The construction of the famous Leaning Tower of that city was begun during his lifetime but was not completed for nearly two centuries. Fibonacci studied in North Africa, where he learned mathematical works available only in Arabic. In 1202, he wrote a book on arithmetic and algebra titled the *Liber Abaci*. In this book, he proposed a problem about rabbits* whose solution was based on the sequence

$$1 \quad 1 \quad 2 \quad 3 \quad 5 \quad 8 \quad 13 \quad \ldots$$

▶ This sequence is called the **Fibonacci sequence.** Its first two terms are 1 and each successive term is the sum of the preceding pair of terms.

$$
\begin{aligned}
1 \quad 1& \\
1 + 1 &= 2 \\
1 + 2 &= 3 \\
2 + 3 &= 5 \\
3 + 5 &= 8 \\
5 + 8 &= 13 \ldots
\end{aligned}
$$

The Fibonacci sequence appears in an amazingly wide variety of creations, both natural and manmade. The family tree of a male

---

* This problem is in the Set I exercises of this lesson.

bee is an interesting example. A male bee has only one parent, his mother, whereas a female bee has both father and mother. The family tree of a male bee, which has a strange pattern as a result, looks like this (each male is represented by the symbol ♂ and each female by the symbol ♀):

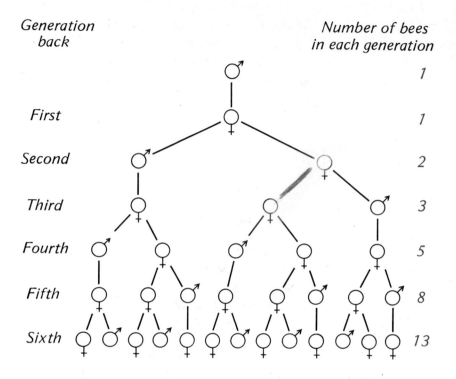

The numbers of bees in successive generations of ancestors form a Fibonacci sequence.

The thirteen ancestors on the bottom row of the tree can be related to the keys of a piano. Each female bee corresponds to a white key and each male bee to a black key.

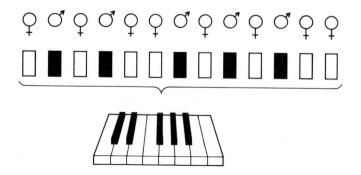

Chapter 2: NUMBER SEQUENCES

The 13 keys shown are one octave of the *chromatic* scale. The preceding number of ancestors, 8, corresponds to the number of white keys in an octave, the notes of a *major* scale. The number before that, 5, is the number of black keys in an octave, the notes of the old *pentatonic* scale.

A fascinating aspect of mathematics is that ideas from one area of study frequently apply to other, apparently unrelated areas. In addition to appearing in the reproduction of bees and musical scales, the Fibonacci sequence has been found in many other subjects, including pine cones, data sorting, Roman poetry, and sunflowers.

Exercises

Set I

Fibonacci's problem about the rabbits went like this.

A pair of rabbits one month old are too young to produce more rabbits, but suppose that in their second month and every month thereafter they produce a new pair. If each new pair of rabbits does the same, and none of the rabbits die, how many pairs of rabbits will there be at the beginning of each month?

The figure below illustrates what happens in the first six months. The lines in color indicate the births of new pairs of rabbits.

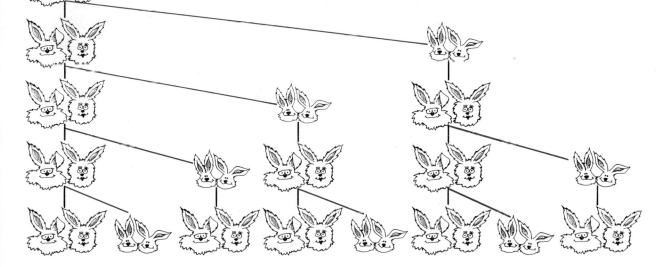

1. Refer to the figure to copy and complete the following table.

Month	Pairs of baby rabbits	Pairs of adult rabbits	Pairs of rabbits altogether
1st	1	0	1
2nd			
3rd			
4th			
5th			
6th			

2. How is the number of pairs of *rabbits altogether* in any given month related to the numbers of pairs of *baby rabbits* and *adult rabbits* in that month?

3. How is the number of pairs of *adult rabbits* in a given month related to the number of pairs of *rabbits altogether* in the preceding month?

4. How is the number of pairs of *baby rabbits* in a given month related to the number of pairs of *adult rabbits* in the preceding month?

5. How is the number of pairs of *baby rabbits* in a given month related to the number of pairs of *rabbits altogether* two months earlier?

Compare the rabbits figure on page 99 with the family tree of a male bee on page 98.

6. What does each pair of baby rabbits correspond to on the bee's family tree?

7. What does each pair of adult rabbits correspond to?

Set II

The first ten terms of the Fibonacci sequence are listed below.

1st	2nd	3rd	4th	5th	6th	7th	8th	9th	10th
1	1	2	3	5	8	13	21	34	55

1. Copy this list and continue it to the 20th term.

Every third term of the Fibonacci sequence is evenly divisible by two.

$$1 \quad 1 \quad 2 \quad 3 \quad 5 \quad 8 \quad 13 \quad 21 \quad 34 \quad 55 \quad \ldots$$

Use the list of terms that you made for exercise 1 to figure out which terms are evenly divisible by

2. three.

3. five.

4. eight.

5. thirteen.

6. Which terms of the sequence do you suppose are evenly divisible by fifty-five?

The following exercises refer to this pattern.

Pattern A

$1 + 1$	$= 2$
$1 + 1 + 2$	$= 4$
$1 + 1 + 2 + 3$	$=$ ▥
$1 + 1 + 2 + 3 + 5$	$=$ ▥
$1 + 1 + 2 + 3 + 5 + 8$	$=$ ▥
$1 + 1 + 2 + 3 + 5 + 8 + 13$	$=$ ▥
$1 + 1 + 2 + 3 + 5 + 8 + 13 + 21$	$=$ ▥

7. Copy it, filling in the missing numbers.

8. Compare the sums with the list that you made for exercise 1. What do you notice?

9. Use what you noticed to guess the sum of the first 12 terms of the sequence without adding them.

Pattern B

$$1^2 + 1^2 = 2$$
$$1^2 + 2^2 = 5$$
$$2^2 + 3^2 = ▥$$
$$3^2 + 5^2 = ▥$$
$$5^2 + 8^2 = ▥$$
$$8^2 + 13^2 = ▥$$
$$13^2 + 21^2 = ▥$$

The following exercises refer to the pattern at the right.

10. Copy it, filling in the missing numbers.

11. What do you notice?

12. Write the next line of the pattern.

The following exercises refer to this pattern.

Pattern C

$$1^2 + 1^2 \qquad\qquad\qquad = \ 2 = 1 \cdot 2$$
$$1^2 + 1^2 + 2^2 \qquad\qquad = \ 6 = 2 \cdot 3$$
$$1^2 + 1^2 + 2^2 + 3^2 \qquad = 15 = 3 \cdot 5$$
$$1^2 + 1^2 + 2^2 + 3^2 + 5^2 \ = 40 = 5 \cdot 8$$

13. Write the next line of the pattern.

14. Use the pattern to guess the sum of the squares of the first 10 terms of the Fibonacci sequence without adding them.

The following exercises refer to this pattern.

Pattern D

$$1^3 + 2^3 - 1^3 = \ 8$$
$$2^3 + 3^3 - 1^3 = 34$$
$$3^3 + 5^3 - 2^3 = \text{▥}$$
$$5^3 + 8^3 - 3^3 = \text{▥}$$

15. Copy it, filling in the missing numbers.

16. Write the next line of the pattern.

17. What do you notice?

18. Which one of the preceding patterns—A, B, C, or D—do these figures illustrate?

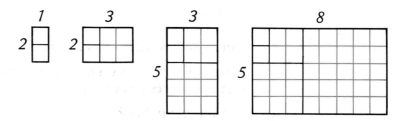

19. Draw the next figure in the pattern.

Chapter 2: NUMBER SEQUENCES

Set III

The arrangement of the leaves on the stems of many trees is related to the Fibonacci sequence in an interesting way. Look, for example, at the drawing of the stem of an elm tree shown here. We find that the *second* leaf up the stem from the bottom leaf is directly above it. If a thread were wound around the stem following the leaves, it would encircle the stem *once* in going from a lower leaf to the first one directly above it.

Look at the drawings below.

Elm

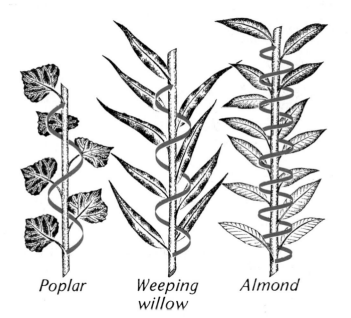

Poplar   *Weeping willow*   *Almond*

1. Which leaf is directly above the bottom one on each stem?

2. How many times does a thread encircle the stem in going from the bottom leaf to the first one directly above it?

3. What do you notice about these numbers?

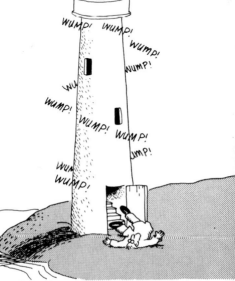

Drawing by Gary Larson; Chronicle Features, 1981

These figures, each showing a square transformed into the profile of a woman and then back into a square, were drawn with the help of a computer. The first was programmed according to an arithmetic sequence and the second according to a geometric sequence.

Courtesy of the Computer Technique Group, Tokyo

Chapter 2 / Summary and Review

In this chapter we have studied the following number sequences:

Arithmetic sequences (*Lesson 1*)   Each successive term is found by adding the same number. An example of an arithmetic sequence is

$$1 \quad 4 \quad 7 \quad 10 \quad 13 \quad \ldots$$

Geometric sequences (*Lessons 2 and 3*)   Each successive term is found by multiplying by the same number. An example of a geometric sequence is

$$2 \quad 6 \quad 18 \quad 54 \quad 162 \quad \ldots$$

The *binary sequence* is a special geometric sequence:

$$1 \quad 2 \quad 4 \quad 8 \quad 16 \quad \ldots$$

Power sequences (*Lessons 4 and 5*)   The terms are found by raising the consecutive counting numbers to the same power. Examples of two power sequences are the *sequence of squares,*

$$1 \quad 4 \quad 9 \quad 16 \quad 25 \quad \ldots,$$

and the *sequence of cubes,*

$$1 \quad 8 \quad 27 \quad 64 \quad 125 \quad \ldots.$$

Chapter 2: NUMBER SEQUENCES

The Fibonacci sequence (*Lesson 6*)   The first two terms of this sequence are 1 and each successive term is the sum of the preceding pair of terms:

$$1 \quad 1 \quad 2 \quad 3 \quad 5 \quad 8 \quad \ldots$$

Exercises

Set I

Find the missing term in each of the following number sequences.

1. 2   6   10 ||||||| 18
2. 9 16 25 ||||||| 49
3. 4 12 36 ||||||| 324
4. 1   8 27 ||||||| 125
5. 8 13 21 ||||||| 55

Whether the following sequence is arithmetic or geometric depends on what numbers are chosen for the missing terms.

$$10 \quad ||||||| \quad 40 \quad ||||||| \quad |||||||$$

6. What should the missing terms be so that it is arithmetic?

7. What should they be in order for it to be geometric?

The following item, used in intelligence testing, is answered incorrectly by more than 20% of the adult population:

John is twice as old as his sister Mary, who is now 5 years of age. How old will John be when Mary is 30 years of age?

Illustrate the correct solution by doing the following.

8. Write a number sequence of Mary's ages at five-year intervals, beginning with her present age and ending with 30.

9. Write a number sequence of John's ages over the same period of time.

10. How old is John when Mary is 30 years of age?

The three figures at the left illustrate part of a table learned in elementary school.

11. What is the table?

12. What kind of number sequences do the numbers in each circled row form?

13. What number sequence do the numbers in the circled diagonal form?

14. What number sequence do the *sums* of the numbers in each circled "L" form?

The first ten terms of the Fibonacci sequence are

$$1 \quad 1 \quad 2 \quad 3 \quad 5 \quad 8 \quad 13 \quad 21 \quad 34 \quad 55$$

15. Copy and complete the following equations.

$$5^2 = \text{▦} \qquad 3 \cdot 8 = \text{▦}$$
$$8^2 = \text{▦} \qquad 5 \cdot 13 = \text{▦}$$
$$13^2 = \text{▦} \qquad 8 \cdot 21 = \text{▦}$$
$$21^2 = \text{▦} \qquad 13 \cdot 34 = \text{▦}$$

16. How does the square of any term of the Fibonacci sequence seem to compare with the product of the term before it and the term after it?

Set II

The three chemical elements lithium, sodium, and potassium are very much alike. They are soft, light metals that will burn your fingers if you touch them. Their atomic weights are 7, 23, and 39.

1. What kind of number sequence is this?

The Rhind Papyrus, written in Egypt in the seventeenth century B.C., contains this list:

Household	7
Cats	49
Mice	343
Barley	2,301
Hekats	16,807

Chapter 2: NUMBER SEQUENCES

2. What kind of number sequence does this seem to be?

3. There is a mistake in the sequence as it appears in the papyrus. What is it?

Among the terms sometimes used by the book industry to indicate the size of a book's pages are *folio*, *quarto*, and *octavo*. These words refer to the number of pages that can be obtained from large printer's sheets by folding them as shown by the brown lines in the figures at the right. Smaller pages obtained from the large sheets are referred to as *16 mo*, *32 mo*, and *64 mo*.

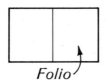
Folio

Quarto

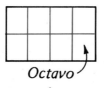
Octavo

4. What kind of number sequence do these terms suggest?

5. Why does the folding of the sheets result in this particular sequence?

The following exercises refer to this pattern.

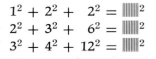

$$1^2 + 2^2 + 2^2 = \text{▥}^2$$
$$2^2 + 3^2 + 6^2 = \text{▥}^2$$
$$3^2 + 4^2 + 12^2 = \text{▥}^2$$

6. Copy it, writing in the missing numbers.

7. Write what you think the next line of the pattern is.

8. Show whether it is true.

9. Write the line of the pattern that starts with 10^2.

10. Show whether it is true.

The brilliant Indian mathematician Srinivasa Ramanujan is pictured on this postage stamp. He specialized in the study of numbers and knew their characteristics in the same way that a baseball fan might know a vast number of statistics about the game. One time a friend went to visit him in a taxi having the number 1,729. When the friend mentioned this number, Ramanujan immediately replied: "1,729 is a very interesting number; it is the smallest number expressible as the sum of two cubes in two *different* ways."

11. Complete the following equations based on this fact. (Refer to the table of cubes on page 93.)

$$1{,}729 = \text{▥}^3 + \text{▥}^3$$
$$1{,}729 = \text{▥}^3 + \text{▥}^3$$

12. The number 50 is the smallest number expressible as the sum of two *squares* in two different ways. What are they?

As the earth's population increases, so does the need for food. In 1798, in his *Essay on the Principle of Population*, the English economist Thomas Malthus used number sequences to represent the rates at which the number of people and the amount of available food increase. He assumed that the population, when unchecked, doubles every 25 years, and said:

> *Supposing the present population equal to a thousand millions, the human species would increase as the numbers 1, 2, 4, 8, 16, 32, 64, 128, 256, and subsistence as 1, 2, 3, 4, 5, 6, 7, 8, 9. In two centuries the population would be to the means of subsistence as 256 to 9.*

13. What type of number sequence did Malthus use to represent the increase in population?

14. What type of number sequence did he use to represent the increase in food?

Malthus went on to predict the increases both in population and in available food for the third century ahead.

15. Write four more numbers in each sequence to find what they were. (Four more numbers are needed in each sequence because the numbers represent the situation every 25 years.)

By permission of Johnny Hart and Field Enterprises, Inc.

Set III

From the use to which the letters were put, it looks as if the "pyramid club" in this cartoon was appropriately named. Notice from the directions in Wiley's letter that each person who receives it is supposed to send 3 copies to him, so that he will first get 3 letters, then 3 more letters, then 3 more letters, and so on.

1. Write the first five terms of the number sequence showing the successive *total* numbers of letters that Wiley has as they start coming in.

2. What kind of sequence is it?

Chapter 2: NUMBER SEQUENCES

The figures below are side and top views of a pyramid similar to the one Wiley built. It has no interior spaces and all of the blocks in it are the same size and shape.

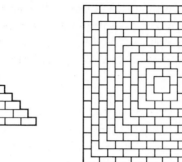

3. Write the first five terms of the number sequence showing the numbers of blocks in the layers of the pyramid, starting from the top.
4. How many blocks are in the pyramid?

Photograph by Christine Simon

Chapter 2 / Problems for Further Exploration

Lesson 1

1. The seventeen-year locust is named for the length of its life-span. The locust is seen only in the last month of its life, the rest of its existence being spent underground. In Oklahoma, Kansas, and Missouri, it last appeared in 1981.

   a) Write a number sequence listing the dates of all of its appearances in the past and present centuries.

   Suppose that the locust had a predator with a life cycle of six years (during most of which it is in larval form), and that the predator appeared as an adult in 1801.

   b) Write a number sequence listing the dates of all of the predator's appearances in the past and present centuries.

   c) In what years did both the locusts and their predators appear?

   d) Do you see a relation between the interval between these years and the life-spans of the locusts and their predators? If so, what is it?

   Suppose that the locust had a life-span of *sixteen* years rather than seventeen and that it appeared in 1981.

   e) Write a number sequence listing the dates of all of its appearances in the past and present centuries.

   f) Which of these years coincide with the years in which the predator appeared?

Suppose that the locust had a life-span of *eighteen* years and that it appeared in 1981.

g) Write a number sequence listing the dates of all of its appearances in the past and present centuries.

h) Which of these years coincide with the years in which the predator appeared?

i) Do you see any relations between the years in parts *f* and *h* and the life-spans of the locusts and their predators? If so, what are they?

2. What is the sum of all of the digits needed to write down all of the whole numbers from 0 to 1,000,000?* Explain how you obtained your answer.

Hint: Look at the pattern that results from pairing the numbers like this:

> 0 and 999,999
> 1 and 999,998
> 2 and 999,997. . .

Lesson 2

1. The rate at which the atoms of a radioactive element break apart into atoms of other elements depends on its *half-life*. Its half-life is the amount of time it takes half of the atoms of the element to disintegrate.

   The half-life of uranium-232 is 70 years. Suppose that a sample of this kind of uranium contained 1,000,000,000 atoms in the year 1000 A.D.

   a) Write a number sequence beginning with this number of atoms and showing the number of atoms left in 1070 A.D., 1140 A.D., 1210 A.D., 1280 A.D., and 1350 A.D.

   b) Approximately how many atoms would be left at the present time?

   c) When would you expect the last atom to disintegrate?

2. The sums of numbers in geometric sequences have interesting patterns. Consider, for example, the sequence

   > 1   3   9   27   81   243   729   . . .

   a) What is the common ratio of the sequence?

---

* This puzzle is similar to one by Pierre Berloquin in his book titled *100 Numerical Games* (Scribner's, 1976).

If we multiply the sums of the terms of the sequence by the common ratio minus 1, we get

Sums	$\times\ (3 - 1) = 2$
$1 + 3 = 4$	8
$1 + 3 + 9 = 13$	26
$1 + 3 + 9 + 27 = 40$	80
$1 + 3 + 9 + 27 + 81 = 121$	242

Compare the products in the table above with the terms of the sequence.

b) What pattern do you see?

c) Find the sum

$$1 + 3 + 9 + 27 + 81 + 243$$

by adding the six numbers.

d) Show how to find the same sum by using the pattern that you found for part b of this exercise.

Now consider the sequence

$$1 \quad 4 \quad 16 \quad 64 \quad 256 \quad 1{,}024 \quad 4{,}096 \quad \dots$$

e) Does a similar pattern exist for it? If you can find one, show what it is.

If the first term of a geometric sequence is 1 and the common ratio is r, the first few terms are

$$1 \quad r \quad r^2 \quad r^3 \quad r^4 \quad r^5 \quad r^6 \quad \dots$$

From your preceding observations, write expressions for the following sums:

f) $1 + r$

g) $1 + r + r^2$

h) $1 + r + r^2 + r^3$

i) $\underbrace{1 + r + r^2 + r^3 + \dots + r^{n-1}}_{n \text{ terms}}$

Lesson 3

1. Experiment: *A Card Sorting System*

One application of the binary sequence is in data processing systems. From the following experiment, you will see how it can be used to sort cards automatically.

Chapter 2: NUMBER SEQUENCES

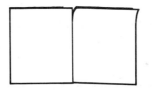

Cut eight file cards in half as shown in the figure above. Take one of the sixteen cards produced and punch a row of four holes below a longer edge as shown in the figure below. The holes should be spaced about 1.5 centimeters apart.

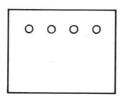

Using this card as a stencil, punch the other fifteen cards to match it. Three cards can easily be punched at a time.

Number the cards from 0 through 15. The four holes represent the first four numbers of the binary sequence in reverse order:

$$8 \quad 4 \quad 2 \quad 1$$

Using the list of binary numerals that you made in the Set I exercise as a guide, write the number 1 above the appropriate holes of each numbered card to represent the matching binary numeral. The first four cards are illustrated below.

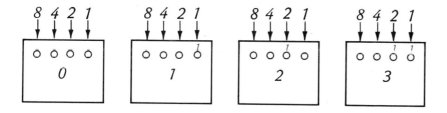

Along the top edge of each card, cut out the space above each hole marked with a 1.

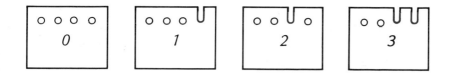

The cards are now ready to use. Shuffle them, being careful not to turn any of them over or upside down.

Make a hook out of a paper clip, something like the one illustrated at the left.

Holding the cards loosely together with one hand, put the hook through the 1-holes and lift up, shaking the hook rapidly so that the cards with notches above the 1-holes remain behind. Slide the cards that came up off the hook and place them on top of the other cards.

Next, put the hook through the 2-holes and carry out the same procedure. Be sure to place the cards that come up on *top* of the cards that remain behind. Repeat with the remaining holes (going from right to left.) When you have done this, the cards should be in correct order from 0 through 15.

After sorting the cards by this method, you might print the following words on them, reshuffle them, and sort them again.

0	This	4	cards	8	sorted	12	of
1	pack	5	has	9	automatically	13	the
2	of	6	just	10	by	14	binary
3	sixteen	7	been	11	means	15	sequence.

a) How many cards could be sorted if each one had five holes rather than four? Explain your reasoning.

b) If each card had ten holes, more than one thousand cards could be sorted in just ten steps. Explain why.

2. In the game of Twenty Questions, one player thinks of an object whose identity the other players try to determine by asking no more than twenty questions, each of which must be phrased so that it can be answered either "yes" or "no." The player who guesses the object correctly is the winner.

It is possible to distinguish between eight objects by asking three questions. If, for example, the object to be guessed is a number from 1 to 8, the three questions and the possible answers might be:

1. Is the number 4 or less?
   Yes, so it is 1, 2, 3, or 4.
2. Is the number 1 or 2?
   No, so it is 3 or 4.
3. Is the number 3?
   Yes.

Notice that this procedure closes in on the answer by cutting the number of possible objects in half with each question.

a) It is possible to distinguish between 32 objects by asking five questions. Show how this might be done by supposing that the object to be guessed is a number from 1 to 32 and listing the five questions that you might ask together with their possible answers.

b) How many different objects can be distinguished by asking twenty questions? Explain your reasoning.

Lesson 4

1. The squares of certain numbers form interesting patterns.
   a) Copy and complete the following equations.

$$33^2 = \text{▌▌▌▌▌▌}$$
$$333^2 = \text{▌▌▌▌▌▌}$$
$$3{,}333^2 = \text{▌▌▌▌▌▌}$$

   b) On the basis of your answers, what do you think $33{,}333^2$ is?
   c) Copy and complete the following equations.

$$66^2 = \text{▌▌▌▌▌▌}$$
$$666^2 = \text{▌▌▌▌▌▌}$$
$$6{,}666^2 = \text{▌▌▌▌▌▌}$$

   Suppose that the ten-digit number, each of whose digits is 6, is squared.
   d) How many digits do you think the resulting number would have?
   e) What would they be?
   f) Find another series of equations that have a similar pattern.

2. The driver of a car glanced at the odometer and saw that it read 15,951 miles.* He said to himself: "That's interesting. The mileage is a palindrome: it reads the same backward as forward. It will be a long time before that happens again."

   Just two hours later, however, the mileage shown on the odometer was a new palindrome. How fast was the car going in those two hours?

Lesson 5

1. A well-known mystery in mathematics is Fermat's Last Theorem. The theorem concerns equations of the form $a^n + b^n = c^n$, in which a, b, c, and n are counting numbers. Some examples of

---

*This puzzle is from a popular Russian book of mathematical puzzles written by Boris A. Kordemsky. Its English edition is titled *The Moscow Puzzles* (Scribner's, 1972).

such equations are $3^2 + 4^2 = 5^2$, $5^2 + 12^2 = 13^2$, and $20^2 + 21^2 = 29^2$.

The seventeenth-century French mathematician Pierre de Fermat made a note in the margin of one of his books that equations of this form can be found for squares only. He wrote that he had discovered a proof of this but that the margin of the book was too narrow to contain it. The mystery is that no one to this day has been able to figure out what Fermat's proof was.

The March 7, 1938, issue of *Time* magazine contained an article reporting that a man had discovered an equation that supposedly disproved the theorem.* The equation was

$$1{,}324^n + 731^n = 1{,}961^n,$$

in which *n* was a counting number larger than 2. A reporter for the *New York Times* proved that the man was mistaken. How did he do it? (*Hint:* What digits can each power in the equation end in?)

2. Look at the following patterns for obtaining squares and cubes.

	1	2	3	4	5	6	7	8	9	10	11	12	...
Form sums:	1		4		9		16		25		36		...

	1	2	3	4	5	6	7	8	9	10	11	12	...
Form sums:	1	3		7	12		19	27		37	48		...
Form sums:	1			8			27			64			...

Find similar patterns for getting fourth and fifth powers. Check your patterns to see whether they work.

Lesson 6

1. This figure shows part of a honeycomb with a bee in the cell numbered 1.† Suppose that the bee moves to the other cells,

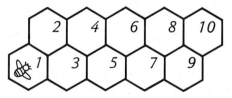

---

*Reported by Martin Gardner in his "Mathematical Games" column, *Scientific American,* July 1971.

†From Martin Gardner's "Mathematical Games" column, *Scientific American,* March 1969.

Chapter 2: NUMBER SEQUENCES

always traveling to either neighboring cell to the right. The possible paths of the bee to cells 2, 3, 4, and 5 are shown in the figures below.

Cell 2

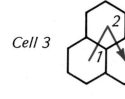

Cell 3

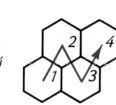 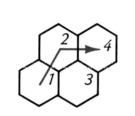

Cell 4

 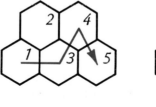 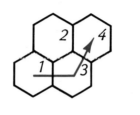

Cell 5

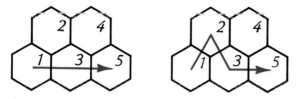

a) How does the number of possible paths to a given cell seem to be related to the number of the cell?
b) How many paths to cell 6 do you think are possible?
c) Trace the figure at the right as many times as necessary and draw all of the possible paths.
d) How many paths to cell 10 do you think are possible?

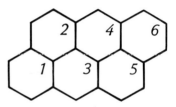

2. The following number trick is based on properties of the Fibonacci sequence. Ask someone to choose two numbers at random and write one number below the other. Have the person add them to get a third number. Add the second and third numbers to get a fourth number. Add the third and fourth numbers to get a fifth number. Ask the person to continue in this fashion as far as he likes.

   Now have him draw a line between any two numbers. You can quickly tell the sum of all of the numbers above the line by simply subtracting the second number in the list from the second number below the line.

   a) Try this trick out.

   b) Represent the two originally chosen numbers by a and b and continue the table below. Then prove that the trick will work for any two numbers and a list of as many as twelve numbers.

Example:

12
7
19
26
45
71
⋮

Numbers	Sums
a	a
b	$a + b$
$a + b$	$2a + 2b$
$a + 2b$	$3a + 4b$

Chapter 3
FUNCTIONS AND THEIR GRAPHS

*A watercolor by Toko Morimoto
featuring two crickets*

Lesson 1

The Idea of a Function

A familiar sound in the country on a warm summer evening is the chirping of crickets. The rate at which crickets chirp depends on the temperature: the warmer it is, the more they chirp in any given time. The table below shows how the rate and temperature are related.

Temperature in degrees Fahrenheit	50	60	70	80	. . .
Number of chirps in fifteen seconds	10	20	30	40	. . .

To each temperature in this table, there corresponds a rate: 10 corresponds to 50, 20 corresponds to 60, and so forth. A mathematician would say that the rate at which crickets chirp is a *function* of the temperature.

120

▶A **function** is a pairing of two sets of numbers so that to each number in the first set there corresponds exactly one number in the second set.

One way to represent a function is with a *table*, as was done on the facing page. Another way is to write a *formula*. From the table, we see that each number on the second line is 40 less than the corresponding number on the first line. If we let F represent the temperature in degrees Fahrenheit and n represent the number of chirps in 15 seconds, we can write

$$n = F - 40.$$

The two letters in this formula are *variables*: as F varies in value, so does n. For example, if $F = 55$, then $n = 55 - 40 = 15$ and, if $F = 90$, then $n = 90 - 40 = 50$.

Because the rate at which crickets chirp depends on the temperature, it is possible to find out how warm it is by using a cricket as a thermometer. By interchanging the two lines of the table on the facing page, we get a table for another function:

n	10	20	30	40	. . .
F	50	60	70	80	. . .

Its formula,

$$F = n + 40,$$

shows how to find the temperature in degrees Fahrenheit. We simply count the number of chirps made in 15 seconds and add 40.

If we know the formula of a function, we can write a table by choosing numbers that we want for the first line and then substituting them into the formula to find the corresponding numbers of the second line. For example, a formula for the temperature in degrees Celsius, C, as a function of the number of chirps in 15 seconds, n, is

$$C = 0.6n + 4.$$

To write a table for this function, we first choose some numbers for n:

n	0	10	20	30	40

Lesson 1: The Idea of a Function

We then substitute these numbers into the formula to find the numbers on the second line:

Substitute $n = 0$:
$C = 0.6(0) + 4 = 0 + 4 = 4$
Find that $C = 4$.

Substitute $n = 10$:
$C = 0.6(10) + 4 = 6 + 4 = 10$
Find that $C = 10$.

Substitute $n = 20$:
$C = 0.6(20) + 4 = 12 + 4 = 16$
Find that $C = 16$.

Substitute $n = 30$:
$C = 0.6(30) + 4 = 18 + 4 = 22$
Find that $C = 22$.

Substitute $n = 40$:
$C = 0.6(40) + 4 = 24 + 4 = 28$
Find that $C = 28$.

The table is

n	0	10	20	30	40
C	4	10	16	22	28

Exercises

Set I

One way to represent a function is with a table. For example, if a function has the formula $y = 5x$, then a partial table for it might be:

x	0	1	2	3	4
y	0	5	10	15	20

Chapter 3: FUNCTIONS AND THEIR GRAPHS

Copy and complete the tables shown for the functions having the following formulas.

1. Formula: $y = x + 4$

x	0	1	2	3	4
y	4				

2. Formula: $y = 7x$

x	0	1	2	3	4
y		7			

3. Formula: $y = 8 - x$

x	0	1	2	3	4
y	8				

4. Formula: $y = \dfrac{12}{x}$

x	1	2	3	4	5
y	12				

5. Formula: $y = 11x + 1$

x	0	1	2	3	4
y	1				

6. Formula: $y = 6x - 5$

x	1	2	3	4	5
y	1				

7. Formula: $y = 0x + 3$

x	1	2	3	4	5
y	3				

8. Formula: $y = x^2$

x	0	1	2	3	4
y					

9. Formula: $y = x^2 + 10$

x	0	1	2	3	4
y					

10. Formula: $y = x^3$

x	1	2	3	4	5
y					

11. Formula: $y = 2^x$

x	2	3	4	5	6
y	4	8			

The second line of numbers in the table shown in color at the beginning of this set of exercises is

y	0	5	10	15	20.

These numbers are part of an arithmetic sequence because each number can be found by adding 5 to the preceding number.

12. In which of the other tables in exercises 1 through 11 are the numbers on the second line in arithmetic sequence?

13. The second line of one of those tables is part of a geometric sequence. Which table is it?

A function is *increasing* if, as one variable increases, the other variable also increases.

14. Which of the functions in exercises 1 through 11 are increasing?

A function is *decreasing* if, as one variable increases, the other variable decreases.

15. Which of the functions in exercises 1 through 11 are decreasing?

Set II

Guess a formula for the function represented by each of the following tables. Begin each formula with $y =$.

1. | x | 1 | 2 | 3 | 4 | 5 |
   |---|---|---|---|---|---|
   | y | 2 | 4 | 6 | 8 | 10 |

2. | x | 0 | 1 | 2 | 3 | 4 |
   |---|---|---|---|---|---|
   | y | 8 | 9 | 10 | 11 | 12 |

3. | x | 7 | 8 | 9 | 10 | 11 |
   |---|---|---|---|---|---|
   | y | 4 | 5 | 6 | 7 | 8 |

4. | x | 3 | 4 | 5 | 6 | 7 |
   |---|---|---|---|---|---|
   | y | 9 | 16 | 25 | 36 | 49 |

5. | x | 9 | 16 | 25 | 36 | 49 |
   |---|---|---|---|---|---|
   | y | 3 | 4 | 5 | 6 | 7 |

6. | x | 2 | 3 | 4 | 5 | 6 |
   |---|---|---|---|---|---|
   | y | 22 | 33 | 44 | 55 | 66 |

7. | x | 2 | 3 | 4 | 5 | 6 |
   |---|---|---|---|---|---|
   | y | 21 | 31 | 41 | 51 | 61 |

8. | x | 0 | 1 | 2 | 3 | 4 |
   |---|---|---|---|---|---|
   | y | 7 | 7 | 7 | 7 | 7 |

9. | x | 1 | 2 | 3 | 4 | 5 |
   |---|---|---|---|---|---|
   | y | 1 | 8 | 27 | 64 | 125 |

10. | x | 1 | 2 | 3 | 4 | 5 |
    |---|---|---|---|---|---|
    | y | 3 | 10 | 29 | 66 | 127 |

(*Hint:* Compare this table with the preceding one.)

11. | x | 1 | 2 | 3 | 4 | 5 |
    |---|---|---|---|---|---|
    | y | 99 | 98 | 97 | 96 | 95 |

12. | x | 1 | 2 | 3 | 4 | 5 |
    |---|---|---|---|---|---|
    | y | 60 | 30 | 20 | 15 | 12 |

13. | x | 2 | 3 | 4 | 5 | 6 |
    |---|---|---|---|---|---|
    | y | 9 | 27 | 81 | 243 | 729 |

The distance that light travels through space is a function of time. Here is a table for this function.

Number of seconds, t	1	2	3	4
Number of kilometers, d	300,000	600,000	900,000	1,200,000

14. What is the speed of light in kilometers per second?

15. Write a formula for this function. Begin the formula with $d =$.

16. How far does light travel in one minute?

Lesson 1: The Idea of a Function

125

The size of a motion picture on the screen is a function of the distance of the projector from the screen. The diagram shows that,

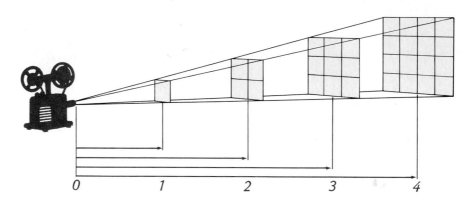

if the distance between the screen and the projector is 3 units, the size of the picture is 9 square units.

17. Copy and complete the following table suggested by the diagram.

Distance from screen, d	1			
Size of picture, s	1			

18. Write a formula for the function, using the letters s and d.

The pulse rate of a child is a function of the child's age. Here is a table for this function.

Age in years, a	0	2	4	6	8	10
Pulse rate in beats per minute, p	135	110	105	95	90	87

19. How does the pulse rate vary with age?

20. Do you think there is a simple formula for this function?

Set III

A person's weight is a function of his or her distance from the center of the earth. Because we are always approximately the same distance from the center of the earth when we weigh ourselves,

Chapter 3: FUNCTIONS AND THEIR GRAPHS

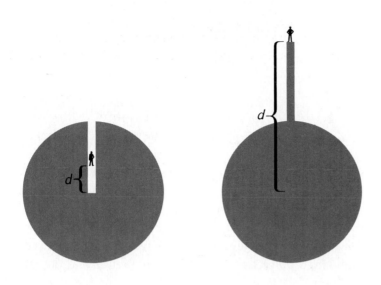

however, our weight at any given time is usually thought of as fixed.

If W represents your weight at the surface of the earth and w represents your weight if you are at a distance of d miles from the center of the earth, then

$$w = \frac{dW}{4{,}000} \text{ if you are below the earth's surface}$$

and

$$w = \frac{16{,}000{,}000\ W}{d^2} \text{ if you are above the earth's surface.}$$

The radius of the earth is 4,000 miles; so whether you are below the earth's surface, at its surface, or above it depends on whether d is less than, equal to, or greater than 4,000.

1. Substituting your own weight for W, show that either formula can be used to find your weight at the surface of the earth.

2. Substituting your own weight for W, use the formulas to complete the following table.

Distance from center of earth in miles, d	0	1,000	2,000	3,000	4,000	5,000	6,000	7,000	8,000
Your weight in pounds, w									

3. How does a person's weight, w, change as his or her distance from the center of the earth, d, increases?

Lesson 2

Descartes and the Coordinate Graph

One of the greatest mathematical achievements of all time was the invention of coordinate geometry by René Descartes. It made possible a new method of studying geometric figures and the relations between them, relations first proved by the ancient Greeks. The new method used algebra, a subject developed many centuries after geometry, and marked the beginning of modern mathematics. With this method, lines, circles, and other figures could be related to equations, so that algebra and geometry were combined into a new subject more powerful than either of its separate parts.

René Descartes was born in France in the late-sixteenth century. In the age in which Descartes lived, Europe was in political and religious turmoil, yet there was also great intellectual progress. In England, Shakespeare was writing his plays; great scientific discoveries were being made by Galileo in Italy; and the French mathematicians Fermat and Pascal were developing another new branch of mathematics called probability theory.

As a young student, Descartes began to question the truth of much of what he was being taught. The subject of mathematics, however, appealed to him because its methods of reasoning seemed universal and without fault. He decided that mathematics, to quote his words, "is a more powerful instrument of knowledge than any other that has been bequeathed to us by human agency." This belief led him to apply deductive reasoning, the mathematical method developed by the Greeks so many centuries earlier, to other areas of study. In 1637, his book titled *A Discourse on the Method of Rightly Conducting the Reason and Seeking Truth in the Sciences* established Descartes as the "father of modern philosophy." The book ended with a section on coordinate geometry, his great contribution to the subject of mathematics.

Descartes's invention was clever yet, like many important discoveries in mathematics, it was very simple. The idea was that the location of a point in a plane can be described by giving its distances from a pair of perpendicular lines. The lines are called the *x*-axis and the *y*-axis, and the point at which they intersect, labeled O, is called the origin. The axes are numbered at equal intervals in each direction from the origin. On the *x*-axis, positive numbers are used to the right of the origin and negative numbers* are used to the left. On the *y*-axis, positive numbers are used above the origin and negative numbers are used below it.

To locate a point, we first move along the *x*-axis until we are directly above or below the point, counting the units as we go. Then we move directly up or down to the point itself, again counting the units along the way. These two numbers are called the *coordinates* of the point and are written in parentheses like this: (3, 2). The first number is the x-*coordinate* and the second number is the y-*coordinate*.

Other examples of how the coordinates of a point are found are shown in the figure below.

———————

*Look on page 628 if you are not familiar with negative numbers.

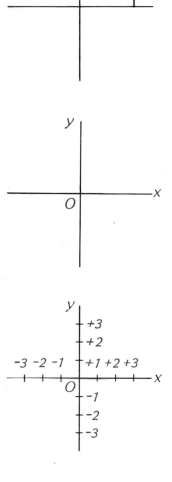

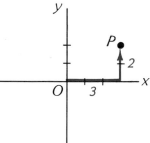

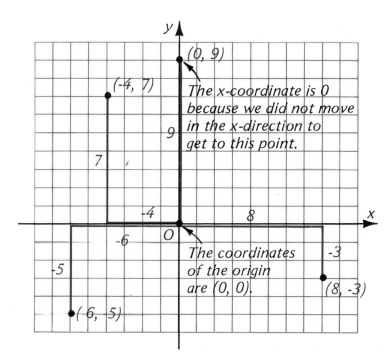

Exercises

Set I

The coordinate graph is valuable in the study of geometric figures because each point in a figure can be located with a pair of numbers.

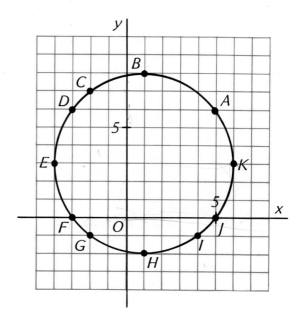

1. The curve in this graph is a circle. The coordinates of point K are (6, 3). Write the coordinates of each of the other points on the circle that is named with a letter.

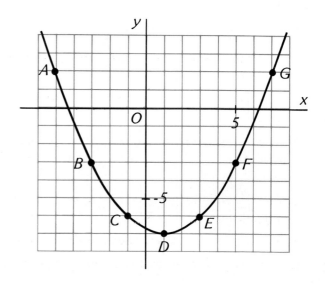

2. The curve in this graph is a parabola. Write the coordinates of each lettered point on the parabola.

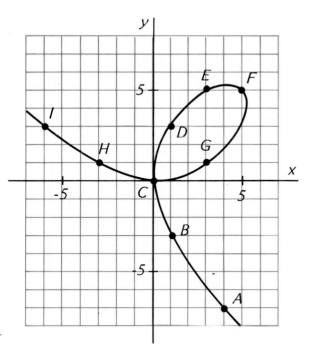

3. This curve is called the "folium of Descartes." Although the coordinates of most of the lettered points on it are not actually whole numbers, write them as if they were.

Set II

On graph paper, draw and label five pairs of axes as shown in the adjoining figure.

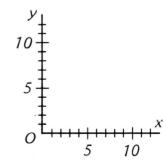

Graphs, Coordinates, and Lines

1. Plot the following points on the first graph: A(0, 2), B(3, 5), C(5, 7), D(6, 8), E(8, 9), and F(9, 11).

2. Which one of the six points in your graph seems to be out of place?

3. The coordinates of the other five points have something in common that the coordinates of the "out of place" point do not. What is it?

4. Plot the following points on the second graph: A(4, 0), B(4, 1), C(4, 3), D(4, 7), and E(4, 10).

5. What do you notice about the five points in your graph?

6. What do the coordinates of the five points have in common?

Graphs, Coordinates, and Midpoints

7. Plot the following points on the third graph: A(1, 6), B(5, 2), C(6, 7), and D(12, 9).

8. Find the point midway between points A and B and label it M. What are the coordinates of point M?

9. How are the coordinates of point M related to the coordinates of points A and B?

10. Find the point midway between points C and D and label it N. What are the coordinates of point N?

11. Are the coordinates of point N related to the coordinates of points C and D in the same way that the coordinates of point M are related to the coordinates of points A and B?

Graphs, Coordinates, and Triangles

12. Plot the following points on the fourth graph and connect them with straight line segments to form the three triangles named.
    Triangle 1: (2, 1) (5, 2) (3, 6)
    Triangle 2: (5, 5) (8, 6) (6, 10)
    Triangle 3: (9, 0) (12, 1) (10, 5)

13. In what ways are the three triangles that you have drawn alike?

14. The x-coordinates of the corners of triangle 2 are 3 more than the x-coordinates of the corresponding corners of triangle 1. How do the y-coordinates of the corners of triangle 2 compare with the coordinates of the corresponding corners of triangle 1?

15. How are the coordinates of the corners of triangle 3 related to the coordinates of the corresponding corners of triangle 1?

16. Plot the following points on the fifth graph and connect them with straight line segments to form the three triangles named.
    Triangle 1: (4, 0) (6, 6) (2, 4)
    Triangle 2: (8, 0) (12, 12) (4, 8)
    Triangle 3: (2, 0) (3, 3) (1, 2)

17. In what way are the three triangles that you have drawn alike?

18. How are the coordinates of the corners of triangle 2 related to the coordinates of the corresponding corners of triangle 1?

19. How are the coordinates of the corners of triangle 3 related to the coordinates of the corresponding corners of triangle 1?

Chapter 3: FUNCTIONS AND THEIR GRAPHS

Set III

Here is an exercise that requires patience but has an amusing result. Draw a pair of coordinate axes extending from −10 to +10 on the x-axis and from −10 to +20 on the y-axis. Connect the points in each list *in order* with straight line segments and a familiar figure will appear. (After the points in one list have been connected in order, start all over again with the next list. In other words, *do not connect* the last point in each list to the first point in the next one.)

A. (−3, 10) (−2, 8) (−5, 2) (−7, 1) (−5, 0) (−2, 0) (−2, −8) (−6, −8) (−7, −9) (−4, −9) (−5, −10) (2, −10) (2, −3)

B. (−4, −9) (0, −9) (0, 5) (−2, 1) (0, 1)

C. (2, 4) (2, 0) (3, −1) (3, −2) (2, −3) (0, 0), (−2, 0)

D. (2, 1) (6, 1) (5, 0) (2, 0)

E. (−5, 2) (−4, 1) (1, 7) (1, 8) (3, 9) (4, 9)

F. (1, 7) (5, 1)

G. (2, 13) (4, 14) (5, 13) (4, 12) (6, 11) (6, 10) (3, 9.5)

H. (−7, 12) (−6, 14) (−5, 15) (−1, 15) (0, 16)

I. (0, 15) (1, 16) (2, 16)

J. (−6, 11) (−5, 14)

K. (−5, 10) (−4, 13)

L. (−4, 9) (−4, 11)

M. (−2, 12) (−3, 12) (−3, 11) (−2, 10)

N. (2, 12) (4, 12)

O. (0, 13) (1, 13)

Lesson 3
Graphing Linear Functions

Kangaroos are the only large mammals that hop. At low speeds, they move somewhat clumsily, using their tails as well as their legs for support. When they hop, however, they can clear 10 meters in a single leap and reach speeds of more than 50 kilometers per hour.

At speeds between 10 and 35 kilometers per hour, kangaroos hop at a fairly constant rate. The table below shows the distance covered in each hop as a function of the speed.

Speed in kilometers per hour	10	15	20	25	30	35
Distance covered in each hop in meters	1.2	1.8	2.4	3.0	3.6	4.2

Notice that the table contains *pairs of numbers,* a distance number for each speed number. These pairs of numbers can be used as coordinates of points to make a picture called the *graph* of the function.

To graph the function, we first plot the points: (10, 1.2), (15, 1.8), (20, 2.4), (25, 3.0), (30, 3.6), and (35, 4.2). The first coordinate of each point is the speed and the second coordinate is the distance, and so we will name the axes s and d. Because the speed coordinates grow much faster than the distance coordinates, we choose scales on the two axes that allow room to show all six points.

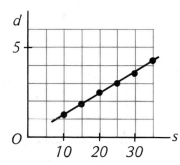

After plotting the points, we see that they lie along a straight line. It makes sense to draw this line because there are other

134

speeds and distances between those listed in the table. The formula of the function is

$$d = 0.12s$$

and the line is called its graph.

To graph a function for which we are given a formula, it is first necessary to make a table. For example, to make a table for the function

$$y = 2x - 1,$$

we first choose some numbers for x: 1, 2, 3, 4, and 5 are convenient. Then we substitute each number into the formula to find the corresponding number for y, as shown below.

Substitute $x = 1$:
$y = 2(1) - 1 = 2 - 1 = 1$
    Find that $y = 1$.

Substitute $x = 2$:
$y = 2(2) - 1 = 4 - 1 = 3$
    Find that $y = 3$.

Substitute $x = 3$:
$y = 2(3) - 1 = 6 - 1 = 5$
    Find that $y = 5$.

Substitute $x = 4$:
$y = 2(4) - 1 = 8 - 1 = 7$
    Find that $y = 7$.

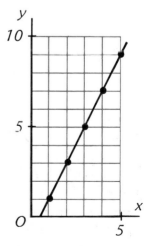

Substitute $x = 5$:
$y = 2(5) - 1 = 10 - 1 = 9$
    Find that $y = 9$.

The resulting table,

x	1	2	3	4	5
y	1	3	5	7	9,

contains the coordinates of five points. Plotting them on a pair of axes and drawing a line through them, we get the graph shown at the right.

Exercises

Set I

A certain function is represented by the table of numbers at the left.

x	0	1	2	3	4
y	3	4	5	6	7

1. What is a formula for this function? Begin your formula with $y =$.

2. How many points are included in the table?

3. Graph the function by drawing a pair of axes, plotting these points, and drawing a line through them.

A certain function has the graph shown here.

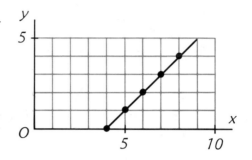

4. Copy and complete the following table for this function.

x	4	5	6	7	8
y					

5. What is a formula for this function?

6. Make tables for the following functions, letting $x = 0, 1, 2, 3, 4$, and 5.

Function A: $y = x$

Function B: $y = \dfrac{1}{2}x$

Function C: $y = 2x$

7. Graph all three functions on one pair of axes. Draw a line through each set of points and write each formula along its line.

8. What do you notice about the three lines?

9. Which line is steepest?

Chapter 3: FUNCTIONS AND THEIR GRAPHS

10. The graphs of the functions $y = \dfrac{1}{3}x$ and $y = 3x$ are shown here. Which line do you think is the graph of each function?

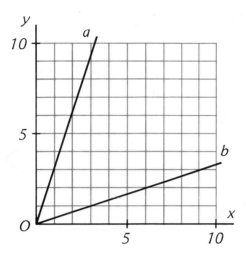

11. Make tables of numbers for the following functions, letting $x = 0, 1, 2, 3, 4,$ and 5.

Function D: $y = x + 1$
Function E: $y = x + 4$
Function F: $y = x + 5$

12. Graph all three functions on one pair of axes. Write the formula for each along its line.

13. What do you notice about these three lines?

14. Where does each line meet the y-axis?

15. Write a formula for the function whose graph is a line in the same direction meeting the y-axis at 8.

16. Make tables of numbers for the following functions, letting $x = 0, 1, 2, 3,$ and 4.

Function G: $y = 2x + 1$
Function H: $y = 2x + 3$

17. Graph both functions on one pair of axes.

18. What do you think is the formula of the function whose graph is the line midway between these lines?

Lesson 3: Graphing Linear Functions   137

19. Make tables of numbers for the following functions, letting $x = 0, 1, 2, 3$, and 4.

$$\text{Function I:} \quad y = 6 + x$$
$$\text{Function J:} \quad y = 6 - x$$

20. Graph both functions on one pair of axes.

21. What angle do the two lines seem to make with each other?

Set II

When lightning strikes, the time between the flash that you see and the thunder that you hear depends on the distance that you are from where the lightning struck.

A table for this function is shown here.

Distance in kilometers, d	1	2	3	4
Time in seconds, t	3	6	9	12

1. Write a formula for this function, using the letters d and t.

2. Graph it, letting the x-axis represent distance and the y-axis represent time.

3. Why does it seem reasonable that the graph of this function should go through the origin?

The amaryllis grows from a bulb to a plant with flowers in just a few weeks. Part of a graph of the height of the plant as a function

of time is shown here.

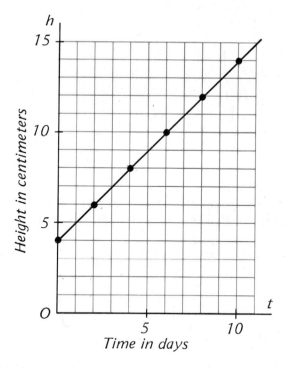

4. Refer to this graph to copy and complete the following table.

t	0	2	4	6	8	10
h	▨	▨	▨	▨	▨	▨

5. Write a formula for this function.

6. If the amaryllis continues to grow at the same rate, how tall would it be after 20 days?

Before the invention of mechanical clocks, candles were sometimes used to measure the passage of time. A formula for the height of such a candle as a function of time is

$$h = 10 - 2t$$

in which h represents the height of the candle in centimeters and t represents the time in hours that the candle has been burning.

7. Use this formula to copy and complete the following table.

t	0	1	2	3	4	5
h	▨	▨	▨	▨	▨	▨

8. Graph the function, letting the x-axis represent time and the y-axis represent height.

9. What does the 10 in the formula represent?

10. What does the 2 in the formula represent?

11. Write a formula for the height of a candle as a function of time that has a height of 18 centimeters at the beginning and burns at the rate of 3 centimeters per hour.

A person's shoe size is a function of the length of his or her foot. Formulas for this function for men's and women's shoes are given below: x represents the length of a person's foot in inches and y represents the corresponding shoe size.

$$\text{Men:} \quad y = 3x - 25$$
$$\text{Women:} \quad y = 3x - 22$$

12. Use these formulas to copy and complete the following tables.

Men

Length of foot in inches, x	9	10	11	12	13	14
Shoe size, y						

Women

Length of foot in inches, x	9	10	11	12	13	14
Shoe size, y						

13. Graph both functions on one pair of axes.

14. What do you notice about their graphs?

15. If a man and a woman have feet of the same length, who has the larger shoe size?

16. If a man and a woman have the same shoe size, who has the longer feet?

Set III

The number of words in a child's vocabulary is a function of the child's age. A formula for the size of the vocabularies of typical children between the ages of 20 months and 50 months is

$$n = 60a - 900$$

in which *a* represents a child's age in months and *n* represents the number of words that the child uses correctly.

1. Copy and complete the following table for this function.

a	20	25	30	35	40	45	50
n	▓	▓	▓	▓	▓	▓	▓

2. Graph this function, letting one unit along the *a*-axis represent 2 years and one unit along the *n*-axis represent 100 words.

3. How many words does a typical 20-month-old child know?

4. How many words does a typical 50-month-old child know?

5. How many new words does a typical child learn each month from the age of 20 months to the age of 50 months?

6. Could the formula be correct for a 10-month-old child? Explain.

By permission of Johnny Hart and Field Enterprises, Inc.

Lesson 4

Functions with Parabolic Graphs

If it takes 16 seconds for a rock to hit the bottom of a well, can anything be concluded about the depth of the well? Suppose, for example, that a heavy rock and a light rock were thrown in at the same time. Would they hit the bottom at the same time?

The Greeks thought that the heavy rock would hit the bottom first. They reasoned that, if one object is heavier than another, it is because it is more strongly attracted to the earth and, the more strongly the earth attracts an object, the faster it will fall. But the Greeks were wrong.

In the seventeenth century, the great Italian scientist Galileo discovered that the speed at which an object falls does not depend

142

on its weight. Drop a small stone and a large rock from the same height and they will hit the ground at the same time.

Galileo knew that the distance an object falls is a function of time. If the time is measured in quarter seconds and the distance in feet, the two variables are related in an especially simple way. This table shows what it is.

Time in quarter seconds, t	0	1	2	3	4	5	...
Distance in feet, d		0	1	4	9	16	25	...

The formula for this function,

$$d = t^2,$$

no doubt delighted Galileo.*

The graph of this function is shown at the right. Unlike those of the graphs of functions that we have already considered, the points in this graph do not lie on a straight line. They can be connected, instead, with a smooth curve to show the times and distances between those listed in the table.

Although the graph does not extend far enough, the formula for the function can be used to find out how many feet deep B.C.'s well is. Noting that 16 seconds is the same as 64 quarter seconds and substituting, we get

$$\begin{aligned} d &= t^2 \\ &= (64)^2 \\ &= 4{,}096. \end{aligned}$$

The well is apparently more than four thousand feet deep!

---

*In 1636, Galileo said: "So far as I know, no one has yet pointed out that the distances traveled during equal intervals of time, by a body falling from rest, stand to one another in the same ratio as the odd numbers beginning with one."

The drawing of the falling rock at the right illustrates this. (Compare it with the diagram on page 83, which shows how the Greeks related square and odd numbers.)

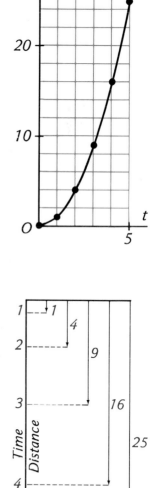

Exercises

Set I

A certain function is represented by this table of numbers.

x	0	1	2	3	4
y	0	1	4	9	16

1. What is a formula for this function? Begin your formula with $y =$.

2. Graph the function by copying the pair of axes shown here, plotting the points included in the table, and drawing a smooth curve through them.

Because the graph is a curve, a more complete picture of it can be obtained by including some points with negative coordinates.

3. Copy and complete the following table for the same function.

x	−4	−3	−2	−1
y	16*			

Add the points to your graph and extend the curve through them.

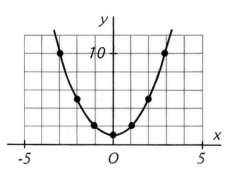

4. Copy and complete the following table for the function pictured in the graph at the left.

x	−3	−2	−1	0	1	2	3
y							

5. What is a formula for this function?

6. Where does its graph cross the y-axis?

7. Make a table for the function

$$y = x^2 + 4,$$

letting $x =$ −3, −2, −1, 0, 1, 2, and 3.

---

*Because the product of two negative numbers is positive, the square of a negative number is always positive. See page 630 if this is not clear.

Chapter 3: FUNCTIONS AND THEIR GRAPHS

8. Graph this function on a pair of axes with the same scales as those in exercise 2.

9. Where does the graph cross the y-axis?

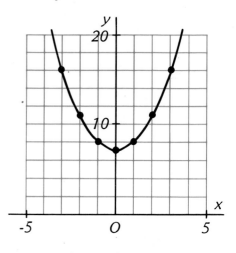

10. Compare the graph shown here with the one that you have just drawn. What do you think is the formula for the function shown?

11. Copy and complete the following table for the function

$$y = 12 - x^2.$$

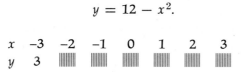

x	−3	−2	−1	0	1	2	3
y	3						

12. Graph it on a pair of axes with the same scales as those in exercise 2.

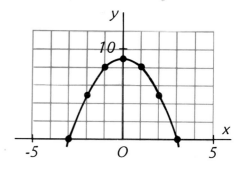

13. Compare the graph shown here with the one that you have just drawn. What do you think is the formula of the function shown?

14. Copy and complete the following table for the function

$$y = (3 - x)^2.$$

x	0	1	2	3	4	5	6
y	9						

15. Graph it on a pair of axes with the same scales as those in exercise 2.

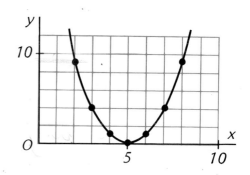

16. Compare the graph shown here with the one that you have just drawn. What do you think is the formula of the function shown?

"Of course you smell pizza pie—
it's right there in front of you!"

Drawing by Tom Henderson

Set II

The area of a circular pizza is a function of its diameter. A graph of this function is shown here.

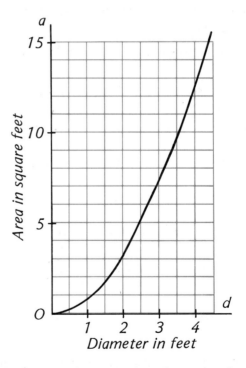

From the graph it can be seen that the area of a pizza with a diameter of 1 foot is a little less than 1 square foot.

1. What is the approximate area of a pizza with a diameter of 2 feet?

2. What is the approximate area of a pizza with a diameter of 4 feet?

Chapter 3: FUNCTIONS AND THEIR GRAPHS

3. If the diameter of a pizza is doubled, is its area doubled?

The collision impact* of an automobile is a function of its speed. For a certain automobile, it is given by the formula

$$I = 2(s^2)$$

in which I represents the collision impact and s represents the speed in kilometers per minute.

4. Use this formula to copy and complete the following table.

s	0	1	2	3	4
I					

5. What happens to the collision impact of the automobile if its speed is doubled?

6. What happens to the collision impact of the automobile if its speed is tripled?

7. Graph this function, labeling the axes as shown here.

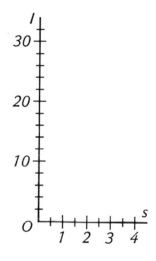

When a ball is thrown straight up in the air, the height at any given moment is a function of time. If the ball is thrown with a speed of 32 feet per second, the formula for this function is

$$h = 8t - t^2$$

in which h represents the height in feet and t represents the time in quarter seconds.

8. Make a table for this function, letting $t = 0, 1, 2, 3, 4, 5, 6, 7,$ and 8.

9. Graph the function.

10. How high does the ball go?

11. How many seconds is the ball in the air? (Remember that the time is measured in quarter seconds.)

12. How does the time that the ball spends going up compare with the time that it spends coming down?

---

*Collision impact is a measure of the damage that a moving automobile can cause when it hits an object.

Set III

On the beach, the distance that you can see out to sea is a function of the height of your eyes above the water. A formula for this function is

$$d = 1.2\sqrt{h}$$

in which h is the height in feet and d is the distance in miles.

1. Use this formula to copy and complete the following table.

h	0	1	4	6	9	16	25
d	▓	▓	▓	2.9*	▓	▓	▓

2. How far can a person 6 feet tall see out to sea?

3. How far can a lifeguard sitting in a tower see out to sea if his eyes are 25 feet above sea level?

4. Graph the function, labeling the axes as shown here.

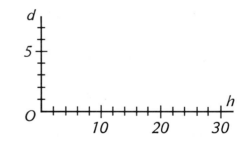

5. How does the curve that results seem to be related to the others in this lesson?

---

* $\sqrt{6} \approx 2.449$ and $1.2(2.449) \approx 2.9$. (The symbol $\approx$ stands for "is approximately equal to.")

Chapter 3: FUNCTIONS AND THEIR GRAPHS

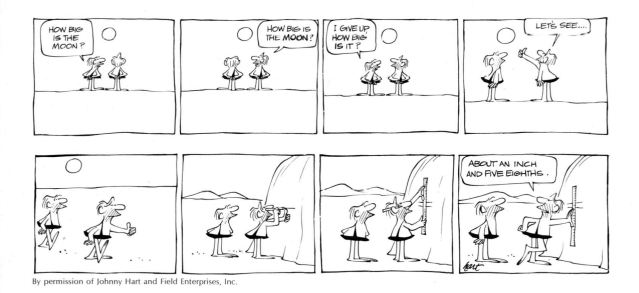

By permission of Johnny Hart and Field Enterprises, Inc.

Lesson 5

More Functions with Curved Graphs

How big is the sun? At certain times, it seems to be the same size as the moon. For example, in a total eclipse of the sun, the moon is just large enough to cover it completely. The Greeks had some strange ideas about the relative sizes and distances of the sun and moon. Democritus, who lived in the fourth century B.C., thought that the sun was smaller than the earth. In fact, a century earlier, the people of Athens were surprised by one astronomer's suggestion that the sun might be as large as the country of Greece. We now know that the sun has a diameter more than 100 times that of the earth; it is about 1,400,000 kilometers across. (Even at that size, however, the sun is only a medium-sized star compared with others in our galaxy.)

The *apparent* size of the sun is a function of the distance from which we see it. If the distance from the earth to the sun were half as great, the sun would appear to be twice as large. A formula for this function is

$$w = \frac{1}{d}$$

in which w represents the apparent width of the sun and d represents the relative distance from the sun.

How large would the sun appear to be from a planet other than the earth? From Mercury, the closest planet to the sun, it would seem the largest and from Pluto, the most distant planet, the smallest. It would be interesting to know just how large and how small. Here is a table for some of the planets.

	Mercury	Venus	Earth	Mars	Jupiter	Saturn
Relative distance of planet from sun	0.4	0.7	1.0	1.5	5.2	9.5
Apparent width of the sun	2.5	1.4	1.0	0.7	0.2	0.1

In the table, the distance of each planet from the sun is given relative to the earth's distance from the sun, chosen as 1 unit. For example, Mars is 1.5 times as far from the sun as the earth is. The apparent widths of the sun are also given relative to its width as seen from the earth. For example, the sun's width as it would be seen from Mercury is 2.5 times its width as we see it.

A graph of this "apparent size of the sun" function is shown below. Disks have been added to the graph to represent the apparent size of the sun as seen from each planet. The parts of the curve between the points from the table represent apparent widths of the sun as it would be seen from positions between the planets. Notice that the curve gets closer and closer to the distance axis as we look toward the right. If the graph were extended to include the remaining three planets, the curve would be very close to the axis by the time we got to Pluto. Thus, the apparent size of the sun as it would be seen from Pluto is smaller than a period on this page.

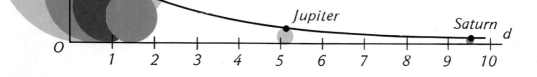

The relative size of the sun as viewed from various planets

Chapter 3: FUNCTIONS AND THEIR GRAPHS

Exercises

Set I

The "apparent size of the sun" function is one member of a family of related functions. A similar function is

$$y = \frac{6}{x}.$$

1. Copy and complete the following table for this function.

x	1	2	3	4	5	6
y	▓	▓	▓	1.5	▓	▓

2. Graph the function by copying the pair of axes shown here, plotting the points included in the table, and drawing a smooth curve through them.

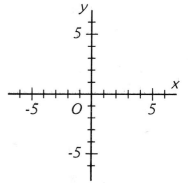

   Notice that this curve is similar in appearance to the graph of the "apparent size of the sun" function.

3. What happens to the values of y as the values of x get larger?

4. What would happen to the curve if the graph were extended to the right?

It is impossible to find a value for y if $x = 0$ because division by zero does not make sense. This means that there is no point on the curve for which $x = 0$. It *is* possible to find values of y corresponding to negative values of x, however.

5. Copy and complete the following table of negative numbers for this function.

x	−6	−5	−4	−3	−2	−1
y	−1	−1.2	▓	▓	▓	▓

Plot the points included in this table on the pair of axes that you have already drawn and connect them with a smooth curve.

6. What would happen to the curve if the graph were extended to the left?

Another function related to the function that you have just graphed is shown in this graph.

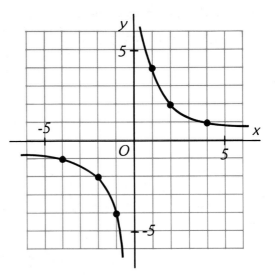

7. Refer to the graph to copy and complete the following table.

x	−4	−2	−1	1	2	4
y	−1					

8. Its formula is similar to the formula of the function that you graphed. What do you think it is?

Another function with a curved graph is

$$y = x^3.$$

9. Copy and complete the following table for this function. (Notice that the cube of a negative number is also negative.)

x	−4	−3	−2	−1	0	1	2	3	4
y	−64								

10. Graph the function, letting 1 unit on the y-axis represent 10.

Compare the graph of the function

$$y = x^3 + 10$$

shown at the right with the graph of the function

$$y = x^3$$

that you have just drawn.

11. In what way are the two graphs alike?

12. In what way are they different?

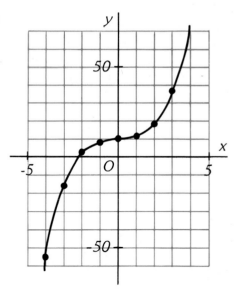

Graphs of the functions $y = x^2$, $y = x^4$, $y = x^5$, $y = x^6$, and $y = x^7$ are shown below.

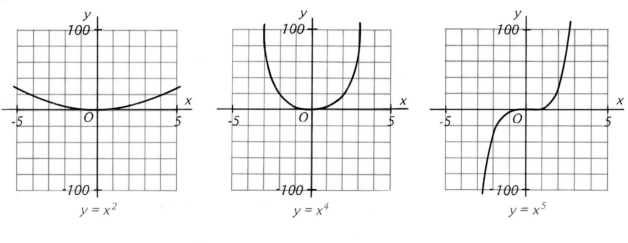

$y = x^2$ $y = x^4$ $y = x^5$

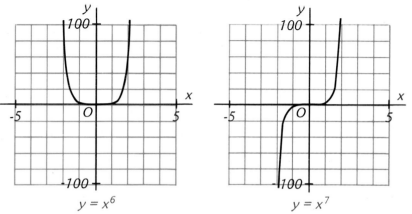

$y = x^6$ $y = x^7$

Lesson 5: More Functions with Curved Graphs

153

13. Which functions have graphs similar to the graph of $y = x^2$?

14. What do their formulas have in common?

15. Which functions have graphs similar to the graph of $y = x^3$? (See exercise 10.)

16. What do their formulas have in common?

Set II

By permission of Johnny Hart and Field Enterprises, Inc.

The wavelength of a sound wave is a function of its frequency. A graph of this function is shown here.

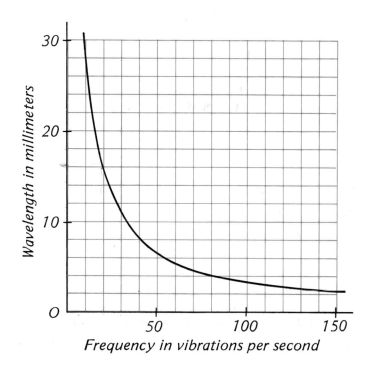

Chapter 3: FUNCTIONS AND THEIR GRAPHS

1. What happens to the wavelength of a sound wave as its frequency increases?

The human ear can hear sound waves with frequencies as low as 20 vibrations per second.

2. What is the approximate wavelength of these sounds?

Some animals can hear sound waves with frequencies of 100,000 vibrations per second.

3. On the basis of the graph, what do you think can be concluded about the wavelengths of such sounds?

The weight of a cube of ice is a function of the length of its edge. A table for this function is shown here.

Length of edge in centimeters, e	2	4	6	8	10
Weight in grams, w	8	64	216	512	1,000

4. Write a formula for this function, using the letters e and w.

5. Graph it, labeling the axes as shown here.

6. Use your graph to estimate the weight of a cube of ice measuring 5 centimeters on each edge.

7. Why does it seem reasonable that the graph of this function should go through the origin?

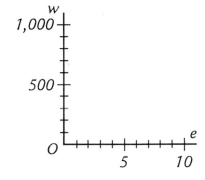

The time that it takes to run 100 meters is a function of the speed of the runner. The formula for this function is

$$t = \frac{100}{s}$$

in which s represents the speed of the runner in meters per second and t represents the time in seconds.

8. Use this formula to copy and complete the following table.

s	2	4	6	8	10
t	50				

9. What happens to the time if the speed of the runner doubles?

10. What happens to the time if the speed of the runner triples?

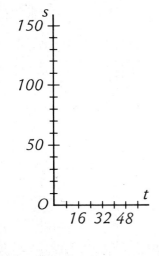

11. Graph the function, letting 1 unit on the t-axis represent 5 seconds.

The size of a fire is a function of time. A fire that is just beginning doubles in size about every eight seconds.

12. Copy and complete the following table for this function.

Time in seconds, t	0	8	16	24	32	40	48	56
Size of fire, s		1	2	4				

13. Graph it, labeling the axes as shown at the left.

14. As time passes, what happens to the rate at which the size of the fire increases?

Set III

The brightness of the light produced by fireflies is a function of temperature. An approximate formula for this function is

$$I = 10 + 0.3t + 0.4(t^2) - 0.01(t^3)$$

in which I is the light intensity and t is the temperature in degrees Celsius.

1. Copy and complete the following table for this function.

t	0	10	20	30	40
I	10	43			

2. Graph the function, letting 1 unit along the t-axis represent 5 degrees.

3. At about what temperature do you think fireflies produce the brightest light?

Chapter 3: FUNCTIONS AND THEIR GRAPHS

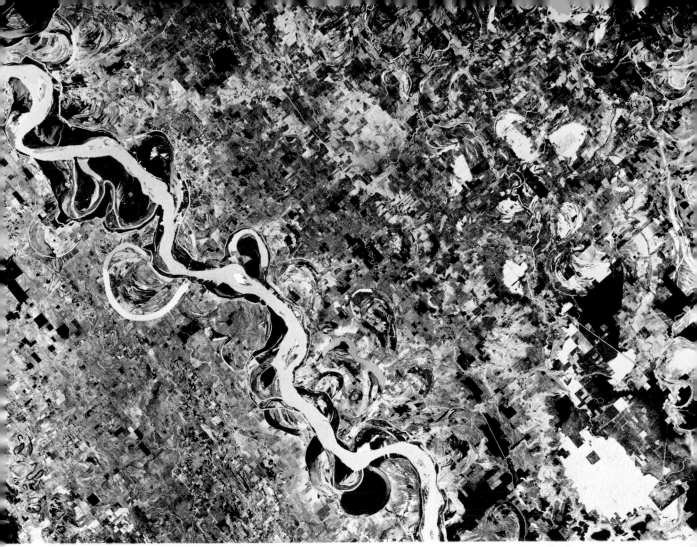

Courtesy of the National Oceanic and Atmospheric Administration

Interpolation and Extrapolation: Guessing Between and Beyond

Mark Twain once remarked that in eternity he planned to spend eight million years on mathematics. In his book *Life on the Mississippi*, Twain used some mathematics to make a strange prediction about the future of the Mississippi River. The river is extremely crooked with many curves, some of which are shown in the photograph above. From time to time, the river changes its course from a wide bend to a more direct path, called a cutoff. As a result, the

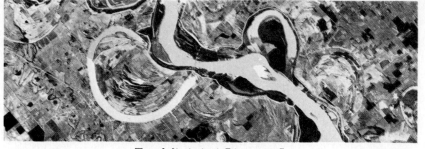

Two Mississippi River cutoffs

length of the Mississippi is becoming shorter and shorter. Twain gave some figures:

> *The Mississippi between Cairo and New Orleans was 1,215 miles long 176 years ago. It was 1,180 after the cutoff of 1722 . . . its length is only 973 miles at present [1875].*

Because the length of the Mississippi is a function of time, a table can be made from these numbers. Rounding them slightly, we get:

Time (the year), t	1700	1720	1875
Length in miles, ℓ	1,215	1,180	975

Although we have no formula for this function, the points in the table can be graphed. The result, shown below, suggests that they might lie on a straight line.

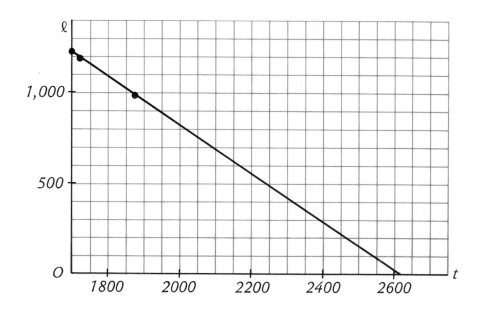

Chapter 3: FUNCTIONS AND THEIR GRAPHS

If a line is drawn through the three points, we can get additional information from it. For example, it appears that in 1800 the river was approximately 1,100 miles long. This information, found by reading between the values that we know, was obtained by *interpolation*.

▶ To **interpolate** is to guess other values of a variable *between* values that are known.

If the line representing the length of the Mississippi is extended until it intersects the time axis, it looks as if the river will disappear sometime about the year 2600. Twain jokingly made this prediction:

Any calm person, who is not blind or idiotic, can see . . . that 742 years from now the Lower Mississippi will be only a mile and three-quarters long. . . .

What's wrong with this reasoning? The trouble is that, although the three points may seem to lie along a straight line, the graph must actually curve later on. The river cannot become any shorter after all of the bends are gone. In drawing the line past the third point, we are *extrapolating*.

▶ To **extrapolate** is to guess other values of a variable *beyond* those that are known.

If a function continues to behave in the same way, then our guess may be very close; otherwise, you can see what may happen.

Exercises

Set I

The time that it takes a planet to travel once around the sun is a function of its distance from the sun. A table for four of the planets is shown here.

	Earth	Jupiter	Uranus	Neptune
Distance in astronomical units, d	1	5	19	30
Time in earth years, t	1	12	84	165

1. Draw a pair of axes as shown here and plot the four points in this table.

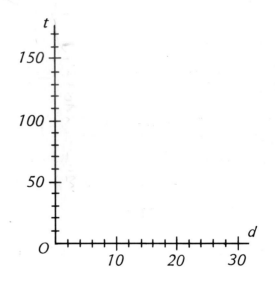

2. Do they seem to lie along a *straight line* or a *curve*? Connect them accordingly.

3. Saturn is 10 astronomical units from the sun. Use your graph to estimate the time in earth years that it takes Saturn to travel once around the sun.

4. Did you *interpolate* or *extrapolate* in making this estimate? Explain.

Chapter 3: FUNCTIONS AND THEIR GRAPHS

In 1787, the French physicist Jacques Charles observed that all gases contract equally when cooled. A table of the volume of a gas as a function of temperature is shown below.

Temperature in °C, t	50	−30	−110	−135	−220
Volume of gas, v	120	90	60	50	20

5. Draw a pair of axes as shown below and plot the five points in this table.

6. What do you notice about the five points?

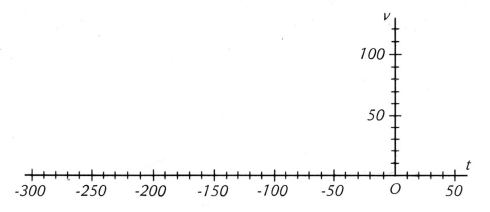

Draw a line through them, extending it until it meets the t-axis. Refer to the graph to answer the following questions.

7. What is the approximate volume of the gas at −60°C?

8. Did you *interpolate* or *extrapolate* to obtain this answer?

9. At approximately what temperature does the volume of the gas appear to become 0?

10. Did you *interpolate* or *extrapolate* to obtain this answer?

Set II

This figure shows a set of weights suspended by a rubber band. The length to which the rubber band is stretched is a function of the weight supported. A table for this function is shown here.

Weight in kilograms, w	0.2	0.3	0.5	1.0	1.2
Length in centimeters, l	15	16	18	23	25

1. Draw a pair of axes as shown here and graph this function.

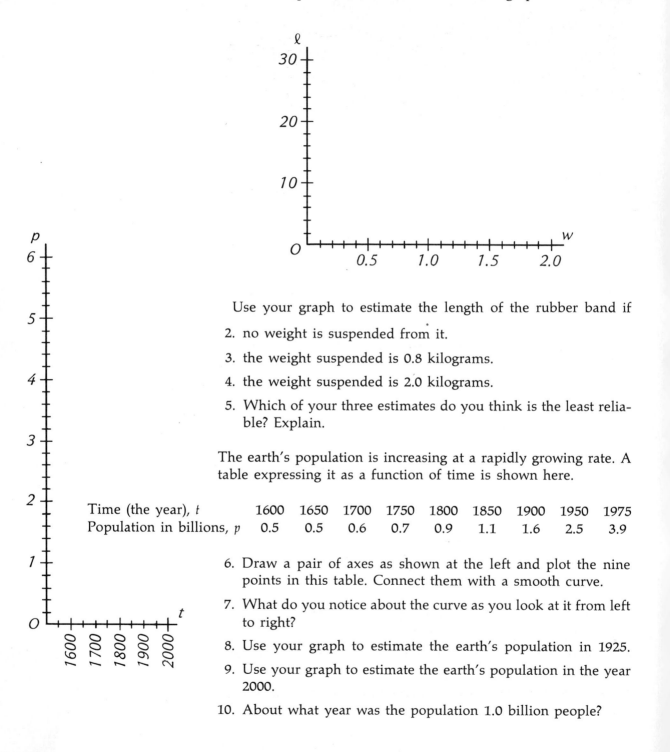

Use your graph to estimate the length of the rubber band if

2. no weight is suspended from it.

3. the weight suspended is 0.8 kilograms.

4. the weight suspended is 2.0 kilograms.

5. Which of your three estimates do you think is the least reliable? Explain.

The earth's population is increasing at a rapidly growing rate. A table expressing it as a function of time is shown here.

Time (the year), t	1600	1650	1700	1750	1800	1850	1900	1950	1975
Population in billions, p	0.5	0.5	0.6	0.7	0.9	1.1	1.6	2.5	3.9

6. Draw a pair of axes as shown at the left and plot the nine points in this table. Connect them with a smooth curve.

7. What do you notice about the curve as you look at it from left to right?

8. Use your graph to estimate the earth's population in 1925.

9. Use your graph to estimate the earth's population in the year 2000.

10. About what year was the population 1.0 billion people?

Record miler Steve Ovett

The first person to run the mile in less than four minutes was Roger Bannister, who did it in 1954 with a time of 3:59. He has predicted that someone will eventually run the mile in three minutes and 30 seconds.

The world records in the mile since 1920 are listed in the table below.

Year, y	1920	1930	1940	1950	1960	1970	1980
Record time, t	4:13	4:10	4:06	4:01	3:55	3:51	3:49

11. Draw a pair of axes as shown here and plot the seven points in

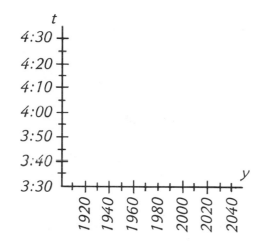

the table. Draw a line through the first and last points and extend it until it meets the y-axis.

12. What do you think the world record in the mile will be in the year 2000?

13. When do you think the mile will be run in three minutes and 30 seconds?

Set III

The temperature of the water in a lake is a function of the depth. A typical table for this function for a lake 20 meters deep at its center is shown here.

Depth in meters, d	0	2	4	6	8	10	12	14	16	18	20
Temperature in °C, t	15	14.5	14	13.5	13	9.5	6	5.5	5	4.5	4

1. Graph this function by plotting the points in this table and connecting them with a smooth curve.

2. If someone saw only the part of the table shown below, graphed it, and were asked to guess the temperature at 20 meters by extrapolation, what would be a reasonable answer?

Depth in meters, d	0	2	4	6	8
Temperature in °C, t	15	14.5	14	13.5	13

3. Near what depths does it seem the most difficult to determine the temperature by interpolation?

4. Why is it meaningless to extrapolate the temperatures beyond 20 meters?

Photograph by David Knudson

Chapter 3 / Summary and Review

In this chapter we have become acquainted with:

The idea of a function (*Lesson 1*)   A function is a pairing of two sets of numbers so that to each number in the first set there corresponds exactly one number in the second set.

A function can be represented by a table of numbers, a formula, or a graph.

The coordinate graph (*Lesson 2*)   The coordinate graph was invented by the French mathematician and philosopher René Descartes.

Each point on a coordinate graph is located by a pair of numbers, called its coordinates, which are the distances of the point from the x- and y-axes.

Functions with linear graphs (*Lesson 3*)   To graph a function for which a formula is known, it is first necessary to write a table for that function. The pairs of numbers in the table are coordinates of points of the graph.

Functions with curved graphs (*Lessons 4 and 5*)   If the graph of a function is a curved line, it is often useful to include points with negative x-coordinates to get a complete picture.

Typical formulas and graphs of functions that we have studied are shown here.

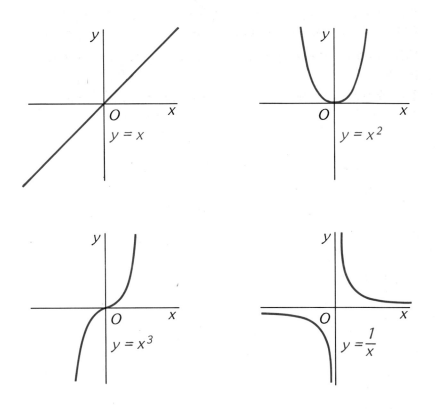

Interpolation and extrapolation (*Lesson 6*) Interpolation is guessing another value of a variable *between* values that are known. Extrapolation is guessing another value *beyond* those that are known.

Exercises

Set I

Guess a formula for the function represented by each of these tables. Begin each formula with $y =$.

1.
x	1	2	3	4	5
y	3	6	9	12	15

2.
x	0	1	2	3	4
y	5	6	7	8	9

3.
x	2	3	4	5	6
y	4	9	16	25	36

4.
x	0	10	20	30	40
y	7	27	47	67	87

5.
x	1	2	3	4	5
y	11	10	9	8	7

Chapter 3: FUNCTIONS AND THEIR GRAPHS

On graph paper, draw and label a pair of axes extending 8 units in each direction from the origin.

6. Plot the following points on your axes: A(1, 2), B(-3, -6), C(3, 3), D(-2, -4), E(0, 0), and F(4, 8).

7. Which one of the six points in your graph seems to be out of place?

8. The coordinates of the other five points have something in common that the coordinates of the "out of place" point do not. What is it?

Copy and complete the tables for the following functions.

9. Formula: $y = x + 6$

x	-3	-2	-1	0	1	2	3
y	3						

10. Formula: $y = x^2 + 6$

x	-3	-2	-1	0	1	2	3
y	15						

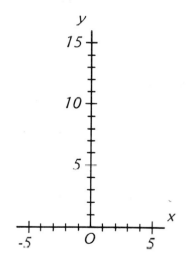

11. Draw and label a pair of axes like the pair shown above. Use the tables that you made in exercises 9 and 10 to graph the functions $y = x + 6$ and $y = x^2 + 6$ on your pair of axes. Write each formula along its line or curve.

12. What do the two graphs have in common?

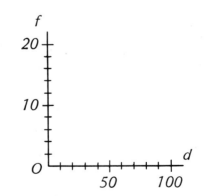

Set II

The lever is a simple machine, which has been known for thousands of years. By using a lever, it is possible to move a heavy weight with a small force.

Suppose that a weight of 30 kilograms is located 20 centimeters from a lever's pivot point. The force necessary to move the weight

is a function of the distance at which it is applied from the other side of the pivot. In this case, the formula for the function is

$$f = \frac{600}{d}$$

in which f is the force in kilograms and d is the distance in centimeters from the pivot.

1. Copy and complete the following table for this function.

d	30	40	50	60	80	100
f	20					

2. Graph it, labeling the axes as shown here.

3. Use your graph to estimate the force at a distance of 75 centimeters from the pivot that would be needed to move the weight.

By permission of Johnny Hart and Field Enterprises, Inc.

Chapter 3: FUNCTIONS AND THEIR GRAPHS

4. In estimating, did you *interpolate* or *extrapolate?* Explain.

The power generated by a windmill is a function of the speed of the wind. A typical formula for this function is

$$P = \left(\frac{s}{10}\right)^3$$

in which P represents the power generated and s represents the speed of the wind in kilometers per hour.

5. Copy and complete the following table for this function.

s	0	10	20	30	40	50
P	▦	▦	8	▦	▦	▦

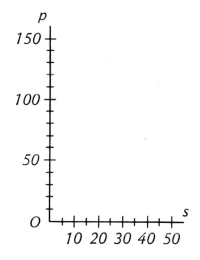

6. What happens to the power generated by a windmill if the speed of the wind doubles?

7. What happens to the power generated by a windmill if the speed of the wind triples?

8. Graph the function, labeling the axes as shown at the right.

Tsunamis are ocean waves produced by earthquakes or underwater volcanic eruptions. The speed of a tsunami as it approaches land is a function of the depth of the water and is approximated by the formula

$$s = 3\sqrt{d}$$

in which d represents the depth of the water in meters and s represents the speed of the tsunami in meters per second.

9. Copy and complete the following table for this function.

d	0	1	4	9	16
s	0	▦	▦	▦	▦

10. Graph it.

The height of a tsunami increases as it slows down.

11. What happens to the *speed* of a tsunami as it approaches land?

12. What happens to its *height?*

Set III

The speed of the hand in a karate strike is a function of time. A table for this function for a hammer-fist strike is given here.*

Time in hundredths of a second, t	0	1	2	3	4	6	8	10	13	14
Speed of fist in meters per second, s	0	2.0	3.0	2.8	2.4	2.7	4.4	6.7	11.0	11.6

Photograph by Bruce Tegner; courtesy of Thor Publishing Company

1. Graph this function.

2. How does the speed of the fist change during a hammer-fist strike?

3. Estimate the speed of the fist at 5 hundredths of a second.

4. Estimate the speed of the fist at 15 hundredths of a second.

5. Which of your estimates do you think is more reliable? Explain.

---

* The data in this table is based on a graph in the article "The Physics of Karate" by Michael S. Feld, Ronald E. McNair, and Stephen R. Wilk, *Scientific American*, April 1979.

Chapter 3
Problems for
Further Exploration

Lesson 1

1. A person lost on a desert sometimes walks a great distance in what he thinks is a straight line, only to return to the place from which he started without realizing it. The reason for this is that his legs are not exactly the same length, so that the steps taken with one foot are slightly longer than those taken with the other. An exaggerated diagram of this is at the right below.

   a) If someone's left leg is longer than his right leg, would you expect him to turn to the *left* or *right*?

   The radius of the circle in which a person walks is a function of the difference between the lengths of his steps. A typical table for this function is shown here.

Difference in millimeters, d	1	2	3	4	5
Radius of circle in meters, r	180	90	60	45	36

   b) What happens to the radius of the circle as the difference between the lengths of the steps decreases?

   c) Write a formula for this function.

   d) If the difference between the lengths of a person's steps is 2.5 millimeters, what do you think would be the radius of the circle in which he walked?

   e) Can the formula be used to find the radius of the circle in which someone would walk if the difference between the lengths of his steps was 0 millimeters? Explain.

Drawing by Jack Tippit;
© 1968 The New Yorker
Magazine, Inc.

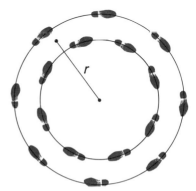

2. The following pattern is based on a table for the function $y = x^2$.

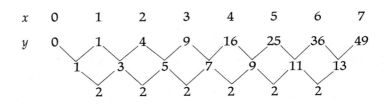

x	0	1	2	3	4	5	6	7
y	0	1	4	9	16	25	36	49

a) How can the numbers in the third and fourth rows of this pattern be obtained from the numbers above them?
b) Use the same procedure to copy and complete the following pattern based on a table for the function $y = x^3$.

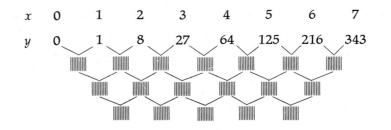

x	0	1	2	3	4	5	6	7
y	0	1	8	27	64	125	216	343

c) Make a similar pattern for the function $y = x^4$.
d) Make a pattern for the function $y = x$.
e) On the basis of the patterns for $y = x$, $y = x^2$, $y = x^3$, and $y = x^4$, what would you predict about the corresponding pattern for $y = x^5$?

Lesson 2
1. In coordinate geometry, to each *geometric line* there corresponds an *algebraic equation*.
   a) To see an example of this, draw a pair of axes, plot the points (1, 3) and (5, 7), and draw a line through them.
   b) Use the line that you drew to copy and complete the following table.

x	0	1	2	3	4	5
y	2	3				7

   c) Write a formula for a function having this table. This formula is the equation of the line you drew.
   d) Draw another pair of axes, plot the points (1, 3) and (3, 9), and draw a line through them.

Chapter 3: FUNCTIONS AND THEIR GRAPHS

e) Use the line that you drew to copy and complete the following table.

x	0	1	2	3	4
y	▓	3	▓	9	▓

f) What is the equation of the line that you drew?

See if you can discover equations for the lines through the following pairs of points by repeating the procedure above.

g) (1, 4) and (3, 10).

h) (2, 5) and (6, 1).

i) (0, −1) and (5, 9).

2. In his book *On Growth and Form*, the great British zoologist D'Arcy Thompson applied Descartes's method of coordinates to the study of the shapes of living things. With a series of examples, he showed how a coordinate system could be applied to the form of an animal and the system then transformed in a certain way to obtain the form of a different animal.

One of the examples concerns a fish of the species *Argyropelecus olfersi*.

a) Draw and label a pair of axes as shown here. Plot the point

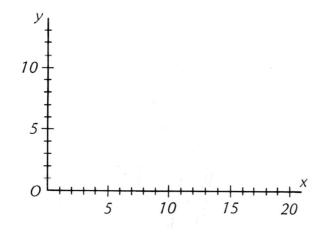

(3, 7) and draw a small circle around it to represent the eye of the fish. Then connect the points in the following list in order with straight line segments to form the outline of the fish.

(0, 6) (1, 4) (3, 2) (8, 0) (8, 1) (10, 2) (10, 4) (11, 3) (12, 5)
(16, 4) (15, 6) (16, 8) (12, 7) (9, 8) (11, 10) (9, 12) (8, 10)
(8, 12) (6, 10) (1, 7) (2, 5) (0, 6)

b) Make another set of coordinates by changing each x-coordinate in the preceding set like this:

$$(x, y) \rightarrow \left(x + \frac{y}{2}, y\right)$$

For example, the eye of the fish would be changed from $(3, 7)$ to $\left(3 + \frac{7}{2}, 7\right)$, or $(6.5, 7)$, and the first point in the outline of the fish would be changed from $(0, 6)$ to $\left(0 + \frac{6}{2}, 6\right)$, or $(3, 6)$.

c) Draw and label another pair of axes and repeat the directions given in part a of this exercise with the list of coordinates that you made in part b. The result is the outline of a fish of a species of a different genus, *Sternoptyx diaphana*.

Lesson 3

1. The following puzzle was given on a radio quiz program*:

You and I have 35 apples together. You have two-fifths of what I have. How many do we each have?

The first person calling in who was able to figure it out would win five dollars, yet no one of the three people given the puzzle could give the correct answer.

One way to solve it is with a graph. If we let x represent the number of apples that I have and y represent the number of apples that you have, we can write the first statement,

You and I have 35 apples together,

as:

$$x + y = 35 \quad \text{or} \quad y = 35 - x.$$

The second statement,

You have two-fifths of what I have,

would be:

$$y = \frac{2}{5}x.$$

---

* "Testing One Two Three," KNX radio, Los Angeles.

Chapter 3: FUNCTIONS AND THEIR GRAPHS

a) Make a table for each of these equations, letting $x = 0, 5, 10,$ and 15.

b) Graph both equations on the same pair of axes, on which 1 unit on each axis represents 2 apples. Extend both lines across the graph.

c) Does the graph of $y = 35 - x$ meet the x-axis in the right place? Explain.

d) How can your graph be used to determine the answer to the puzzle?

e) What is the answer?

2. In about 450 B.C., the Greek philosopher Zeno made up a well-known puzzle about a race between Achilles and a tortoise on which the following exercise is based. Achilles runs ten times as fast as the tortoise, and so the tortoise is given a head start. If they run at the rates of 500 meters per minute and 50 meters per minute, respectively, and the tortoise is given a head start of 1,000 meters, when and where will Achilles overtake the tortoise?

a) Make a table showing the distances Achilles covers in 1, 2, 3, and 4 minutes.

b) Write a formula for the distances run by Achilles, d, as a function of time, t.

c) Make a table showing the tortoise's distances in 1, 2, 3, and 4 minutes, supposing that its distance at the beginning of the race is 1,000 meters.

d) Write a formula for the distances of the tortoise along the track, d, as a function of time, t.

e) Draw and label a pair of axes as shown here and graph both functions on it.

f) Approximately when and where does Achilles overtake the tortoise?

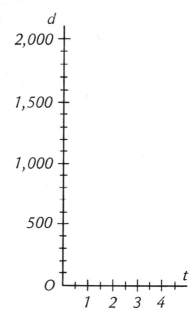

Lesson 4

1. The number of pieces into which a pancake can be cut by a series of straight cuts, each of which crosses the others, is a function of the number of cuts. One cut across a pancake divides it into two pieces. A second cut crossing the first one results in four pieces, and a third cut crossing the first two cuts results in seven pieces.

a) Draw two figures representing pancakes that have been cut with four cuts and five cuts respectively, in which each cut crosses each of the others.
b) Refer to the figures above and the figures that you have drawn to copy and complete the following table.

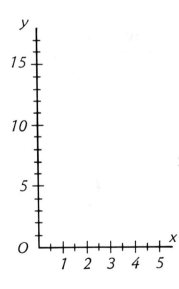

Number of cuts, x	1	2	3	4	5
Number of pieces, y					

c) Graph it, labeling the axes as shown at the left.
d) What do you think the value of y would be if $x = 0$? Does your answer make any sense? Explain.
e) Use your graph to decide the value of y when $x = 3.5$. Does your answer make any sense? Explain.

2. As a swimming pool is drained for cleaning, the volume of water remaining in the pool is a function of the time that the water has been draining. A formula for this function for a certain pool is

$$v = 20(30 - t)^2$$

in which v represents the number of gallons of water in the pool and t represents the time in minutes.

a) Copy and complete the following table for this function.

t	0	5	10	20	25	30
v	18,000					

Chapter 3: FUNCTIONS AND THEIR GRAPHS

b) Graph it, labeling the axes as shown here.

c) Use your graph to estimate the time that it takes to drain half the water from the pool.

d) Does the water drain from the pool at a steady rate? Explain.

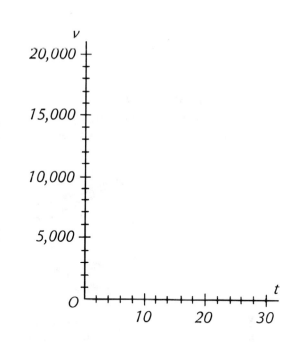

Lesson 5

1. Experiment: *The Volume of a Tray*

On a sheet of graph paper ruled 4 units per inch (or 2 units per centimeter), draw four rectangles, each having a length of 20 units and a width of 12 units. Cut the rectangles out and draw squares having sides of lengths 1, 2, 3, and 4 units in their corners as shown here. Cut out the squares and fold up the

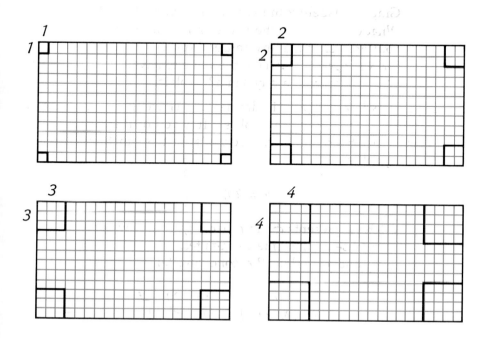

sides of each rectangle to form four trays. The shallowest tray is shown here.

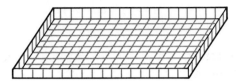

The volume of each tray is the number of cubes measuring 1 unit on each edge that are needed to completely fill it. This number can be found by multiplying its length, width, and height.

a) Refer to your four trays to copy and complete the following table.

Sides of squares cut from corners	1	2	3	4
Length of tray	18			
Width of tray	10			
Height of tray	1			
Volume of tray	180			

Use the patterns in the table to add a column for the tray that would be produced if squares with sides 5 units long were cut from the corners of a 20 × 12 rectangle.

The volume of the tray formed from a 20 × 12 rectangle is a function of the length of the sides of the squares cut from its corners.

b) Refer to your table to copy and complete the following table for this function.

Length of side of squares cut from corners, s	0	1	2	3	4	5
Volume of tray, v	0	180				

c) Draw and label a pair of axes as shown at the left, plot the five points from your table, and draw a smooth curve through them.

d) How do you think a tray having the largest possible volume could be formed by cutting squares from the corners of a 20 × 12 rectangle?

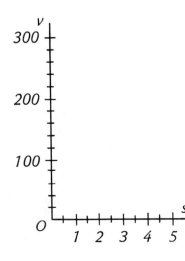

Chapter 3: FUNCTIONS AND THEIR GRAPHS

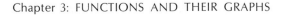

2. The number of people who catch the flu during an epidemic is a function of time. A typical formula for a small town is

$$n = \frac{170t^2}{t^2 + 1}$$

in which n represents the number of people who have caught the flu and t represents the number of weeks from the beginning of the epidemic.
a) Make a table for this function, letting $t = 0, 1, 2, 3$, and 4.
b) Graph the function, labeling the axes as shown here.
c) What happens to the number of cases as time passes?
d) What happens to the *rate* at which the number of cases changes as time passes?
e) The number of cases, n, gets closer to a certain number as t gets larger. What do you think the number is?

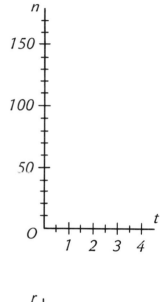

Lesson 6
1. The sex ratio of a population is the number of males for every 100 females. The following table shows its value for the United States as determined by the last eight censuses.

Year, y	1910	1920	1930	1940	1950	1960	1970	1980
Sex ratio, r	106	104	103	101	99	98	96	95

a) Graph the function, labeling the axes as shown here. (The scale on the r-axis jumps from 0 to 95 to save space.)
b) What do you notice about the graph?
c) If the sex ratio continued to change at the same rate, when would it become 0? Explain your reasoning.

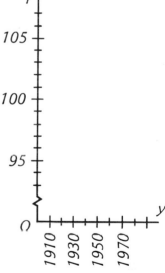

2. An application of extrapolation about which there is still much uncertainty concerns the age of the universe. That the universe is expanding seems certain, but whether it will expand forever is not known.
   According to an article in *Scientific American,*

   *If the motions of the galaxies are extrapolated into the past as far as possible, a state is eventually reached in which all the galaxies were crushed together at infinite density. That state represents the big bang, and it marks the origin of the universe and everything in it.**

---

* "Will the Universe Expand Forever?" by J. Richard Gott, III, James E. Gunn, David N. Schramm, and Beatrice M. Tinsley in *Scientific American,* March 1976.

The graph below, from the article, shows three possible ways in which the scale of the universe may be changing with the passage of time.

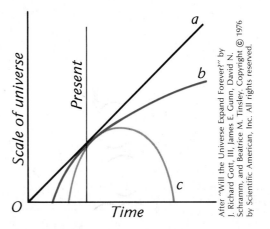

a) What does line *a* suggest as one possible history of the universe?

b) What does curve *b* suggest? (Assume that it continues to rise as time passes.)

c) What does curve *c* suggest?

d) Which of these three possibilities suggests the greatest age for the universe?

e) What is it about the line and two curves at the line marked "present" that makes it difficult to decide which one represents the actual history of the universe?

Chapter 4

LARGE NUMBERS AND LOGARITHMS

Lesson 1

Large Numbers

What is the largest number that you can think of? A billion? A trillion? If you wrote down a one and started adding zeros, it would not take you very long to write a number so big that it would be larger than most of the numbers whose names are commonly known. Schroeder has thought of "googol," which, strange though it may seem, is the accepted name for the number 1 followed by a hundred zeros.* Although it has a name, this number is far too big to be comprehended in ordinary terms. In fact, it is difficult for us to grasp the size of a number even as "small" as a million, because we have little personal experience with such a number.

Do you know how large a million is? It is easy to write 1,000,000, but how big is that? If you counted to one million and could name a number every second without stopping, it would take you nearly 12 days. And how long is a million days? A million days ago was in the eighth century B.C. How far is a million inches? Almost 16 miles.

If our idea of the size of a million is vague, our notions of larger numbers must be even vaguer. To take another example, how large is a billion? Oddly enough, it depends on where you live. In the United States a billion is 1,000,000,000, whereas in England it is 1,000,000,000,000. Perhaps the reason that the difference remains is that until recently there was very little need for such a large number. Using the smaller version, a billion seconds is still a

---

* The name "googol" was invented by a young nephew of the American mathematician Edward Kasner when the boy was asked to make up a name for a very large number.

long time; a billion seconds from now is in the twenty-first century.

Is there much difference in the size of two numbers such as

$$100000000 \quad \text{and} \quad 1000000000?$$

At a glance they look about the same. Yet, when we add a zero to the end of the first number to give the second, we have multiplied it by ten. Do not be fooled into thinking that, because zero stands for nothing, adding a zero to the end of a number makes little difference.

Each number in the sequence

$$1 \quad 10 \quad 100 \quad 1{,}000 \quad 10{,}000 \quad 100{,}000 \quad 1{,}000{,}000 \quad \dots$$

is ten times as large as the number preceding it. Another way to write this sequence is

$$1 \quad 10 \quad 10^2 \quad 10^3 \quad 10^4 \quad 10^5 \quad 10^6 \quad \dots$$

and this suggests the use of exponents as a way to write large numbers compactly. Notice that the *exponent* of each 10 is also the *number of zeros* that follow the 1 if the number is written the long way. This pattern suggests that it makes sense to write

$$10 \text{ as } 10^1$$

and even

$$1 \text{ as } 10^0 \text{ (1 followed by 0 zeros).}$$

The beginning of the sequence can then be written as

$$10^0 \quad 10^1 \quad 10^2 \quad 10^3 \quad \dots$$

Here is a list of the names and exponential forms of some large numbers.

10^2	hundred	10^{18}	quintillion
10^3	thousand	10^{21}	sextillion
10^6	million	10^{24}	septillion
10^9	billion	10^{27}	octillion
10^{12}	trillion	10^{30}	nonillion
10^{15}	quadrillion	10^{33}	decillion

Exercises

Set I

Refer to the information in this lesson to write each of the following numbers as a power of 10.

1. One.

2. Ten.

3. One thousand.

4. Ten billion.

5. One hundred quadrillion.

6. One thousand octillion.

7. A googol.

The small figure at the left below contains 10^2 dots.

8. Estimate the number of dots in the large figure as a power of 10.

Chapter 4: LARGE NUMBERS AND LOGARITHMS

A tourist from England visited Yankee Stadium to see a baseball game. He did not understand the game and left when the scoreboard read

1 0 0 0 0 0 0 0 0
1 0 0 0 0 0 0 0 0

When a boy outside the gate asked him the score, he answered: "It's up in the millions."

9. How did the man get so confused?

The human brain contains about 10^{10} cells, called neurons, and is far more complex than any computer that has ever been built.

10. Write the number of cells in decimal form.

11. Name the number in words.

If this number was represented by a row of dots like this,

. . . . ,

it would stretch more than 6,000 miles.

The Milky Way galaxy, of which our solar system is a part, is estimated to contain approximately one hundred billion stars.

12. Write one hundred billion in decimal form.

Star clouds in the Sagittarius region of our galaxy, the Milky Way

Courtesy of the Mount Wilson and Palomar Observatories

13. Write it as a power of 10.

The universe is estimated to contain one trillion galaxies.

14. Write one trillion in decimal form.

15. Write it as a power of 10.

16. How many stars are there in the universe if each galaxy contains the same number as our own?

A glass of water contains approximately

$$10,000,000,000,000,000,000,000,000$$

water molecules.

17. Write this number as a power of 10.

18. Name it.

19. Write the approximate number of molecules contained in 10 glasses of water as a power of 10.

The Atlantic Ocean contains approximately

$$10,000,000,000,000,000,000,000,000,000,000,000,000,000,000$$

molecules of water.

20. Write this number as a power of 10.

The Pacific Ocean contains approximately

$$20,000,000,000,000,000,000,000,000,000,000,000,000,000,000$$

molecules of water.

21. How does this number compare with the number of water molecules in the Atlantic Ocean?

22. Write it in a more compact way.

You know that 10^2 means "1 followed by 2 zeros," or 100. The following exercises deal with powers of numbers larger than 10.

23. Copy and complete the following table of squares.

$$100^2 = 100 \times 100 = \text{▦}$$
$$1,000^2 = \text{▦} \times \text{▦} = \text{▦}$$
$$10,000^2 = \text{▦} \times \text{▦} = \text{▦}$$

Because $100 = 10^2$, we can write the numbers in the first line of the table as powers of 10:

$$(10^2)^2 = 10^2 \times 10^2 = 10^4$$

24. Using the fact that $1,000 = 10^3$, write the numbers in the second line of the table as powers of 10.

25. Write the numbers in the third line of the table as powers of 10.

26. Copy and complete the table of cubes at the right.

$$100^3 = \text{▥} \times \text{▥} \times \text{▥} = \text{▥}$$
$$1,000^3 = \text{▥} \times \text{▥} \times \text{▥} = \text{▥}$$
$$10,000^3 = \text{▥} \times \text{▥} \times \text{▥} = \text{▥}$$

Writing the numbers in the first line of this table as powers of 10, we get

$$(10^2)^3 = 10^2 \times 10^2 \times 10^2 = 10^6$$

27. Write the numbers in the second line of the table as powers of 10.

28. Write the numbers in the third line of the table as powers of 10.

29. Which number is larger: 10^{100} or 100^{10}? Show your reasoning.

A billion seconds is approximately 32 years.

30. Use this information to copy and complete the following table.

Seconds ago	Years ago	Date
0	0	(Fill in the present year here.)
10^8	▥	▥
10^9	32	▥
10^{10}	▥	▥
10^{11}	▥	▥
10^{12}	▥	▥

Set II

The Greek mathematician Archimedes, who lived in the third century B.C., devised a method for forming large numbers. He used it in a book called *The Sand Reckoner* to calculate the number of grains of sand that would be needed to fill the entire universe.

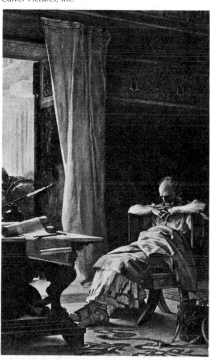

Archimedes

The largest number given a name by the Greeks was a "myriad," which we call "ten thousand."

1. Write this number as a power of 10.

Archimedes began by thinking of a "myriad of myriads."

2. What did he mean by this?

3. What would we call this number?

He referred to the numbers from 1 to 100,000,000 as numbers of "the first order."

4. Write 100,000,000 as a power of 10.

Archimedes referred to the numbers from 100,000,000 to $100,000,000^2$ as numbers of "the second order."

5. Write $100,000,000^2$ out the long way.

6. Write it as a power of 10.

The numbers from $100,000,000^2$ to $100,000,000^3$ were considered to be the numbers of "the third order."

7. Write $100,000,000^3$ out the long way.

8. Write it as a power of 10.

Archimedes continued in this way until he got to the numbers of "the 100,000,000th order": the numbers from $100,000,000^{99,999,999}$ to $100,000,000^{100,000,000}$.

9. Write $100,000,000^{100,000,000}$ as a power of 10.

This number is extraordinarily large.

10. If it were written out in full, what would its digits consist of?

11. Do you think you could write it out the long way in a reasonable length of time?

Archimedes calculated that, in contrast with this immense number, the number of grains of sand needed to fill the entire universe was no more than 10^{63}.

12. To which of Archimedes' orders of numbers does 10^{63} belong?

According to current estimates of the size of the observable universe, the number of grains of sand of the size used by Archimedes needed to fill it completely is no more than 10^{93}.

Set III

Most of the world's major countries are linked by submarine telephone cables. Some of them are shown in this map.

Amplifiers, called repeaters, are spaced evenly along each cable to keep the sound loud enough to hear. Each repeater multiplies the signal strength about a million times. One of the longest links in the submarine cable system is the one connecting California and Hawaii. It contains 57 repeaters in each direction.

The first repeater on the cable from California to Hawaii amplifies the signal 10^6 times. After it passes through the second repeater, it has been boosted $10^6 \times 10^6$ times.

1. How many times is that? Express it as a power of 10.

Upon passing through the third repeater, the signal has been strengthened $10^6 \times 10^6 \times 10^6$ times.

2. Altogether how many times has it been boosted by the time it passes through the last repeater just before reaching Hawaii?

3. Why is the speaker's voice not deafening after having been amplified by all those repeaters?

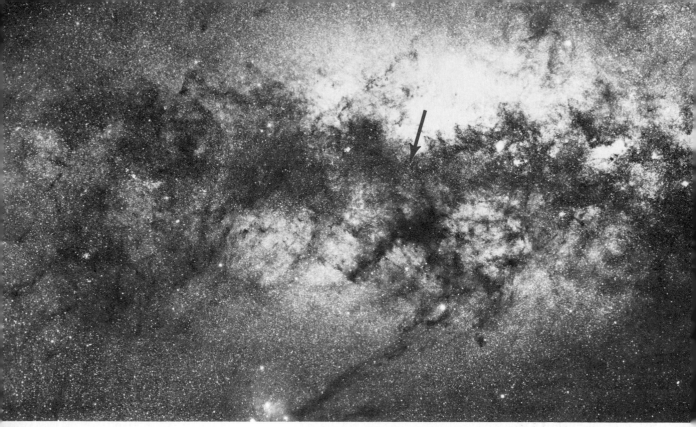

Palomar Observatory, California Institute of Technology; courtesy of Hale Observatories

Lesson 2 Scientific Notation

The center of our Milky Way galaxy lies hidden in dust clouds in the direction of the constellation Sagittarius. Marked with an arrow in the photograph above, it is surrounded by millions of stars and may contain a black hole.*

Astronomers have determined that the distance between the earth and this point is approximately

$$300,000,000,000,000,000$$

kilometers. This number can be written in a more convenient form by observing that

$$300,000,000,000,000,000 = 3 \times 100,000,000,000,000,000.$$

---

* "The Central Parsec of the Galaxy" by Thomas R. Geballe, *Scientific American*, July 1979, and "The Center of the Galaxy" by R. H. Sanders and G. T. Wrixon, *Scientific American*, April 1974.

Writing 100,000,000,000,000,000 as a power of 10, we have

$$3 \times 10^{17}.$$

Written in this form, the number is in *scientific notation.*

▶ A number is in **scientific notation** if it is written in the form

$$a \times 10^b$$

in which a is a number that is at least 1 but less than 10.

Compare 300,000,000,000,000,000 with its form in scientific notation, 3×10^{17}. In the scientific form, the decimal point has been moved from the end of the number to just after its first digit. The point has been moved 17 decimal places and the 17 has become the exponent of the 10.

$$3\,0\ 0\ ,\ 0\ 0\ 0\ ,\ 0\ 0\ 0\ ,\ 0\ 0\ 0\ ,\ 0\ 0\ 0 = 3 \times 10^{17}$$
17 16 15 14 13 12 11 10 9 8 7 6 5 4 3 2 1

The mass of the Milky Way is thought to be about

400,000,000,000,000,000,000,000,000,000,000,000,000,000

kilograms. Following the procedure described above, we can write this number as

$$4 \times 10^{41}.$$

It is easy to see that the larger the number the greater the advantage of writing it in scientific notation.

When we look at the center of the Milky Way, we are looking into the past because it takes 3.2×10^4 years for light from it to reach us. How many years is that? We can reverse the procedure described above to find out.

$$3.2 \times 10^4 = 3\,2\,0\,0\,0$$
1 2 3 4

Writing in a comma, we get 32,000. When we look at the center of our galaxy, we are seeing it as it existed 32,000 years ago.

Exercises

Set I

It would take 3,000,000,000,000,000,000,000,000,000 candles to give as much light as the sun.

1. Write this number in scientific notation.

2. What advantages does this way of writing the number have over the decimal form used above?

The earth picks up approximately 1.2×10^7 kilograms of dust from outer space each day.

3. Write this number in decimal form.

4. Name it.

Signals have been sent from the radio telescope in Arecibo, Puerto Rico, in the hope of making contact with an extraterrestrial civilization.* They were sent on a radio frequency of 1,420,000,000 cycles per second.

5. Write this number in scientific notation.

6. Name it.

Q. How many beans are in the box?

A. About twice as many beans as you'd find in a normal station wagon.

The Volkswagen company once claimed in one of its ads that "exactly 1,612,462 beans" can be put into a Volkswagen station wagon.

7. Write this number in scientific notation.

8. Is this way of writing the number any shorter than writing it in decimal form?

If the number of beans were rounded off, we might say that a VW wagon will hold *about* 1,600,000 beans.

9. Write this number in scientific notation.

10. Is this way of writing the number any shorter than writing it in decimal form?

* See the photograph on page 1.

Chapter 4: LARGE NUMBERS AND LOGARITHMS

11. Which kind of numbers are simplified the most by scientific notation: exact numbers or approximate numbers?

The number of hairs on a person's head varies with the color of his or her hair. A brunette has about 1.05×10^5 hairs, a blond has about 1.4×10^5, and a redhead has about 9×10^4.

12. Write each of these numbers in decimal form.

13. Which one is biggest?

14. If two numbers are written in scientific notation, how can you tell which one is bigger without changing them to decimal form?

The nearest star beyond the sun, Proxima Centauri, is 42 trillion kilometers away. One way of writing this distance is

$$42 \times 10^{12} \text{ kilometers.}$$

15. This number is not in scientific notation. Refer to the definition of scientific notation given in this lesson to explain why not.

16. Write this number in decimal form.

17. Write it in scientific notation.

According to John Scarne, "the reason Bridge surpasses most other card games in strategy is due to the fact that . . . the number of different card combinations that face each player is virtually infinite."*

The number of different possible bridge hands is about 635,000,000,000.

18. Name this number.

19. Write it in scientific notation.

Inflation is a serious economic problem. In 1946, inflation of the currency in Hungary was so great that the gold pengo was worth one hundred thirty quintillion paper pengos.

20. Write this number in decimal form.

21. Write it in scientific notation.

A bill worth 100 quintillion pengos

_____

* *Scarne's New Complete Guide to Gambling* by John Scarne (Simon and Schuster, 1974).

Set II

It is often more convenient to make calculations with large numbers when they are written in scientific notation than when they are written in decimal form.

Consider, for example,

$$30{,}000 \times 15{,}000{,}000.$$

1. Find the answer by multiplying the numbers in the form shown.

2. Rewrite both numbers in scientific notation.

3. Rewrite the answer to the problem in scientific notation.

4. How could this answer have been obtained by multiplying the numbers as written in scientific notation?

Use this shortcut to find the following products.

5. $(2 \times 10^{10}) \times (3 \times 10^5)$

6. $(5 \times 10^8) \times (1.6 \times 10^{12})$

Multiplying two numbers by this method may not immediately give the result in scientific notation. If this is the case, an additional step is required, as illustrated in the example below.

$$
\begin{aligned}
(4 \times 10^3) \times (5 \times 10^4) &= 20 \times 10^7 \\
&= 2 \times 10 \times 10^7 \\
&= 2 \times 10^8
\end{aligned}
$$

Write the results in each of these problems in scientific notation.

7. $(5 \times 10^6) \times (9 \times 10^2)$

8. $(8 \times 10^3) \times (7.5 \times 10^{10})$

One of the largest swarms of locusts ever seen was estimated to contain 4×10^{10} insects.

9. Write this number in decimal form.

A locust weighs 2 grams.

10. What is the total weight (in decimal form) of such a swarm?

11. Write it in scientific notation.

A locust is capable of consuming as much as 60 grams of grain in a month.

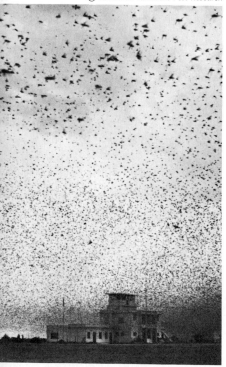

Photograph by A.J. Wood;
© Centre for Overseas Pest Research

A swarm of locusts

Chapter 4: LARGE NUMBERS AND LOGARITHMS

12. How many grams of grain (in decimal form) could be consumed in a month by a swarm of the size mentioned above?

13. Write it in scientific notation.

The crystals used in quartz-crystal clocks make 10^5 vibrations in one second.

14. How many vibrations do they make in one minute? Give your answer in scientific notation.

There are approximately 3.2×10^7 seconds in a year.

15. How many vibrations do such crystals make in a year? Give your answer in scientific notation.

One ounce of gold contains approximately 9.5×10^{22} atoms. Each bar of gold kept in Fort Knox weighs 400 ounces.

Fort Knox, Kentucky

16. Approximately how many atoms of gold are in one bar?

The amount of gold recovered from the earth since the beginning of time is estimated to be 2×10^9 ounces.

17. Approximately how many atoms of gold have been recovered from the earth?

Light travels approximately 3×10^8 meters per second. Moonlight takes 1.25 seconds to reach the earth.

18. Approximately how many meters is the moon from the earth?

It takes light from the sun 500 seconds to reach the earth.

19. Approximately how many meters is the sun from the earth?

It takes light from the Great Galaxy in Andromeda, the most distant object in the universe that can be seen without a telescope, approximately 7×10^{13} seconds to reach the earth.

20. Approximately how many meters is this galaxy from the earth?

Set III

The photograph below shows a large number of Peruvian booby birds on South Guanape Island, Peru.

1. Make an estimate of the number of birds in the picture and write it in scientific notation.

2. Explain how you made your estimate.

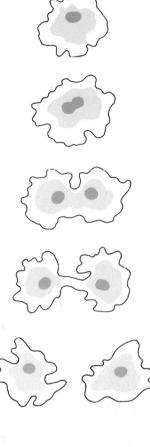

Lesson 3

An Introduction to Logarithms

An amoeba can do an unusual mathematical trick: it multiplies by dividing. After it has grown to a certain size, the amoeba's single cell divides in half to produce two amoebas. In about a day, the two amoebas have grown to the point at which they are ready to divide and form four; the day after that, there are eight amoebas, and so forth. How many amoebas will there be at the end of a week?

The number of amoebas is a function of the time that has passed. A table for this function looks like this:

Time in days	0	1	2	3	4	5	6	7	8	9	10	...
Number of amoebas	1	2	4	8	16	32	64	128	256	512	1,024	...

The answer is in the table: at the end of 7 days there will be 128 amoebas.

Notice that the second line of numbers in this table is the binary sequence, a geometric sequence in which each term is twice the one it follows. The first line is an arithmetic sequence in which each term is one more than the preceding term.

Something remarkable about this table was discovered by the Scottish mathematician John Napier in the early seventeenth century. Suppose that we choose two numbers from the second sequence, say 4 and 32. If we multiply these numbers,

$$4 \times 32 = 128,$$

we get another number of the sequence. The numbers in the first sequence that correspond to 4, 32, and 128 are 2, 5, and 7, and here is the remarkable part:

$$2 + 5 = 7$$

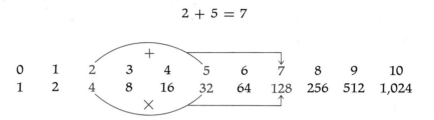

We have discovered that a *multiplication* in the second sequence corresponds to an *addition* in the first sequence.

Here is another example:

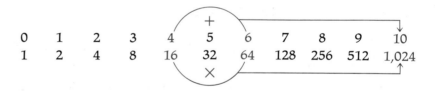

These examples suggest that, if we do not want to *multiply*, we can *add* instead by using the table. For example, what is 16 × 64? Finding these numbers in the second sequence, we look above them and see 4 and 6; adding 4 and 6, we get 10; looking below 10 we see 1,024.

So, without doing any multiplying, we have found that 16 × 64 = 1,024 by simply adding 4 and 6.

The numbers in the first sequence are the *logarithms* of the corresponding numbers of the second sequence:

Logarithms, x	0	1	2	3	4	5	6	7	8	9	10 ...
Numbers, y	1	2	4	8	16	32	64	128	256	512	1,024 ...

A formula for this table is

$$y = 2^x$$

Chapter 4: LARGE NUMBERS AND LOGARITHMS

in which y represents a number in the binary sequence and x represents its logarithm. Notice that the logarithm, x, is the *exponent* of the 2. The logarithm of 8, for example, is 3 because $8 = 2^3$; the logarithm of 16 is 4 because $16 = 2^4$, and so forth.

Exercises

Set I

Here is a longer version of our logarithm table. It is arranged in columns with the terms of the binary sequence listed in the first column and their logarithms listed in the second.

Number	Loga-rithm	Number	Loga-rithm	Number	Loga-rithm
1	0	2,048	11	4,194,304	22
2	1	4,096	12	8,388,608	23
4	2	8,192	13	16,777,216	24
8	3	16,384	14	33,554,432	25
16	4	32,768	15	67,108,864	26
32	5	65,536	16	134,217,728	27
64	6	131,072	17	268,435,456	28
128	7	262,144	18	536,870,912	29
256	8	524,288	19	1,073,741,824	30
512	9	1,048,576	20		
1,024	10	2,097,152	21		

1. Find the answer to this problem by multiplying: 32×128. Show your work.

2. Referring to the table, show the corresponding addition problem that can be used to find the answer without multiplying.

3. Find the answer to this problem by multiplying: $512 \times 1,024$. Show your work.

4. Referring to the table, show the corresponding addition problem that can be used to find the answer without multiplying.

Use the table to find answers to the following problems without doing any multiplying.

5. $16 \times 65,536$

6. $8,192 \times 16,384$

7. $64 \times 256 \times 2,048$

Logarithms can also be used to raise numbers to powers, as the example below illustrates.

$$\text{Numbers} \quad 16 \times 16 \times 16 = (16)^3 = 4{,}096$$
$$\text{Logarithms} \quad 4 + 4 + 4 = 3 \times 4 = 12$$

Use this method and the table to find answers to the following problems.

8. $(128)^2$

9. $(64)^3$

10. $(32)^4$

11. $(8)^{10}$

12. $(512)^0$

Set II

Why the logarithm table works as it does can be easily understood by looking at it in the following form.

Number	Logarithm
1	0
2 = 2	1
4 = 2 × 2	2
8 = 2 × 2 × 2	3
16 = 2 × 2 × 2 × 2	4
32 = 2 × 2 × 2 × 2 × 2	5
64 = 2 × 2 × 2 × 2 × 2 × 2	6
128 = 2 × 2 × 2 × 2 × 2 × 2 × 2	7
256 = 2 × 2 × 2 × 2 × 2 × 2 × 2 × 2	8
512 = 2 × 2 × 2 × 2 × 2 × 2 × 2 × 2 × 2	9
1,024 = 2 × 2 × 2 × 2 × 2 × 2 × 2 × 2 × 2 × 2	10
⋮	⋮

1. What does the fact that the logarithm of 32 is 5 tell us about 32?

2. What does the fact that the logarithm of 4,096 is 12 tell us about 4,096?

3. What does the fact that the logarithm of 1 is 0 tell us about 1?

The figure below shows why we can solve the multiplication problem $8 \times 16 = 128$ by doing the addition problem $3 + 4 = 7$.

Numbers $\underline{8}$ $\times$ $\underline{16}$ $=$ $\underline{128}$
           $\underbrace{2 \times 2 \times 2}$ $\times$ $\underbrace{2 \times 2 \times 2 \times 2}$ $=$ $\underbrace{2 \times 2 \times 2 \times 2 \times 2 \times 2 \times 2}$

Logarithms   3 $+$ 4 $=$ 7

A second example is shown below.

Numbers $\underbrace{2 \times 2}$ $\times$ $\underbrace{2 \times 2 \times 2 \times 2 \times 2 \times 2}$ $=$ $\underbrace{2 \times 2 \times 2 \times 2 \times 2 \times 2 \times 2 \times 2}$

Logarithms $+$ $=$

4. What multiplication problem is illustrated?

5. What addition problem is illustrated?

The next figure shows why we can solve the raising-to-a-power problem 4^3 by doing the multiplication problem $3 \times 2 = 6$.

Numbers $\underline{4}$ $\times$ $\underline{4}$ $\times$ $\underline{4}$ $=$ 4^3 $=$ $\underline{64}$
           $\underbrace{2 \times 2}$ $\times$ $\underbrace{2 \times 2}$ $\times$ $\underbrace{2 \times 2}$ $=$ $\underbrace{2 \times 2 \times 2 \times 2 \times 2 \times 2}$

Logarithms   2 $+$ 2 $+$ 2 $= 3 \times 2 =$ 6

A second example is shown below.

Numbers $\underbrace{2 \times 2 \times 2 \times 2 \times 2}$ $\times$ $\underbrace{2 \times 2 \times 2 \times 2 \times 2 \times 2}$ $=$

Logarithms $+$ $=$

                                  $=$ $\underbrace{2 \times 2 \times 2 \times 2 \times 2 \times 2 \times 2 \times 2 \times 2 \times 2 \times 2}$

                        $\times$ $=$

6. What raising-to-a-power problem is illustrated?

7. What multiplication problem is illustrated?

Lesson 3: An Introduction to Logarithms

The patterns that we have been considering can be expressed more compactly with exponents.

Number	Logarithm
1	0
$2 = 2^1$	1
$4 = 2^2$	2
$8 = 2^3$	3
$16 = 2^4$	4
$32 = 2^5$	5
$64 = 2^6$	6
$128 = 2^7$	7
$256 = 2^8$	8
$512 = 2^9$	9
$1{,}024 = 2^{10}$	10

The following example illustrates the multiplication pattern.

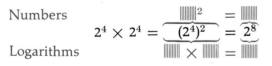

8. What multiplication problem is illustrated?

9. What addition problem is illustrated?

10. How do the logarithms appear in the equation in the middle?

The next example illustrates the raising-to-a-power pattern.

Numbers

$$2^4 \times 2^4 = \underbrace{(2^4)^2} = \underbrace{2^8}$$

Logarithms

11. What raising-to-a-power problem is illustrated?

12. What multiplication problem is illustrated?

Set III

Biologists refer to the growth period of bacteria as their "log phase" because of the connection between their repeated doublings and the logarithms introduced in this lesson. The time that it takes a population of bacteria to double varies with each species. *Escherichia coli,* one of the most rapidly growing bacteria, has a doubling time of about 15 minutes.

This means that, starting with 1 cell, there would be 2 cells after 15 minutes, 4 cells after 30 minutes, 8 cells after 45 minutes, and 16 cells after one hour.

1. If this process continued, how many cells would there be after two hours?

2. How many cells would there be after three hours?

Although it would be impossible for the cells to continue multiplying in this way for 24 hours, it is interesting to realize what would happen if they could.

3. Approximately how many cells would there be after 24 hours?

About 100,000,000,000,000 such cells weigh 1 kilogram (2.2 pounds).

4. Approximately how much would the cells in existence after 24 hours weigh?

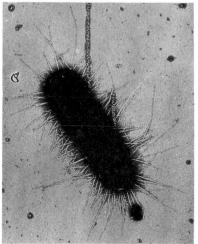

Escherichia coli

Culver Pictures, Inc.

John Napier

Lesson 4

Logarithms and Scientific Notation

Logarithms were given their name by John Napier, the Scottish mathematician credited with their invention. His book, *A Description of the Wonderful Law of Logarithms*, was published in 1614. On the three hundredth anniversary of its publication, Lord Moulton wrote:

> *The invention of logarithms came on the world as a bolt from the blue. No previous work had led up to it, foreshadowed it or heralded its arrival. It stands isolated, breaking in upon human thought abruptly without borrowing from the work of other intellects or following known lines of mathematical thought.* *

---

* "Inaugural Address: The Invention of Logarithms," *Napier Tercentenary Memorial Volume* (London, 1915).

Napier was fascinated by the properties of the binary sequence and the table below appears in his writings:

I	II	III	IV	V	VI	VII	...
1	2	4	8	16	32	64	128 ...

By rewriting it in the following form, we recognize Napier's table as the logarithm table of the preceding lesson.

Logarithms	0	1	2	3	4	5	6	7	...
Numbers		2^0	2^1	2^2	2^3	2^4	2^5	2^6	2^7 ...

This table is based on powers of *two*. However, a logarithm table built on powers of ten would be more convenient for us because our number system is based on the number *ten*. Such a table would look like this:

Logarithms	0	1	2	3	4	5	6	7	...
Numbers	10^0	10^1	10^2	10^3	10^4	10^5	10^6	10^7 ...

These logarithms are referred to as *base ten*, or *decimal*, logarithms to distinguish them from the *base two*, or *binary*, logarithms studied earlier. A formula for this table is

$$y = 10^x$$

in which y represents a number and x represents its base-ten logarithm. The connection between base-ten logarithms and our number system is a very simple one: the logarithm of 1,000,000, for example, is 6 because $1,000,000 = 10^6$.

Unfortunately, in the form in which it is written above, the base-ten table is relatively useless. Consider, for example, the problem

$$100 \times 1,000.$$

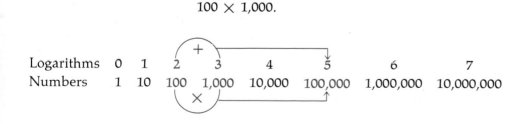

Logarithms	0	1	2	3	4	5	6	7
Numbers	1	10	100	1,000	10,000	100,000	1,000,000	10,000,000

There is little point in using logarithms to solve this problem, because 100 and 1,000 are so easy to multiply in the first place.

The value of the base-ten table becomes apparent only by filling in the gaps between the numbers on the second line. If we rewrite the beginning of the table, filling in the numbers 2 through 9 on the second line, we get

Logarithms	0	▦	▦	▦	▦	▦	▦	▦	▦	1
Numbers	1	2	3	4	5	6	7	8	9	10

The missing logarithms are evidently numbers between 0 and 1. The method for finding out what they are is quite complicated, and so no attempt will be made to explain it here.*

When the missing logarithms are computed and listed with their corresponding numbers, we have the table at the left. How this table can be used will be illustrated in the exercises.

Number	Logarithm
1	0.000
2	0.301
3	0.477
4	0.602
5	0.699
6	0.778
7	0.845
8	0.903
9	0.954
10	1.000

Exercises

Set I

From the preceding lesson we learned that *multiplying* numbers corresponds to *adding* their logarithms. This is true regardless of the number on which the logarithms are based. An example illustrating this property with base-ten, or decimal, logarithms is shown here.

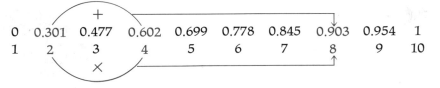

Logarithms	0	0.301	0.477	0.602	0.699	0.778	0.845	0.903	0.954	1
Numbers	1	2	3	4	5	6	7	8	9	10

---

* The easiest current method is to push the *log* button on an electronic calculator (every scientific calculator has such a button) and let the calculator figure the
logarithms out. The values had been accurately determined long before the invention of the calculator, however.

Chapter 4: LARGE NUMBERS AND LOGARITHMS

The logarithms of many numbers larger than 10 can be found in a similar way.

1. Find the logarithm of 12 by adding the logarithms of 2 and 6. Show your work.

2. Check your answer by adding the logarithms of 3 and 4. Show your work.

3. Find the missing logarithms in the table below.

Logarithms	1.041	1.079	1.114				1.230		1.279	
Numbers	11	12	13	14	15	16	17	18	19	20

4. Find the missing logarithms in this table.

Logarithms	1	1.301								
Numbers	10	20	30	40	50	60	70	80	90	100

5. Find the missing logarithms in this table.

Logarithms	1	2						
Numbers	10	100	1,000	10,000	100,000	1,000,000	10,000,000	100,000,000

The insect population of the earth is estimated to be about 1,000,000,000,000,000,000. Of this population, 10^{15} are ants. What is the logarithm of

6. 1,000,000,000,000,000,000?

7. 10^{15}?

The name of the number whose logarithm is 33 is one decillion.* What is the name of the number whose logarithm is

8. 3?

9. 9?

10. 100?

---

* See the table on page 183.

"If it's true that the world ant population is 10^{15}, then it's no wonder we never run into anyone we know."

Set II

The advantage of decimal logarithms is that they are closely related to the way in which we write numbers in scientific notation.

1. Copy and complete the last two lines of this table.

Number in decimal form	Number in scientific notation	Logarithm of number
20	2×10^1	$0.301 + 1 = 1.301$
200	2×10^2	$0.301 + 2 = 2.301$
2,000		
20,000		

What is the logarithm of

2. 2×10^7?

3. 2,000,000,000,000?

4. Copy and complete the last two lines of this table.

Number in decimal form	Number in scientific notation	Logarithm of number
7,000	7×10^3	$0.845 + 3 = 3.845$
70,000	7×10^4	$0.845 + 4 = 4.845$
700,000		
7,000,000		

What number has the logarithm

5. 9.845?

6. 16.845?

7. Copy and complete the following table.

Number in decimal form	Number in scientific notation	Logarithm of number
30,000,000		
1,000,000,000		
	8×10^2	
	5×10^0	
		1.954
		6.602

Chapter 4: LARGE NUMBERS AND LOGARITHMS

The color of light depends on its frequency. Examples are listed in the table below.

8. What are the missing numbers?

Frequency in cycles per second	Logarithm of frequency	Color
4×10^{14}	‖‖‖‖	Red
5×10^{14}	‖‖‖‖	Yellow
6×10^{14}	‖‖‖‖	Blue

When Alaska and Hawaii became states, the coastline of the United States more than doubled in length, increasing from about 8,000 kilometers to almost 20,000 kilometers.
What is the logarithm of

9. 8,000?

10. 20,000?

The CRAY-1 is one of the world's fastest computers. It can perform 2×10^8 operations in a single second and deal with numbers as large as 1×10^{2500}.
What is the logarithm of

11. 2×10^8?

12. 1×10^{2500}?

Set III

Counting on our fingers may have led to our use of ten as a base for our number system and our logarithms. Because mathematics is a universal language, it is likely that, if intelligent beings exist on a planet somewhere else in space, they also might have developed the idea of logarithms even though their number system might be very different from our own.

Suppose that we established communication with another civilization, translated its numbers into our own, and learned some of its logarithms as shown in the following table:

Number	1	2	3	4	5	6	7	8	9	10
Logarithm	0	0.39	0.61	‖‖‖	0.90	‖‖‖	1.09	‖‖‖	‖‖‖	‖‖‖

1. On the assumption that the logarithms of this other civilization behave just as ours do, determine the missing logarithms in this table.

2. What number do you think is the base of this number system?

3. What number would have 2 for its logarithm in this system?

Chapter 4: LARGE NUMBERS AND LOGARITHMS

Part of a table of logarithms
published in China in 1713

From *Science and Civilisation in China*, volume 3, by Joseph Needham, Cambridge University Press

Computing with Logarithms

Scientists and mathematicians all over the world recognized the usefulness of logarithms immediately after their invention. This recognition was due in great part to the work of the English mathematician Henry Briggs, who wrote:

*Logarithms are numbers invented for the more easy working of questions in arithmetic and geometry. By them all troublesome multiplications are avoided and performed only by addition. . . . In a word, all questions not only in arithmetic and geometry but in astonomy also are thereby most plainly and easily answered.**

It was Briggs who suggested the number ten as the most practical base for logarithms and Briggs who constructed the first table

---
* *Arithmetica Logarithmica*, 1624.

of base-ten logarithms. Tables derived from Briggs's work were soon published in many countries. Part of one such table, shown on the preceding page, is from a book published at the request of the Chinese Emperor K'anghsi in 1713. The book, printed from wooden plates carved by hand, listed the logarithms of the numbers from 1 to 100,000, each to ten decimal places.

The table on the next page lists the logarithms of numbers from 1 to 9.9. Even though this table is very brief, it can be used in solving problems with numbers larger than 9.9 because of the relation between numbers written in scientific notation and their base-ten logarithms. For example, the table below illustrates the simple way in which the logarithms of 2, 20, 200, and 2,000 are related.

Number in decimal form	Number in scientific notation	Logarithm of number
2	2×10^0	$0.301 + 0 = 0.301$
20	2×10^1	$0.301 + 1 = 1.301$
200	2×10^2	$0.301 + 2 = 2.301$
2,000	2×10^3	$0.301 + 3 = 3.301$

▶ The *whole number* part of the logarithm *is the exponent of the 10* when the number is written in scientific notation. The *fraction* part of the logarithm is looked up in the table. It *is determined by the number by which the power of 10 is multiplied* in the scientific form of the number.

When a number is written in scientific notation, the number in front is always less than 10, and so the logarithm table does not need to include numbers larger than 10. More examples of how the table can be used to find the logarithms of numbers larger than 10 are shown below.

Number in decimal form	Number in scientific notation	Logarithm of number
121,000	1.21×10^5	5.083
35	3.5×10^1	1.544
	6.0×10^{23}*	23.778

* This is Avogadro's number, a large number used in chemistry, whose decimal form is 600,000,000,000,000,000,000,000.

Chapter 4: LARGE NUMBERS AND LOGARITHMS

Table of Logarithms

No.	Log.	No.	Log.	No.	Log.	No.	Log.	No.	Log.
1.00	.000	1.40	.146	1.80	.255	3.00	.477	6.0	.778
1.01	.004	1.41	.149	1.81	.258	3.05	.484	6.1	.785
1.02	.009	1.42	.152	1.82	.260	3.10	.491	6.2	.792
1.03	.013	1.43	.155	1.83	.262	3.15	.498	6.3	.799
1.04	.017	1.44	.158	1.84	.265	3.20	.505	6.4	.806
1.05	.021	1.45	.161	1.85	.267	3.25	.512	6.5	.813
1.06	.025	1.46	.164	1.86	.270	3.30	.519	6.6	.820
1.07	.029	1.47	.167	1.87	.272	3.35	.525	6.7	.826
1.08	.033	1.48	.170	1.88	.274	3.40	.531	6.8	.833
1.09	.037	1.49	.173	1.89	.276	3.45	.538	6.9	.839
1.10	.041	1.50	.176	1.90	.279	3.50	.544	7.0	.845
1.11	.045	1.51	.179	1.91	.281	3.55	.550	7.1	.851
1.12	.049	1.52	.182	1.92	.283	3.60	.556	7.2	.857
1.13	.053	1.53	.185	1.93	.286	3.65	.562	7.3	.863
1.14	.057	1.54	.188	1.94	.288	3.70	.568	7.4	.869
1.15	.061	1.55	.190	1.95	.290	3.75	.574	7.5	.875
1.16	.064	1.56	.193	1.96	.292	3.80	.580	7.6	.881
1.17	.068	1.57	.196	1.97	.294	3.85	.585	7.7	.886
1.18	.072	1.58	.199	1.98	.297	3.90	.591	7.8	.892
1.19	.076	1.59	.201	1.99	.299	3.95	.597	7.9	.898
1.20	.079	1.60	.204	2.00	.301	4.0	.602	8.0	.903
1.21	.083	1.61	.207	2.05	.312	4.1	.613	8.1	.908
1.22	.086	1.62	.210	2.10	.322	4.2	.623	8.2	.914
1.23	.090	1.63	.212	2.15	.332	4.3	.633	8.3	.919
1.24	.093	1.64	.215	2.20	.342	4.4	.643	8.4	.924
1.25	.097	1.65	.217	2.25	.352	4.5	.653	8.5	.929
1.26	.100	1.66	.220	2.30	.362	4.6	.663	8.6	.934
1.27	.104	1.67	.223	2.35	.371	4.7	.672	8.7	.940
1.28	.107	1.68	.225	2.40	.380	4.8	.681	8.8	.944
1.29	.111	1.69	.228	2.45	.389	4.9	.690	8.9	.949
1.30	.114	1.70	.230	2.50	.398	5.0	.699	9.0	.954
1.31	.117	1.71	.233	2.55	.407	5.1	.708	9.1	.959
1.32	.121	1.72	.236	2.60	.415	5.2	.716	9.2	.964
1.33	.124	1.73	.238	2.65	.423	5.3	.724	9.3	.968
1.34	.127	1.74	.241	2.70	.431	5.4	.732	9.4	.973
1.35	.130	1.75	.243	2.75	.439	5.5	.740	9.5	.978
1.36	.134	1.76	.246	2.80	.447	5.6	.748	9.6	.982
1.37	.137	1.77	.248	2.85	.455	5.7	.756	9.7	.987
1.38	.140	1.78	.250	2.90	.462	5.8	.763	9.8	.991
1.39	.143	1.79	.253	2.95	.470	5.9	.771	9.9	.996

Exercises

Set I

Using the table and the examples given in this lesson as a guide, find the logarithms of the following numbers.

1. 1.15
2. 1.15×10^4
3. 1,150,000
4. 8.2
5. 8.2×10^3
6. 82

Find the numbers having the following logarithms.

7. .562
8. 1.562
9. 7.562
10. .021
11. .210
12. 2.100
13. 21.000

Microwave ovens use radio waves with frequencies of 900,000,000 and 2,450,000,000 cycles per second.

14. Write 900,000,000 in scientific notation.
15. Write its logarithm.
16. Write 2,450,000,000 in scientific notation.
17. Write its logarithm.

The numbers on the Richter scale for measuring earthquakes are logarithms. Earthquakes with readings of 0 have been detected by seismographs.

18. What number has a logarithm of 0?

It is unusual for an earthquake measuring 2 on the Richter scale to be felt.

19. What number has a logarithm of 2.000?

An earthquake measuring 7.25 would be a major disaster if it occurred near a populated area.

20. What number has a logarithm of 7.250?

Set II

Compare the following methods of multiplying numbers in scientific notation without and with logarithms.

Without logarithms

Numbers	In decimal form

$$
\begin{array}{r}
5.2 \times 10^6 \rightarrow \quad 5,200,000 \\
3.75 \times 10^2 \rightarrow \times \quad\quad 375 \\
\hline
26\,000\,000 \\
36\,400\,000 \\
15\,600\,000 \\
\hline
1.95 \times 10^9 \leftarrow 1,950,000,000
\end{array}
$$

With logarithms

Numbers	Logarithms

$$
\begin{array}{r}
5.2 \times 10^6 \rightarrow \quad 6.716 \\
3.75 \times 10^2 \rightarrow + 2.574 \\
\hline
1.95 \times 10^9 \leftarrow \quad 9.290
\end{array}
$$

Use similar formats to multiply

1. 2.25×10^3 and 7.6×10^4 *without* logarithms.

2. the same numbers *with* logarithms.

Showing your work, multiply

3. 7.5×10^1, 9×10^2, and 2.8 *without* logarithms.

4. the same numbers *with* logarithms.

Our solar system revolves around the center of the Milky Way once each 225,000,000 years, traveling about 6,800,000,000 kilometers each year.

5. Write 225,000,000 in scientific notation.

6. Write 6,800,000,000 in scientific notation.

7. Use logarithms to determine the distance traveled by our solar system in making one revolution around the center of the Milky Way. Show your work.

The first animated feature film was Walt Disney's *Snow White and the Seven Dwarfs*. Its length was 82 *minutes* and, as is true of modern motion pictures, 24 pictures were projected on the screen each *second*. The number of pictures that appeared on the screen was

$$82 \times 60 \times 24.$$

8. Use logarithms to determine the approximate number of pictures drawn for *Snow White and the Seven Dwarfs*. Show your work.

In one year, approximately 4.2×10^5 kilograms of plants is produced beneath each square kilometer of ocean surface. About 3.6×10^8 square kilometers of the earth is covered by ocean.

9. Use logarithms to determine the approximate weight of the plants that grow in the earth's oceans in one year. Show your work.

In one year, each square kilometer of the earth's land produces approximately 2.8×10^5 kilograms of plants.* About 1.5×10^8 square kilometers of the earth is land.

10. Use logarithms to determine the approximate weight of the plants that grow on land in one year. Show your work.

Set III

The "big idea" is Einstein's famous formula relating the amount of energy contained in an object to its mass. The energy, *e*, is found by multiplying the mass, *m*, by the square of the speed of light, *c*.

Suppose that the anthill in the cartoon has a mass of 4 kilograms. The speed of light is 300,000,000 meters per second.

Use logarithms to find the amount of energy contained in the mass of the anthill. Show your work and express the answer in scientific notation.

This amount of energy, measured in units named *joules*, would be enough to run the entire United States for several days.

---

* These figures, as well as those given for plants produced beneath each square mile of ocean surface, are averages.

By permission of Johnny Hart and Field Enterprises, Inc.

Chapter 4: LARGE NUMBERS AND LOGARITHMS

Logarithmic Scales

Light is a form of electromagnetic radiation. So are radio and television signals, x-rays, and the waves produced by alternating electric current.

The alternating current used in North America has a frequency of 60 cycles per second. The stations on the AM radio dial transmit signals with frequencies varying between 540,000 and 1,600,000 cycles per second. The television channels numbered 2 through 13 and all FM radio stations operate on frequencies ranging from 54,000,000 to 216,000,000 cycles per second. The light spectrum ranges from 390,000,000,000,000 (red) to 770,000,000,000,000 (violet) cycles per second. X-rays used in dentist's offices and airport security stations have frequencies varying between 30,000,000,000,000,000 and 100,000,000,000,000,000,000 cycles per second.

These examples show that the frequencies of electromagnetic waves range from relatively small numbers to incredibly large ones. As a result, it is not practical to represent them on an ordinary linear scale (one in which numbers in arithmetic sequence are evenly spaced) such as the one shown here.

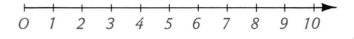

$$0 \quad 1 \quad 2 \quad 3 \quad 4 \quad 5 \quad 6 \quad 7 \quad 8 \quad 9 \quad 10$$

There is another sort of scale, however, on which numbers ranging in size from very small to very large can easily be represented. It is called a *logarithmic scale* and is formed by replacing each number in the preceding scale with the corresponding power of 10.

$$10^0 \quad 10^1 \quad 10^2 \quad 10^3 \quad 10^4 \quad 10^5 \quad 10^6 \quad 10^7 \quad 10^8 \quad 10^9 \quad 10^{10}$$

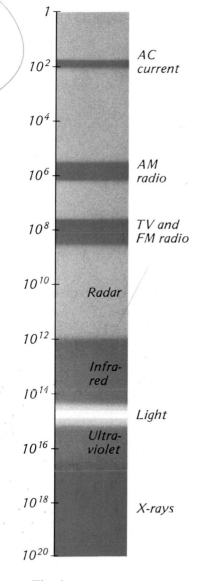

The electromagnetic spectrum

It is easy to see why this scale is called a *logarithmic scale* by looking at our original table of base-ten logarithms:

Logarithms	0	1	2	3	4	5	6	7	...
Numbers	10^0	10^1	10^2	10^3	10^4	10^5	10^6	10^7	...

▶ The numbers on a **logarithmic scale** are arranged so that numbers in geometric sequence are evenly spaced.

Because of this arrangement, each successive unit of length on a logarithmic scale covers a wider range of numbers. The first unit of length, from 10^0 to 10^1, includes the numbers from 1 to 10; the second unit, the numbers from 10 to 100, the third unit, the numbers from 100 to 1,000, and so on. Consequently, such a scale is ideally suited for representing a wide range of numbers.

To locate numbers on a logarithmic scale, we write them in scientific notation and determine their logarithms. The numbers are then located on the scale according to their logarithms.

Type of wave	Frequency	Logarithm of frequency*
Alternating current	$60 = 6 \times 10^1$	1.8
AM radio waves	$540,000 = 5.4 \times 10^5$ to	5.7 to
	$1,600,000 = 1.6 \times 10^6$	6.2
Television and	$54,000,000 = 5.4 \times 10^7$ to	7.7 to
FM radio waves	$216,000,000 = 2.16 \times 10^8$	8.3
Light	$390,000,000,000,000 = 3.9 \times 10^{14}$ to	14.6 to
	$770,000,000,000,000 = 7.7 \times 10^{14}$	14.9
X-rays	$30,000,000,000,000,000 = 3 \times 10^{16}$ to	16.5 to
	$100,000,000,000,000,000,000 = 1 \times 10^{20}$	20.0

*Rounded to the nearest tenth.

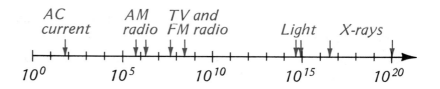

Chapter 4: LARGE NUMBERS AND LOGARITHMS

Exercises

Set I

Here is a table of logarithms that have been rounded to the nearest hundredth.

Number, x	1	2	3	4	5	6	7	8	9	10
Logarithm, y	0	.30	.48	.60	.70	.78	.85	.90	.95	1

1. Draw and label a pair of axes as shown at the left below. Plot the points in the table and connect them with a smooth curve.

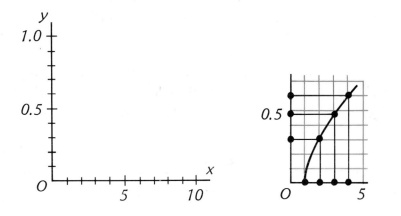

Draw a vertical line from each point down to the x-axis and a horizontal line across to the y-axis. Mark the point at which each line meets the axis with a dot. The figure at the right above shows how the first part of your graph should look.

2. What happens to the direction of the curve as you look at the graph from left to right?

Notice that the points on the x-axis are evenly spaced.

3. What do you notice about the points on the y-axis as you look upward from the origin?

They are arranged in a logarithmic scale.

4. What do you notice about the arrangement of the frets (the ridges to guide the fingers) in this photograph of a guitar?

5. On what kind of a scale (mathematical, not musical) do you think the frets of a guitar are arranged?

The amount of light that enters a camera is determined by the f-stop setting of the lens. The commonly used f-stop numbers are: 1, 1.4, 2, 2.8, 4, 5.6, 8, 11, 16, 22, 32, and 45.

6. Draw and label a scale as shown below and indicate each f-stop setting on it with an arrow. (The first three, 1, 1.4, and 2, have been marked as examples.)

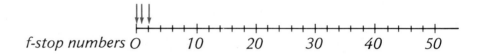

f-stop numbers   O   10   20   30   40   50

7. What do you notice about the spacing of the f-stops on this scale as you look at it from left to right?

8. Use the logarithm table on page 213 to copy and complete the following table. Round each logarithm to the nearest hundredth.

f-stop number	1	1.4	2	2.8	4	5.6	8	11	16	22	32	45
Logarithm of f-stop number	0	1.5 ≈15						1.04*				

9. Draw and label a logarithmic scale as shown below and indicate each f-stop setting on it with an arrow. (The first three have been marked as examples.)

10. What do you notice about the spacing of the f-stops on this scale as you look at it from left to right?

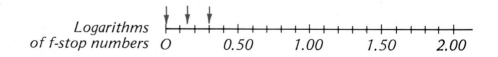

Logarithms of f-stop numbers   O   0.50   1.00   1.50   2.00

---

*Remember that, to find the logarithm of a number greater than 10, it is necessary to first write the number in scientific notation: $11 = 1.1 \times 10^1$, and so its logarithm is $1.041 \approx 1.04$.

Chapter 4: LARGE NUMBERS AND LOGARITHMS

Set II

Even though a second is a very brief interval of time, a billion seconds is a long time: it is more than 31 years.

Here is a table relating seconds to other units of time.

Interval of time	Number of seconds
One minute	60
One hour	3,600
One day	86,000*
One year	32,000,000
One century	3,200,000,000
One thousand years	32,000,000,000
One million years	32,000,000,000,000
One billion years	32,000,000,000,000,000

*The numbers beginning with this entry are approximate.

1. Write each of these numbers in scientific notation.

2. Use the logarithm table on page 213 to write the logarithm of each number.

3. Draw and label a scale as shown below and indicate each interval of time on it with an arrow.

Time in seconds $\quad 1 \quad 10^2 \quad 10^4 \quad 10^6 \quad 10^8 \quad 10^{10} \quad 10^{12} \quad 10^{14} \quad 10^{16} \quad 10^{18}$

Logarithm of time $\quad 0 \quad 2 \quad 4 \quad 6 \quad 8 \quad 10 \quad 12 \quad 14 \quad 16 \quad 18$

Human beings can hear sounds varying over an incredibly wide range of loudness: from sounds that can barely be heard to sounds a *hundred trillion* times as loud. The loudness of sounds is measured on the decibel scale, explained in this chart from *Time* magazine. The least audible sound is said to be at the "threshold of hearing" and is rated as 0 decibels.

4. Refer to the explanation at the top of the chart to copy and complete the following table in which 0 decibels has been assigned a relative loudness of 1 unit.

Relative loudness	1	10	10^2						
Logarithm of relative loudness	0	1	2						
Decibels	0	10	20	30	40	50	60	70	80

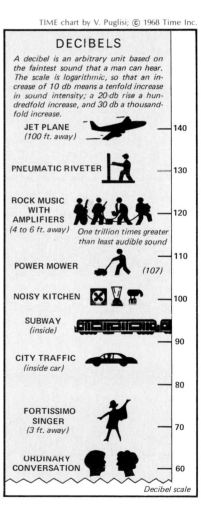

DECIBELS

A decibel is an arbitrary unit based on the faintest sound that a man can hear. The scale is logarithmic, so that an increase of 10 db means a tenfold increase in sound intensity; a 20-db rise a hundredfold increase, and 30 db a thousandfold increase.

JET PLANE (100 ft. away) — 140

PNEUMATIC RIVETER — 130

ROCK MUSIC WITH AMPLIFIERS (4 to 6 ft. away) — 120 *One trillion times greater than least audible sound*

POWER MOWER (107) — 110

NOISY KITCHEN — 100

SUBWAY (inside) — 90

CITY TRAFFIC (inside car) — 80

FORTISSIMO SINGER (3 ft. away) — 70

ORDINARY CONVERSATION — 60

Decibel scale

5. How is the decibel rating of a sound related to the logarithm of its relative loudness?

The sound of leaves rustling in a breeze has a loudness of 20 decibels.

6. How many times as loud as the least audible sound is the sound of leaves rustling in a breeze?

According to the chart, an amplified rock group is a trillion times as loud as the least audible sound.

7. Write a trillion as a power of 10.

8. What is the logarithm of a trillion?

9. How does the loudness of rock music in decibels compare with this logarithm?

The loudest sound that has been created in a laboratory had a rating of 210 decibels.

10. How many times as loud as the least audible sound was this sound?

Set III

The "zone system" is a scale ranging from black to white through eight shades of gray. Invented by the landscape photographer Ansel Adams, it is illustrated in the figure below.

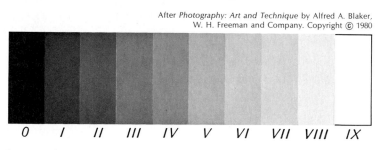

After *Photography: Art and Technique* by Alfred A. Blaker, W. H. Freeman and Company. Copyright © 1980

The numbers of the zones are keyed to units of exposure time as shown in the table below.

Number of zone	0	I	II	III	IV	V	VI	VII	VIII	IX
Exposure time	0.5	1	2	4	8	16	32	64	128	256

Compare the figure and the table.

1. What exposure time results in dead black (no exposure of the negative)?

Chapter 4: LARGE NUMBERS AND LOGARITHMS

2. What exposure time results in pure white (complete exposure)?

3. In what kind of number sequence are the exposure times of the ten zones?

4. Copy and complete the following table by determining the decimal logarithms of the exposure times in the table on page 213. Round each logarithm to the nearest tenth.

Exposure time	0.5	1	2	4	8	16	32	64	128	256
Logarithm	—	0	0.3	▦	▦	▦	▦	▦	▦	2.4

5. In what kind of number sequence are the logarithms of the exposure times?

A dash was written in the space for the logarithm of 0.5 because we have not studied logarithms of numbers less than 1.

6. According to the pattern in the other logarithms, what number do you think seems appropriate as the logarithm of 0.5?

"First you forget logarithms.
Then you forget how to do long division.
Then the multiplication table begins to go. . ."

Chapter 4 / Summary and Review

In this chapter we have become acquainted with:

Large numbers and exponents (*Lesson 1*) In writing a large number as a power of 10, we use an exponent to indicate the number of zeros that would follow the 1 if the number was written the long way. $10^1 = 10$; $10^0 = 1$.

The names and exponential forms of some large numbers are:

10^2	hundred	10^{18}	quintillion
10^3	thousand	10^{21}	sextillion
10^6	million	10^{24}	septillion
10^9	billion	10^{27}	octillion
10^{12}	trillion	10^{30}	nonillion
10^{15}	quadrillion	10^{33}	decillion

Scientific notation (*Lesson 2*) A number is in scientific notation if it is written in the form

$$a \times 10^b$$

in which a is a number that is at least 1 but less than 10.

Logarithms (*Lessons 3, 4, and 5*) Logarithms are exponents of numbers that are in a geometric sequence. Decimal logarithms are based on powers of 10:

Numbers	10^0	10^1	10^2	10^3	10^4	10^5	10^6	10^7	. . .
Logarithms	0	1	2	3	4	5	6	7	. . .

To find the logarithm of a number, first write the number in scientific notation: $a \times 10^b$. The whole-number part of the logarithm is the power of 10, b, and the fraction part of the logarithm can be obtained from a logarithm table in which it appears beside the number a.

Numbers can be multiplied by adding their logarithms to get the logarithm of the answer.

Logarithmic scales (*Lesson 6*) The numbers on a logarithmic scale are arranged so that numbers in geometric sequence are evenly spaced.

Exercises

Set I

There are approximately 10^{14} grains of sand in Malibu Beach, California.

1. Write this number in decimal form.

2. Name it.

Black holes are tremendously massive regions in space from which light cannot escape. Although such holes eventually evaporate, one having the mass of our sun would last for about

1,000,000,000,000,000,000,000,000,000,000,000,000,000,000,000,000,000,000,000,000,000,000,000 years.*

3. Write this number as a power of 10.

4. What advantages does this way of writing the number have over the decimal form used above?

---

* "The Quantum Mechanics of Black Holes" by S. W. Hawking, *Scientific American*, January 1977.

Photographs courtesy of *Coin World*

The first U.S. pennies were minted in Philadelphia in 1793.

5. Copy and complete the following table relating various numbers of pennies to their values in dollars.

Number of pennies	Value in dollars
1,000	▥
1,000,000	▥
1,000,000,000	▥

6. What is the value of 10^9 pennies in dollars? Express the answer as a power of 10.

7. What is the value of 10^n pennies in dollars? Express the answer as a power of 10.

8. Use your answer to exercise 7 to find the value of 10^2 pennies in dollars. Does the answer make sense? Explain.

The Hindus consider "one day in the life of God" to be 4,320,000,000 years.

9. Write this number in scientific notation.

"One day in the life of God" is approximately equal to 1.6×10^{12} human days.

10. Write this number in decimal form.

11. Name it.

Here is a table of logarithms based on a number other than 2 or 10.

Number	Logarithm
1	0
3	1
9	2
27	3
81	4
243	5
729	6
2,187	7
6,561	8
19,683	9
59,049	10

12. On what number is this logarithm table based?

13. Find the answer to this problem by multiplying: 81×729. Show your work.

14. Show the corresponding addition problem that can be used with the table to find the answer without multiplying.

Set II

Since 1900, the burning of fossil fuels has added about 3.5×10^{14} kilograms of carbon dioxide to the earth's atmosphere. One kilogram of carbon dioxide contains approximately 1.37×10^{25} molecules.

1. What is the logarithm of 3.5×10^{14}?

2. What is the logarithm of 1.37×10^{25}?

3. Use logarithms to determine the approximate number of molecules of carbon dioxide that were added to the earth's atmosphere in the past century by the burning of fossil fuels.

As explained in this newspaper article, the Richter scale used to measure the intensity of earthquakes is logarithmic.

4. Copy and complete this table in which an earthquake rated 4 on the Richter scale has been assigned a relative intensity of 1 unit.

Relative intensity	1	10	▓	▓	▓	▓
Richter scale	4	5	6	7	8	9

Someone who does not understand the Richter scale might think that an earthquake that measures 8 is twice as strong as one that measures 4.

5. How do the two actually compare in intensity?

6. How does an earthquake that measures 9 compare in intensity with an earthquake that measures 7?

HOW RICHTER SCALE GAUGES EARTHQUAKES

The Richter scale, used by seismologists to measure the magnitude of earthquakes, operates on a logarithmic basis so there is a 10-fold increase from one unit or number to the next.

A magnitude of 6, which was about the strength of Monday night's earthquake, would be 10 times greater than a magnitude 5 earthquake and 100 times a magnitude 4 earthquake.

Thus the Tehachapi earthquake in 1952 with a magnitude of 7.5 was more than 30 times stronger than Monday night's earthquake and the 1906 San Francisco earthquake with a magnitude of 8.25 was more than 110 times stronger.

Courtesy of the *Los Angeles Times*

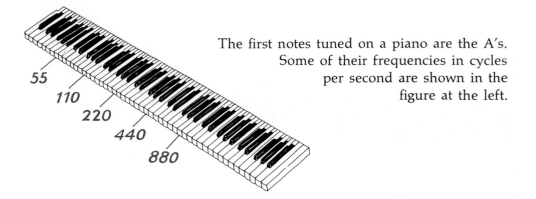

The first notes tuned on a piano are the A's. Some of their frequencies in cycles per second are shown in the figure at the left.

7. Refer to the table of logarithms on page 213 to copy and complete the following table.

Frequency	55	110	220	440	880
Logarithm of frequency	1.740	▓▓▓	▓▓▓	▓▓▓	▓▓▓

8. What kind of number sequence do the frequencies form?

9. What is their common ratio?

10. What kind of number sequence do the logarithms of the frequencies form?

11. What is their common difference?

12. Explain why you would expect this common difference from the common ratio of the frequencies.

Set III

The effects of an electric shock depend on the strength of the current.* Some of them are shown on the scale below.

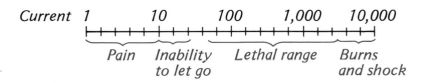

Current 1   10   100   1,000   10,000

         Pain   Inability   Lethal range   Burns
                to let go   and shock

---

*Strange as it may seem, very strong electric currents can be less lethal than smaller currents.

Chapter 4: LARGE NUMBERS AND LOGARITHMS

1. On what kind of scale are the currents shown?

2. How can you tell?

Currents less than 1 milliampere in strength produce a tingling effect and, as the scale shows, currents between 1 and 10 milliamperes create a sensation of pain.

Use the table on page 213 to estimate the range of electric currents corresponding to

3. the inability to let go.

4. the lethal range.

5. burns and shock.

Chapter 4 / Problems for Further Exploration

Lesson 1

1. The word "million" is thought to have originated in Italy in the fourteenth century and may have been invented by Marco Polo. It, together with names for larger numbers, appeared in a book written by the French mathematician Nicolas Chuquet in 1484. The list below shows the names used by Chuquet and the Latin names of the numbers from 1 through 10.

Word used by Chuquet	*Meaning*	*Latin words for numbers*	*Meaning*
Million	10^6	Unum	1
Byllion (billion)	10^9	Duo	2
Tryllion (trillion)	10^{12}	Tria	3
Quadrillion	10^{15}	Quattour	4
Quyllion (quintillion)	10^{18}	Quinque	5
Sixlion (sextillion)	10^{21}	Sex	6
Septyllion (septillion)	10^{24}	Septem	7
Ottyllion (octillion)	10^{27}	Octo	8
Nonyllion (nonillion)	10^{30}	Novem	9
(decillion)*	10^{33}	Decem	10

a) At what place do the names in the two lists first seem to be related?

   Since Chuquet made his list, it has been extended as far as a "centillion." The related Latin word is "centum," meaning 100.

b) What number do you think "centillion" names? Explain your reasoning.

---

* Chuquet did not name this number.

According to *The Random House Dictionary of the English Language*, "zillion" is an informal term meaning "an extremely large, indeterminate number." In other words, it does not name a specific number.

c) How do you suppose the name "zillion" may have originated?

2. *Writing* a large number in decimal form and *counting* to that number are very different things. Suppose that, ignoring the commas, you wrote at the rate of one digit per second. How long would it take you to write

a) "one million" in decimal form?

b) "one billion" in decimal form?

Suppose that you counted at the rate of one number per second. (In other words, counting to "ten" would take ten seconds.) How long would it take you, in *days*, to *count* to

c) one million?

d) one billion?

As you know, the name *googol* refers to the number 10^{100}. Do you think a computer could

e) print this number in decimal form? Explain.

f) "count" to this number by printing a list of all of the numbers from one to a googol? Explain. (The number of atoms in the universe is thought to be no more than 10^{80}.)

At the same time that the word googol was invented, the word *googolplex* was suggested as the name of an even larger number. At first, a goolgolplex was considered to be "1, followed by writing zeros until you got tired."[*] Then it was decided that this was too vague, so a goolgolplex was defined to be 1 followed by a googol of zeros.

g) Do you think that a computer could print a googolplex in decimal form? Explain.

Lesson 2

1. Scientific notation is useful in representing not only numbers that are very large but also numbers that are very small. Nuclear physicists, for example, have discovered that particles having the shortest lifetimes exist for only

0.0000000000000000000000000016

second. To discover how this incredibly small number can be

---

[*] Edward Kasner and James Newman in *Mathematics and the Imagination* (Simon and Schuster, 1940).

written in scientific notation, look at the following table of powers of 10.

10,000	1,000	100	100	1				
10^4	10^3	10^2	10^1	10^0				

a) In what kind of number sequence are the numbers in the first line of this table?

b) In what kind of number sequence are the *exponents* of the 10's in the second line?

c) What do you think the next four numbers of the first line of the table should be? Explain your reasoning.

d) What do you think the next four powers of 10 in the second line of the table should be? Explain your reasoning.

The average lifetime of a meson is 0.0000022 second.

e) How do you think this number would be written in scientific notation? (Observe that $0.0000022 = 2.2 \times 0.000001$.)

f) How do you think the lifetime of the shortest lived particles, 0.000000000000000000000000016 second, would be written in scientific notation?

2. The universe seems to be expanding. In the 1920s, the astronomer Edwin P. Hubble discovered that the other galaxies in space seem to be moving away from us. This suggests that the universe began with a "big bang," an explosion in which the outward motion of the galaxies began.

The farther another galaxy is from ours, the faster it seems to be moving. Galaxies 10,000,000 light-years away are estimated

"Now, if we run our picture of the universe backwards several billion years, we get an object resembling Donald Duck. There is obviously a fallacy here..."

to be traveling 170 kilometers per second; galaxies twice as far away are estimated to be traveling twice as fast. A light-year is the distance that light travels in one year: about 9.46×10^{12} kilometers.

a) How many kilometers away from us is a galaxy that is 10,000,000 light-years away?
b) On the assumption that this galaxy has been moving away from us at a steady rate, determine the approximate age of the universe. Explain your reasoning. (One year is approximately 3.2×10^7 seconds.)

Lesson 3

1. In addition to using logarithms for multiplying and raising numbers to powers, we can use them to divide and to find roots of numbers.

a) Find the answer to this problem by dividing:

$$\frac{65,536}{512}$$

Show your work.
b) Show how logarithms from the table on page 199 could be used to find the answer without dividing.
c) Find the answer to this problem by dividing:

$$\frac{134,217,728}{32}$$

Show your work.
d) Show how to find the answer without dividing.
The square root of 16,384 is 128 because

$$128^2 = 16,384.$$

e) Show how logarithms could be used to discover this.
f) Find the square root of 1,073,741,824 by using logarithms.

2. Imagine that you are driving on a road along which signs are posted indicating the logarithms of the distances that you have driven in meters. The first ten meters of the road are shown in the figure below.

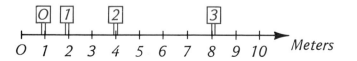

Compare this figure with the logarithm table on page 199. Notice that you would drive 1,024 meters in order to reach the sign numbered 10. Because one kilometer is equal to 1,000 meters, you would arrive at the sign numbered 10 after roughly 1 kilometer, or 0.6 mile.

Estimate, both in kilometers and miles, the distance that you would have to drive to reach the sign numbered

a) 20.
b) 30.
c) 40.
d) 50.

The distances to the last three posts are astonishingly great. By way of comparison, the diameter of the solar system is about 10,000,000,000 kilometers.

Lesson 4

1. One way to figure out the logarithms of the numbers between 1 and 10 is to draw a graph and interpolate. To draw the graph, we can use the following facts:

$$\text{Fact 1:} \quad 3.2 \times 3.2 \approx 10$$
$$\text{Fact 2:} \quad 1.8 \times 1.8 \times 1.8 \times 1.8 \approx 10$$

Table from Fact 1

Numbers	$3.2 \times 3.2 \approx 10$
Logarithms	▦ + ▦ = 1

a) What is the logarithm of 3.2?

Table from Fact 2

Numbers	$1.8 \times 1.8 \times 1.8 \times 1.8 \approx 10$
Logarithms	▦ + ▦ + ▦ + ▦ = 1

b) What is the logarithm of 1.8?
c) Copy and complete the following table.

Number	1	1.8	3.2	10
Logarithm	0	▦	▦	▦

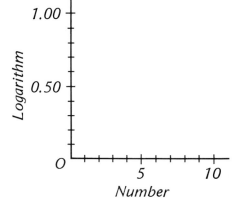

d) Draw and label a pair of axes as shown here, plot the four points corresponding to your table, and draw a smooth curve through them.

e) Use the graph to estimate the logarithms of the numbers from 2 through 9, each to the nearest hundredth.

When you are finished, you should find that your estimates are in fairly close agreement with the values in the table on page 206.

2. It has been suggested that, in performing age-related experiments with animals, the ages of the animals should be in geometric sequence.*

Suppose that an experiment is performed with five groups of mice, the youngest group being 16 days old and the next older group being 24 days old.

a) Copy and complete the second line of the following table so that the logarithms of the ages form an *arithmetic* sequence.

Age	16	24	▦	▦	▦
Logarithm of age	1.204	1.380	▦	▦	▦

Then copy and complete the first line of the table by finding the numbers corresponding to the logarithms on the second line. Refer to the table on page 213 to do this.

b) Are the numbers in the first line in geometric sequence? Explain.

Lesson 5

1. In 1626, Peter Minuit paid the Indians of New York $24 for Manhattan Island. Suppose that they had invested the money in a savings account paying 5% interest compounded annually.

At the end of the first year, their money would have been worth

$$\$24 \times 1.05 = \$25.20.$$

At the end of the second year, it would have been worth

$$\$24 \times (1.05)^2 = \$26.46;$$

after the third year,

$$\$24 \times (1.05)^3 = \$27.78,$$

and so on.

---

*N. O. Calloway, "Ages of Experimental Animals," *Science*, December 31, 1965.

If the Indians had left all of the money in their account until 1776, how much would it have been worth? To find out, we first observe that the money would have been in the account for 150 years. Evidently the amount would be

$$\$24 \times (1.05)^{150}.$$

To find this amount using logarithms, we will use a slightly more precise value of the logarithm of 1.05 than the one given in the table on page 213. It is .0212.

a) Use the method illustrated below to show that the value of $\$24 \times (1.05)^{150}$ is about $36,000.

Numbers $\qquad 24 \times (1.05)^{150} = \qquad 24 \times \underbrace{1.05 \times 1.05 \times 1.05 \times \cdots \times 1.05}_{150 \text{ times}}$

Logarithms $\qquad\qquad\qquad\qquad 1.380 + \underbrace{.0212 + .0212 + .0212 + \cdots + .0212}_{150 \text{ times}}$

Use logarithms to find the approximate value of the money

b) in the year 1876.

c) in the year 1976.

d) in the year 2076.

2. The largest number that can be written using only two digits is

$$9^9.$$

Its approximate value in decimal form can be found by using logarithms as shown below.

Numbers $\qquad 9^9 = \qquad \underbrace{9 \times 9 \times \cdots \times 9}_{9 \text{ times}} \qquad\qquad \approx 3.85 \times 10^8$

Logarithms $\qquad \underbrace{.954 + .954 + \cdots + .954}_{9 \text{ times}} = 9(.954) = \qquad 8.586$

$$3.85 \times 10^8 = 385{,}000{,}000$$

The largest number that can be written using three digits is

$$9^{9^9}.$$

When two exponents are used in this way, they are interpreted to represent

$$9^{(9^9)}$$

a) Use logarithms to find the approximate value of this number in scientific notation.
b) If this number were written in decimal form, approximately how many digits would it contain?

Lesson 6
1. Experiment: *A Slide Rule*

From *Science and Civilisation in China*, volume 3, by Joseph Needham, Cambridge University Press

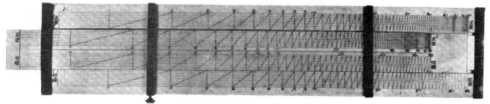

A Chinese slide rule dated at 1660

The slide rule was invented by the English mathematician William Oughtred in 1622, soon after the invention of logarithms. In the following experiment, you will construct a simple slide rule and discover how it works.

First, use the table of logarithms below to graph the logarithm function on graph paper that has 10 squares to the inch (or 5 squares per centimeter).

Number	1	2	3	4	5	6	7	8	9	10
Logarithm	0	.30	.48	.60	.70	.78	.85	.90	.95	1.00

Let the x-axis represent the numbers, with 5 small units on it standing for 1. Let the y-axis represent the logarithms, with 5 small units on it standing for 0.1. Plot the ten points in the table and connect them with a smooth curve. The lower left-hand corner of your completed graph should look like this.

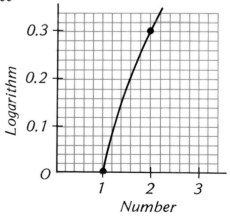

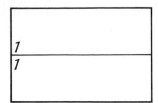

Next, draw an accurate line on a 4-by-6-inch card to divide it in half as shown at the left, and cut it into two pieces along the line. Write the number 1 in the two corners indicated.

Turn the graph sideways, so that the logarithm axis is at the top and the number axis is at the left. Take the lower piece of your card and place it on the graph so that the left edge falls along the number axis and the upper left-hand corner is at 2, as shown in the figure below. Put a mark on the upper edge of the card where you see the curve crossing it and write the number 2 below the mark.

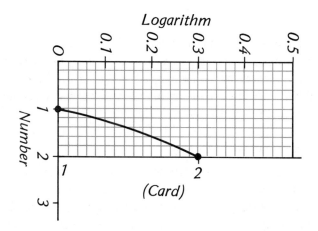

Now slide the card down the graph so that the corner comes to 3 on the number axis. Mark the upper edge of the card where you see the curve crossing it and write the number 3 below the mark, as shown in the figure below.

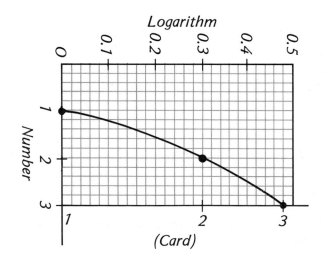

Chapter 4: LARGE NUMBERS AND LOGARITHMS

Slide the card down to 4, 5, 6, 7, 8, 9, and 10 on the number axis and do the same thing: put a mark on the edge of the card at the point where the curve crosses it and write the corresponding number below the mark.

Your finished card should look like the one shown in this figure.

| 1 | 2 | 3 4 5 6 7 8 9 10 |

Now put it below the other half of the card and copy the scale and the numbers from the finished card onto the bottom of the other card, as shown in the figure at the right. The two scales are logarithmic and, when you have finished making them, your slide rule is ready to use.

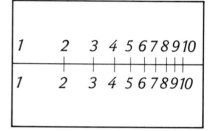

Using a simple example like 2 × 3, explain how the slide rule can be used to multiply.

2. The great Greek mathematician and astronomer Hipparchus, who lived in the second century B.C., classified the stars that he could see according to six classes of brightness. The brightest stars were assigned to the first class and the faintest stars that he could see were assigned to the sixth class.

In the nineteenth century, his system was refined as shown in the following table.

Magnitude	1	2	3	4	5	6
Brightness	100	40	16	6.3	2.5	1

As the table shows, a star of the first magnitude is 100 times as bright as a star of the sixth magnitude.

a) Refer to the table of logarithms on page 213 to copy and complete the following table, which is based on the table above. Round each logarithm to the nearest tenth.

Magnitude	6	5	4	3	2	1
Brightness	1	2.5	6.3	16	40	100
Logarithm of brightness	▓	▓	▓	▓	▓	▓

b) What do you notice about the logarithms of the brightnesses?

Lesson **1**

Symmetry

This remarkable alphabet, designed by Scott Kim, has been arranged and drawn so that its left and right halves are mirror images of each other. To see that this is so, hold a mirror perpendicular to the page so that the edge of the mirror lies on the vertical black line. The letter A at the upper left is reflected in the mirror as the letter F at the upper right. The letter B is reflected as the letter E, the letter C becomes part of the letter D, the letter G becomes the letter J, part of the letter H becomes the letter I, and so on. The line down the center is called a mirror line, and it is the line of symmetry of the figure.

▶ A figure has **line symmetry** if there is a line along which a mirror can be placed to reflect either half of the figure so that it reproduces the other half.

Another way to see if a figure has a line of symmetry is to fold the figure along the line to see if the two halves coincide (fit together exactly). Look, for example, at the flag of Turkey shown here. It is symmetrical with respect to a horizontal line through its center. If the flag is folded along this line, the pattern on one side of the fold coincides with the pattern on the other side.

A figure can be symmetrical with respect to more than one line. A square, for example, has four lines of symmetry. It is easy to see that the horizontal and vertical lines in the figure below are lines of symmetry; the two lines through the opposite corners of the square are less obvious.

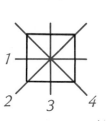

The paper windmill shown here is symmetrical, yet it does not have any lines of symmetry. It is impossible to fold it along a line so that any two halves coincide. The windmill can be *rotated*, however, into other positions that look exactly the same—that is, when the windmill is rotated one-quarter of a circle (90°), one-half of a circle (180°), or three-quarters of a circle (270°).* Therefore, it has *rotational symmetry.*

▶ A figure has **rotational symmetry** if it can be rotated through an angle of less than 360° so that it coincides with its original position.

One-quarter
circle
(90°)

One-half
circle
(180°)

Three-quarters
circle
(270°)

*Angles and their measures are discussed on pages 624–626.

Set I

ABCDEFGHI

KLMNOPQR

STVXYZ

*A fragment of an inscription on
a Roman column of about 260 B.C.*

Courtesy of the Musei Capitolini, Roma

The Roman alphabet, which we use today, has existed since about 700 B.C. For a long time, it consisted of the 23 capital letters shown at the left above. The letters J, U, and W were not added until the Middle Ages.

The adjoining figure shows that the first letter of the Roman alphabet is symmetric with respect to a vertical line.

1. What letters in addition to A have vertical lines of symmetry?

The second letter of the Roman alphabet is symmetric with respect to a horizontal line.

2. What letters in addition to B have horizontal lines of symmetry?

The last letter of the Roman alphabet has rotational symmetry because it looks the same if it is rotated 180°.

3. What letters in addition to Z have rotational symmetry?

Many playing cards are symmetrical.

Chapter 5: SYMMETRY AND REGULAR FIGURES

4. What kind of symmetry do they have?

5. Why do you suppose they have this symmetry?

The photograph at the right is of a microscopic sea urchin.

6. How many lines of symmetry does it have?

The two highly symmetric figures below were drawn by means of a computer. The first figure took five minutes to produce and the second took one hour.

From *Art Forms in Nature* by Ernst Haeckel, 1974, Dover Publications, Inc., New York

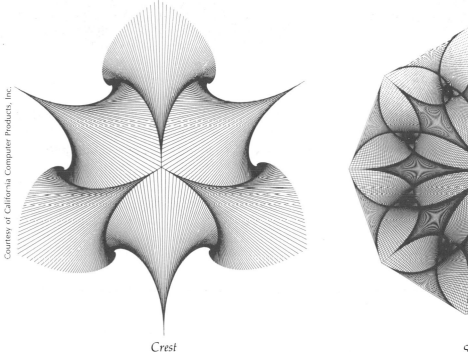

Courtesy of California Computer Products, Inc.

Crest

Symplexity

7. How many lines of symmetry does *Crest* have?

8. Does *Crest* have rotational symmetry? Explain.

9. How many lines of symmetry does *Symplexity* have?

10. What is the smallest angle through which *Symplexity* can be turned so that it coincides with its original position?

Trace the following figures and draw every line that you think is a symmetry line for each. If you do not think that the figure has any lines of symmetry, write "none."

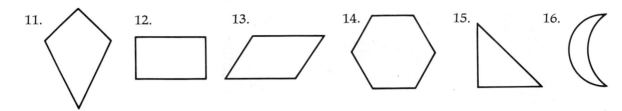

11.   12.   13.   14.   15.   16.

17. Which of the figures in exercises 11 through 16 have rotational symmetry?

The butterfly in this photograph is symmetrical with respect to a vertical line through its center.

Describe the kinds of symmetry, if any, possessed by the objects in the following pictures.

18.

19.

Photograph courtesy of the Jet Propulsion Laboratory

20.

21.

Courtesy of the National Oceanic and Atmospheric Administration; photograph by W. A. Bentley

22.

Courtesy of Barbara Ferenstein

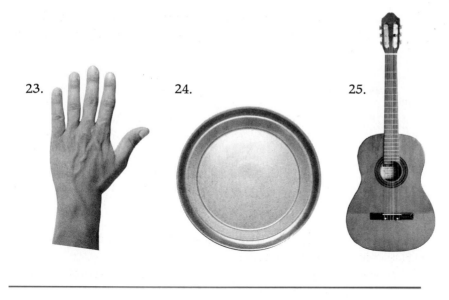

23.

24.

25.

Set II

Experiment: SYMMETRIES WITH A KALEIDOSCOPE

The kaleidoscope, invented in 1816, uses mirrors to produce symmetrical patterns. A simple one can be made by hinging two mirrors together with tape, as shown in the photograph at the right.
Stand the mirrors on figure A so that they form the angle shown.

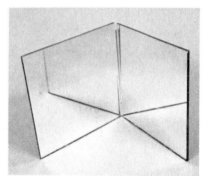

1. What does the figure formed in the kaleidoscope look like?

Stand the mirrors on figures B, C, D, E, and F as indicated. What

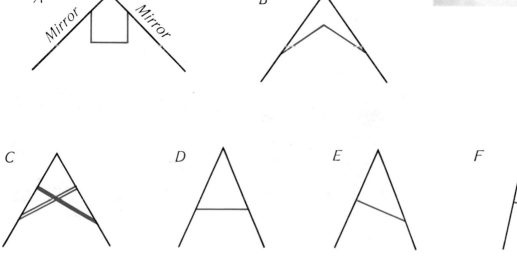

A

Mirror   Mirror

B

C   D   E   F

does the figure formed from

2. figure B look like?   5. figure E look like?

3. figure C look like?   6. figure F look like?

4. figure D look like?

  Look again at the figures formed in the kaleidoscope to answer the following questions.

7. How many lines of symmetry does each figure have?

8. Which figures have rotational symmetry?

  The type of figure formed by the kaleidoscope depends on the angle between the mirrors.

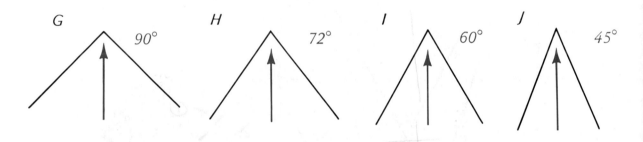

  Stand the mirrors on Figure G.

9. How many arrows do you see?

  Stand the mirrors on figures H, I, and J as indicated, adjusting them as necessary so that the arrows are evenly spaced in each figure.

10. What happens to the number of arrows seen as the angle between the mirrors gets smaller?

11. Copy and complete the following table.

Angle of mirrors, a	90°	72°	60°	45°
Number of arrows, n	▓▓▓	▓▓▓	▓▓▓	▓▓▓

  Stand the mirrors on figure G again.

12. What is the sum of the measures of the four angles that surround the point at the center?

  Stand the mirrors on figure H again.

13. What is the sum of the measures of the five angles that surround the point at the center?

Chapter 5: SYMMETRY AND REGULAR FIGURES

Look at the table that you completed in exercise 11.

14. What do the pairs of numbers in the table have in common?

15. Write a formula for n, the number of arrows, in terms of a, the angle of the mirrors.

16. What angle do you think the mirrors should form to make 10 equally spaced arrows appear?

Set III

This photograph was taken as a gag.

Photograph by Peter Renz

1. How might the scene be changed to make it look more convincing?

2. Explain by means of symmetry why the board would still look wrong, even with this change.

By permission of Johnny Hart and Field Enterprises, Inc.

Lesson **2**

Regular Polygons

A square is certainly not the ideal shape for a wheel and, even though a wheel in the shape of an equilateral triangle "eliminates one bump," it is clearly even less practical. The equilateral triangle and the square are the simplest geometric figures among the *regular polygons*.

*Equilateral
triangle*

▶ A **regular polygon** is a figure of which all sides are the same length and all angles are equal.

Square

The names of the rest of the regular polygons are taken from their numbers of sides. These figures were studied extensively by the early Greek mathematicians and the names they gave them have been used ever since. The origin of the names is apparent in the table of Greek names for numbers below.

5	6	7	8	9	10	12
pente	hex	hepta	octo	ennea*	deca	dodeca

The set of regular polygons, like the set of counting numbers, goes on indefinitely. The greater the number of sides that a regular polygon has, the more symmetrical it is and the more circular in shape. A regular polygon with 100 sides, for example, looks so

---

* The Greek name for the regular polygon having nine sides was *enneagon*. We now use the word "nonagon," from the Latin word *nonus*, which translates into "nine."

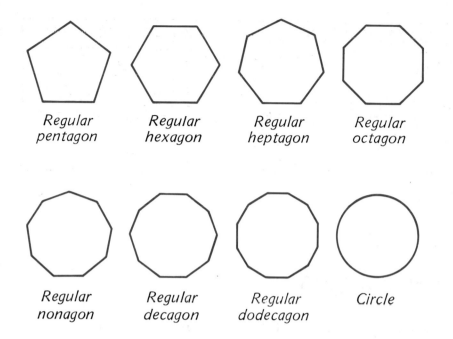

| Regular pentagon | Regular hexagon | Regular heptagon | Regular octagon |

| Regular nonagon | Regular decagon | Regular dodecagon | Circle |

much like a circle that it might very well serve as the shape of a wheel. For this reason, some of the properties of the circle were discovered by comparing it with a regular polygon having a large number of sides.

In the *Elements*, Euclid showed how to draw some of the regular polygons by locating their corners on a circle. The tools that he used to do so were the *straightedge*, for drawing straight lines, and the *compass*, for drawing circles and parts of circles, called arcs. The Greek philosopher Plato is said to have chosen these tools because of their simplicity, and they have been used ever since to make *geometric constructions.*

▶**Geometric constructions** are drawings made with only two tools: a straightedge and a compass.

Of all the regular polygons, the equilateral triangle and regular hexagon are the easiest to construct, followed closely by the square. Drawing a regular pentagon with straightedge and compass is rather tricky and the construction of a regular heptagon or regular nonagon with these tools is *impossible.**

---

* It *is* possible to draw these figures if you use a protractor or some other instrument.

You might think that it should be possible to solve every problem in mathematics if a person is clever enough and keeps trying. But there are mathematical problems that are impossible now and always will be, no matter what someone may think of in the future.

Exercises

Set I

This photograph, taken through a microscope, is of ice crystals.

1. What shape do ice crystals have?

2. Use a straightedge and compass to construct this polygon by doing the following.
   a) Draw a circle as shown in the first figure below and label its center O.* Choose a point A on the circle and, leaving the

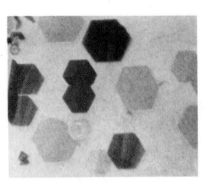

Courtesy of B. J. Mason

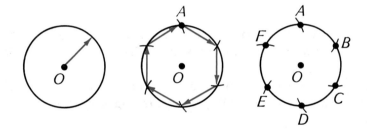

compass adjusted to the radius of the circle, mark off equal arcs around the circle as shown in the second figure. If you do this accurately, the last arc should intersect the circle in point A.
   b) Label the points at which the arcs intersect the circle B, C, D, E, and F as shown in the third figure. Connect the points in order to form a regular polygon.

3. Draw another circle and repeat the same procedure. Rather than connecting all six points in order, however, connect only points A, C, and E to form a regular polygon.

4. What regular polygon is formed?

Napoleon Bonaparte had a surprising interest in mathematics. In 1797, he told two mathematicians in Paris about new discoveries in

---

*In these figures, the arrows show where to put the compass on the paper. Each arrow begins at the point at which the metal point of the compass should be placed and points to the point at which the pencil point should be placed.

Chapter 5: SYMMETRY AND REGULAR FIGURES

geometry that he had learned of in Italy. One of the mathematicians is said to have remarked: "General, we expect everything of you, except lessons in geometry!"

One of the problems that interested Napoleon was how to locate the corners of a square without using a straightedge. The solution is described in the following exercise.

5. The construction begins in the same way as those in exercises 2 and 3.

a) Draw a circle, mark a point on the top of it, and, leaving the compass adjusted to the radius of the circle, draw three arcs like those in the first figure below. Label the points A, B, C, and D.

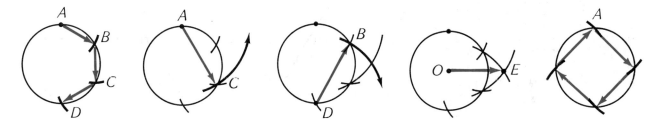

b) Adjust the compass so that the metal point is on A and the pencil point is on C and draw an arc like the one shown in the second figure.

c) Now put the metal point on D and the pencil point on B and draw another arc crossing the preceding one as shown in the third figure.

d) Label the point at which the arcs cross E. Adjust the compass so that the metal point is at the center of the circle, O, and the pencil point is on E. The distance between these two points is equal to the length of the side of the square.

e) Now pick the compass up and, starting at A, draw four arcs around the circle, as shown in the fifth figure. The points at which they intersect the circle are the corners of the square, and we used only a compass to find them. Of course, to draw the sides of the square you will need to use your straightedge.

Perhaps the most common type of traffic sign is the one shown in this photograph.

6. What shape does it have?

7. To simplify the construction of this polygon, we will use the square constructed in exercise 5.

a) First, draw a circle having the same radius as the circle with the square. Then adjust your compass to draw the four arcs

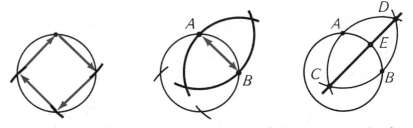

whose intersections locate the corners of the square in the first circle, as shown in the first figure above.

b) Label two of the points A and B and draw two equal arcs with centers at A and B, as shown in the second figure.

c) Label the two points at which the arcs intersect C and D and draw a line through them. Label the point at which this line intersects the circle E; this point divides arc AB of the circle into two equal arcs.

d) The distance between A and E is equal to the length of the side of the polygon. If you continue drawing arcs equal in length to arc AE around the circle, you should get eight equal arcs. Connect the points with line segments and the polygon is complete.

Set II

Several methods for constructing a regular pentagon are known. We will use a method taken from Ptolemy's great work on astronomy, the *Almagest*. Ptolemy lived and worked in Alexandria about 150 A.D.

1. First, draw a circle with a radius of about 4 centimeters and draw a diameter in it, as shown in the first figure below.

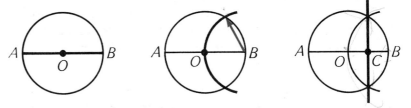

a) Label the center of the circle O and the end points of the diameter A and B. Place the metal point of your compass at B and the pencil point on O and draw an arc like the one shown in the second figure.

b) Draw a line through the two points at which the arc crosses the circle, and label the point at which it intersects the diameter C. This point is midway between points O and B.

To make the remaining diagrams easier to understand, some

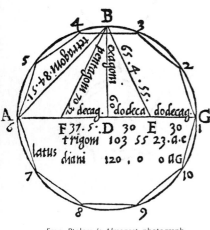

From Ptolemy's *Almagest*; photograph by Owen Gingerich

Chapter 5: SYMMETRY AND REGULAR FIGURES

of the lines have been omitted, but do not erase any lines on your own paper.

c) Put the metal point of the compass at A and the pencil point on B, and draw an arc as shown in the first figure below.

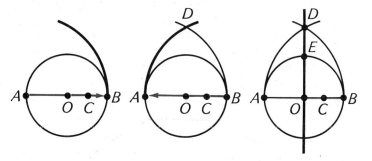

d) Keeping your compass at the same length, put the metal point at B. Draw another arc intersecting the preceding arc at a point above the circle, as shown in the second figure. Label the point D.

e) Draw a line through points D and O; notice that it is perpendicular to the diameter AB. Label the point at which the line crosses the top of the circle E.

f) Put the metal point at C and the pencil point at E. Draw an arc that crosses the diameter of the circle as shown in the first figure at the right. Label the point where it meets the diameter F.

g) The distance between E and F is equal to the length of a side of the pentagon. Put the metal point at E and adjust the compass so that the pencil falls on F. Now, if your work has been very accurate, you should be able to draw five equal arcs around the circle, starting and ending at E. Connect these points with line segments to form a regular pentagon.

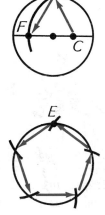

Ptolemy not only gave directions for the construction of the regular pentagon in the *Almagest,* but also explained why the method works. The explanation can be found in most advanced geometry books.

2. We have constructed regular polygons of 3, 4, 5, 6, and 8 sides. As mentioned in this lesson, it is impossible to construct a regular polygon of 7 or 9 sides using only a straightedge and compass. This was proved by Carl Friedrich Gauss in 1796 when he was nineteen years old.

There is, however, a method for constructing a regular polygon having 10 sides. Look again at the method for constructing the regular octagon from the square described in exercise 7 of Set I. Use a similar procedure to construct the regular decagon from the regular pentagon.

Set III

Here is an interesting puzzle. It is called a *dissection* puzzle and the problem is to cut a regular dodecagon into pieces that can be rearranged to form a square.

Use a ruler to accurately trace the figure shown here. Notice that it consists of six pieces, one of which looks like an equilateral triangle.

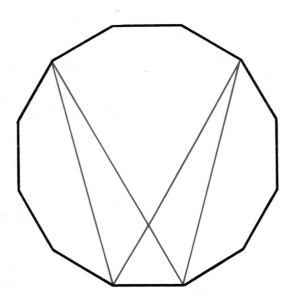

Cut the six pieces apart with scissors and try to rearrange them to form a square. If you solve the puzzle, either make a sketch to show what the arrangement of the pieces looks like or tape them together on your paper.

Some dissection puzzles were discovered by the Greeks and, in the tenth century, the Persian astronomer Abul Wefa wrote an entire book about them. The most complete book on them at the present was written by Harry Lindgren, an Australian patent examiner.*

---

* *Recreational Problems in Geometric Dissections and How to Solve Them* by Harry Lindgren (Dover, 1972).

Photograph from the Moody Institute of Science

Mathematical Mosaics

Bees are expert mathematicians when it comes to building honeycombs. The fourth-century Greek mathematician Pappus said:

> *Though God has given to men the best and most perfect understanding of wisdom and mathematics, He has allotted a partial share to some of the unreasoning creatures as well. . . . This instinct is specially marked among bees. They prepare for the reception of the honey the vessels called honeycombs, with cells all equal, similar and adjacent, and hexagonal in form.**

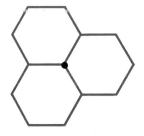

Pappus observed that bees use the regular hexagon exclusively for the shape of the cells in the honeycomb. The photograph shows how the cells are arranged. Each corner point in the honeycomb is surrounded by exactly three hexagons, and there is no wasted space between cells.

Is this hexagonal arrangement of cells the most efficient one? Would other regular polygons work just as well?

---

* *Selections Illustrating the History of Greek Mathematics* by Ivor Thomas (Harvard University Press, 1939).

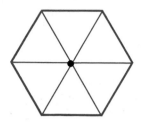

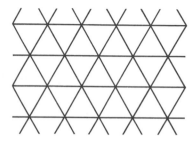

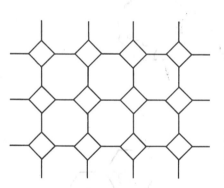

The regular polygon with the least number of sides is the equilateral triangle. The angles of an equilateral triangle are smaller than those of a regular hexagon, and so more than three triangles will fit around a point. In fact, there is room for exactly six.

If bees used equilateral triangles, their honeycomb would look like the second figure at the left. This arrangement, however, would require more wax to form the cells than does the hexagonal honeycomb, if the cells in both patterns are equal in size.

What if bees used a regular polygon having *more* sides than a hexagon rather than fewer sides? How would, say, regular octagons work? The angles of a regular octagon are larger than those of a hexagon; so there is room for only two around a point, with some space remaining unfilled. However, there is just the right amount of room for a square in this space; so two regular octagons and one square surround a point exactly. If this pattern were repeated, the

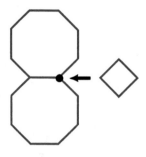

honeycomb would look like the third figure at the left. This design has the disadvantage of being more complicated than the simple hexagonal pattern and it, too, would require more wax.

Each of the designs that we have considered is a *mosaic*. Although a mosaic is any arrangement of shapes fitted together to cover a surface with no gaps or overlapping, we will consider only mosaics of the following type in this lesson.

▶ A **mathematical mosaic** is a set of regular polygons arranged so that the polygons share their sides and the polygons at each corner point are alike in number, kind, and order.

Chapter 5: SYMMETRY AND REGULAR FIGURES

Exercises

Set I

Experiment: MATHEMATICAL MOSAICS

The pattern used by bees in building
a honeycomb is also used in the construction
of doors. The doors, made by gluing thin sheets
of wood to both sides of a paper honeycomb, have the
advantage of being light in weight yet exceptionally strong.

1. What regular polygon is used in the shape of the cells of the honeycomb?

2. How many of them surround each corner point of the honeycomb?

Look at the regular polygons shown below.

3. What happens to the measure of each angle of a regular polygon as the number of sides of the polygon increases?

A set of polygons will surround a point exactly only if the sum of the measures of the angles at the point is 360°. Three hexagons surround a point because

$$120° + 120° + 120° = 360°.$$

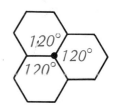

4. How many squares surround a point exactly?

5. How many equilateral triangles surround a point exactly?

6. No number of regular pentagons will surround a point exactly. Explain why not.

To discover other ways to surround a point exactly, we will experiment with regular polygons of several different shapes. If you are not given materials for making the necessary polygons, you can make them by following the directions in the next paragraph.

Lesson 3: Mathematical Mosaics

259

The figure below contains the five polygons that you will need. Copy it accurately on tracing paper. (All that you need to do to make the tracing is to mark the corners of each polygon.) Put a sheet of stiff paper underneath the tracing paper and poke a hole with the metal point of a compass at each of the twelve corners of the dodecagon. Then remove the paper, use a straightedge to draw the sides of the polygon on the stiff paper, and carefully cut it out. Repeat this procedure until you have:

> 2 regular dodecagons,
> 2 regular octagons,
> 2 regular hexagons,
> 3 squares, and
> 4 equilateral triangles.

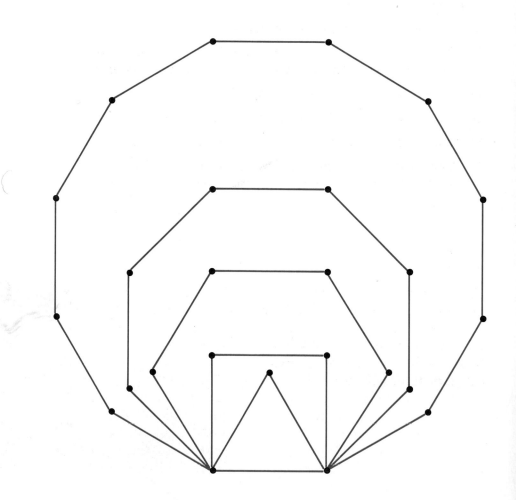

Chapter 5: SYMMETRY AND REGULAR FIGURES

On another sheet of paper, draw a point and surround it with a square and two octagons as shown at the right.

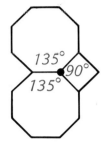

7. What is the sum of the measures of the three angles surrounding the point?

Two dodecagons together with another polygon will exactly surround a point.

8. What is the polygon? (Use your polygons to find out which one it is.)

9. Use arithmetic to figure out the measure of one of the angles of a dodecagon.

Use your polygons to discover whether a point can be surrounded by the following combinations of shapes. If it is possible to surround the point, *indicate how many of each polygon are used.* If it is not possible, say so.

10. One or more triangles and squares.

11. One or more triangles and hexagons. (Two different combinations are possible in this case. Tell what both of them are.)

12. One or more squares and hexagons.

13. One or more triangles, squares, and hexagons.

14. One or more triangles, squares, and octagons.

15. A dodecagon and polygons of two other shapes. (Two different combinations are possible in this case. Tell what both of them are.)

Set II

This photograph shows ceramic tiles available commercially in the form of a mosaic. The mosaic can be represented by the symbol 4-8-8 to show that one square and two octagons surround each point.

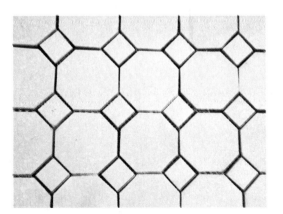

Write symbols to indicate the polygons that surround each point in the mosaics illustrated in the following figures.

1.

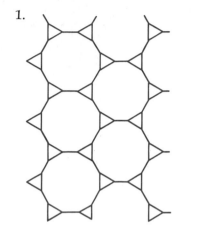

2.

3.

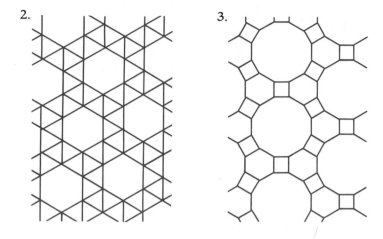

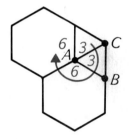

Although two triangles and two hexagons exactly surround a point, it is easy to show that they will not form a mosaic in the order 3-3-6-6.

4. What is the sum of the measures of the four angles surrounding point A in the figure at the left?

5. Trace the figure on your paper and then add a triangle and a hexagon to your drawing so that point B is also surrounded in the order 3-3-6-6.

It is now impossible to surround point C in the same way.

6. Why?

In exercises 5 and 6, you showed that a 3-3-6-6 mosaic cannot exist, yet in the mosaic illustrated below, each corner point is surrounded by two triangles and two hexagons.

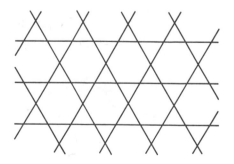

7. How is this possible?

Chapter 5: SYMMETRY AND REGULAR FIGURES

A triangle, two squares, and a hexagon can be used to surround a point in two different orders, as shown in the figures at the right. Only one of these orders can be used to make a mosaic, however. It is shown in the figure below.

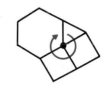

3-4-4-6

3-4-6-4

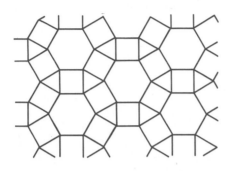

8. Which order is it?

In the first figure below, point A is exactly surrounded by two triangles, a square, and a dodecagon in that order. In the second figure, point B is exactly surrounded by polygons of the same type in the same order.

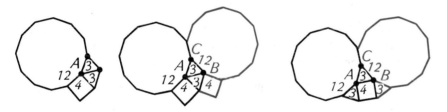

9. What does the second figure reveal about the possible existence of a 3-3-4-12 mosaic? Explain.

10. What does the third figure reveal about the possible existence of a 3-4-3-12 mosaic?

Three different patterns consisting of equilateral triangles and squares are shown below. One of them is not a mathematical

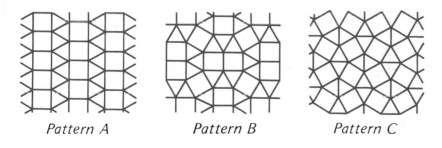

Pattern A   *Pattern B*   *Pattern C*

mosaic because every corner point is not surrounded by the same regular polygons in the same order.

11. Which one?

12. Write appropriate symbols for the other two patterns.

Set III

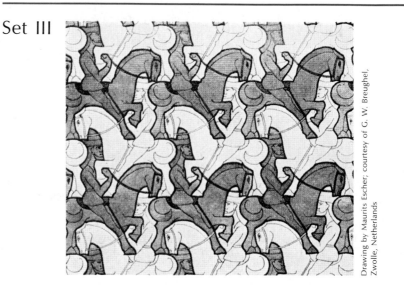

Drawing by Maurits Escher; courtesy of G. W. Breughel, Zwolle, Netherlands

We have defined a mathematical mosaic as an arrangement of *regular polygons* that completely cover a surface. It is possible to fill a surface with a repeating pattern of other shapes as well.

The Dutch artist Maurits Escher created many mosaics using picture shapes.* A remarkable example of these drawings is the one of knights on horseback shown above.

Another of Escher's drawings is based on the 6-6-6 mosaic. It consists of identical hexagons filled with the design at the left.

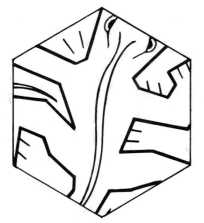

After a periodic drawing by M. C. Escher;
© Beeldrecht, Amsterdam/V.A.G.A., New York;
Collection Haags Gemeentemuseum—The Hague, 1981

1. Place a sheet of tracing paper over the design and make an accurate copy of it. Then see if you can figure out how Escher fitted the hexagons together to form the mosaic. (You will need to turn the tracing paper in different directions as you make additional tracings of the design.) Continue the mosaic until you have at least seven hexagons in your drawing.

2. What is pictured in the mosaic?

To make them easier to see, you might shade them in different colors.

---

* Forty-two of them are included in the book *Fantasy and Symmetry: The Periodic Drawings of M. C. Escher* by Caroline H. MacGillavry (Abrams, 1976).

Chapter 5: SYMMETRY AND REGULAR FIGURES

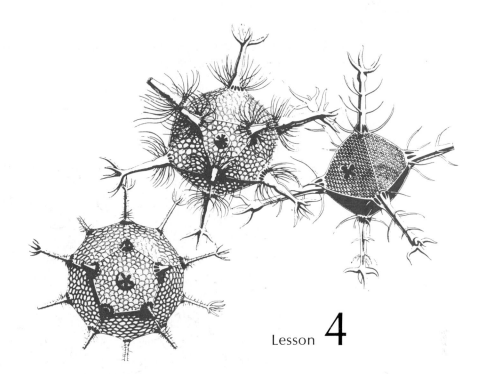

Regular Polyhedra: The Platonic Solids

The skeletons of sea creatures so small that their structure can be seen only under a microscope are shown above. Called radiolarians, these one-celled animals are found floating near the surface of warm ocean water. The skeletons of radiolarians appear to be built of equilateral triangles and regular pentagons; more than that, each one has the shape of a *regular polyhedron.**

▶ A **regular polyhedron** is a solid having faces (surfaces) in the shape of a regular polygon. All of its faces, edges, and corners are identical.

The simplest regular polyhedron is the regular **tetrahedron,** a solid with four equilateral triangles for its faces. Three triangles meet at each corner of a tetrahedron. The adjoining figures show a tetrahedron and the three triangles that meet at one of its corners.

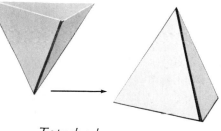

Tetrahedron

---

* The plural of *polyhedron* is *polyhedra.*

If there were four triangles at each corner instead of three, we would have a solid with eight faces, called a regular **octahedron.**

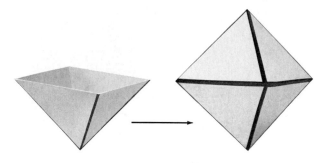

Octahedron

Five triangles at each corner result in a solid called a regular **icosahedron.** An icosahedron has twenty faces in all.

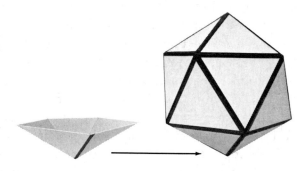

Icosahedron

The most familiar regular polyhedron is the one whose faces are squares: the hexahedron, or **cube.** Three squares meet at each corner of a cube and a cube has six faces altogether.

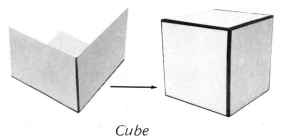

Cube

You may recall that when three regular pentagons are put to-gether around a point, they leave a gap. If the pentagons are folded upward to close the gap, we get a corner of a regular **dodecahedron,** a solid having twelve faces in all.

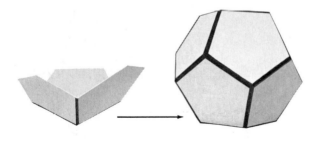

Dodecahedron

The regular polyhedra were first studied by a group of Greek mathematicians under the leadership of Pythagoras in the sixth century B.C. They are sometimes referred to as the **Platonic solids,** after Plato, who gave instructions for making models of them. Plato claimed that the atoms of the four elements of ancient science had the shapes of regular polyhedra. Atoms of earth were supposed to have the shape of cubes; atoms of air, octahedrons; atoms of fire, tetrahedrons; and atoms of water, icosahedrons. The universe itself was thought to be in the shape of a dodecahedron.

Although these ideas now seem naive, it is true that many crystals grow in the shape of regular polyhedra as a result of the arrangement of the atoms in them. The cube and tetrahedron, for example, appear in the shape of sodium chlorate crystals; the octahedron appears in crystals of chrome alum. So Plato was on the right track in relating the shape of matter to the regular polyhedra.

From *Crystals and Crystal Growing* by Alan Holden and Phylis Singer. Copyright © 1960 by Educational Services Inc. Reprinted by the permission of Doubleday & Company, Inc., and Heinemann Educational Books Ltd.

Exercises

Set I

Experiment: MODELS OF THE REGULAR POLYHEDRA

To do the first four parts of this experiment, you will need 20 drinking straws,* 120 paper clips, and a pair of scissors. Cut the straws into 60 equal lengths (6.5 centimeters is convenient).

---

* The straws should be of the standard size ($\frac{1}{4}$ inch diameter).

Part 1. The Tetrahedron Take three paper clips and link them together as shown in the picture at the left. The narrow end of each clip should be at the center. We will refer to this linkage as a "3-clip corner assembly."

Make three more 3-clip corner assemblies. Now take six of the short straws that you cut and put the tetrahedron together by inserting the clips of the corner assemblies into them. The finished model is shown in the first photograph at the left. (Be sure to insert the wide end of each clip into the straw; otherwise, the straw and clip will not hold together.)

Look at your model to answer the following questions.

1. How many triangles does the tetrahedron contain?

2. How many corners does it have?

3. How many edges (straws) does it have?

4. How many edges meet at each of its corners?

Part 2. The Octahedron Make six 4-clip corner assemblies. Take twelve of the short straws and put the octahedron together by inserting the clips of the corner assemblies into them. The finished model is shown in the second photograph at the left.

Look at your model to answer the following questions.

5. How many triangles does the octahedron contain?

6. How many corners does the octahedron contain?

7. How many edges does it have?

8. How many edges meet at each of its corners?

9. What regular polygon other than equilateral triangles do the straws form?

10. How many of these polygons do you see?

Part 3. The Cube Make eight 3-clip corner assemblies. These corner assemblies and twelve short straws are needed to put the cube together. A photograph of the finished model is not shown because the model is not rigid and therefore loses its shape.

Look at your model to answer the following questions.

11. How many squares does the cube contain?

12. How many corners does it have?

13. How many edges does it have?

14. How many edges meet at each of its corners?

15. Why does this model, unlike the models of the other polyhedra, lack rigidity?

16. The model could be braced with straws as shown in the drawing at the right. What polyhedron do the bracing straws form?

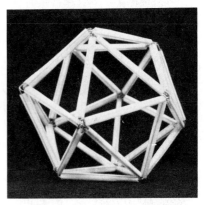

Part 4. The Icosahedron Make twelve 5-clip corner assemblies. These assemblies and the rest of the short straws are needed to put the icosahedron together. The finished model is shown in the photograph at the right.

Look at your model to answer the following questions.

17. How many triangles does the icosahedron contain?

18. How many corners does it have?

19. How many edges does it have?

20. How many edges meet at each of its corners?

21. What regular polygon other than equilateral triangles do the straws form?

22. How many of these polygons do you see?

Part 5. The Dodecahedron A straw model of a dodecahedron would be even less rigid than that of a cube; so we will construct a dodecahedron by another method. The large figure below is the pattern for the model. Make two copies of this figure on tracing

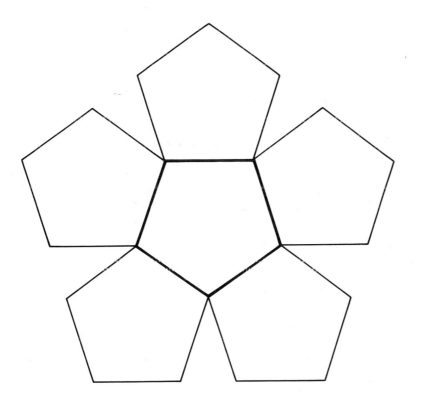

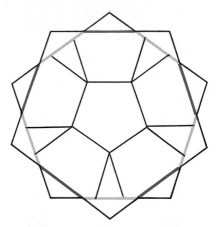

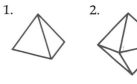

paper. (All that you need to do is to mark the corners of the pentagons.) Put a piece of stiff paper underneath the tracing paper and poke a hole with the metal point of a compass at each of the twenty points of your tracing. Remove the tracing paper, use a straightedge to draw all of the line segments between the twenty holes in the stiff paper, and cut the two figures out. Fold the five pentagons in each figure one at a time onto the pentagon in the center, making a sharp crease along each of the heavy line segments in the pattern. The result will be two "bowls."

Hold the "bowls" so that they face each other and press them together so that they flatten out. Take a rubber band (size 19) and weave it alternately above and below the corners as shown in the first figure at the left, while holding the two pieces flat. If you carefully let go, a dodecahedron will pop into shape.

Look at your model to answer the following questions.

23. How many pentagons does the dodecahedron contain?

24. How many corners does it have?

25. How many edges does it have?

26. How many edges meet at each of its corners?

Set II

4-4-4   4-4-4-4

The regular polyhedra are comparable to mathematical mosaics in that the regular polygons that meet at each corner point are alike in number, kind, and order. In a cube, for example, three squares meet at each corner. In a mathematical mosaic consisting of squares, four squares meet at each corner point. The appropriate symbols for these patterns are shown in the adjoining figures.

Write the appropriate symbols for the following mosaics and regular polyhedra.

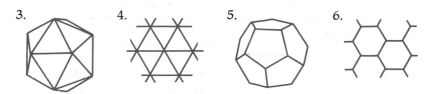

7. What is the fewest number of polygons that can meet at one corner point of a mosaic or regular polyhedron?

Look at the measures of the angles of the regular polygons shown below.

Because each corner of a cube is surrounded by three squares, the sum of the measures of the angles at each corner is

$$90° + 90° + 90° = 270°.$$

What is the sum of the measures of the angles that surround each corner of

8. a tetrahedron? (See exercise 1.)

9. an octahedron? (See exercise 2.)

10. an icosahedron? (See exercise 3.)

11. a dodecahedron? (See exercise 5.)

In general, what is true about the sum of the measures of the angles that surround each corner point

12. of a mosaic?

13. of a regular polyhedron?

Lewis Carroll once referred to the regular polyhedra as being "provokingly few in number." The answer to exercise 13 is part of the reason for the fact that there are only five of them.

A regular tetrahedron has four triangular faces. One way to determine how many corners and edges it has is to imagine putting it together from the four triangles.

Four triangles have
4 × 3 = 12 corners and
4 × 3 = 12 sides.

Three triangles meet at each corner; so a tetrahedron has $\frac{12}{3} = 4$ corners.

Two triangles meet at each edge; so a tetrahedron has $\frac{12}{2} = 6$ edges.

Four separate triangles

A tetrahedron

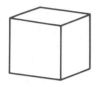

A cube

An octahedron

A dodecahedron

An icosahedron

A cube has six square faces.

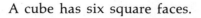

Six separate squares

14. Use the same method of reasoning to determine how many corners and edges it has. Show the equations used.

An octahedron has eight triangular faces.

Eight separate triangles

15. Use the same method of reasoning to determine how many corners and edges it has. Show the equations used.

A dodecahedron has twelve pentagonal faces.

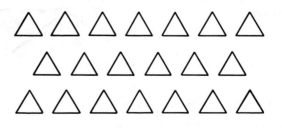

Twelve separate pentagons

16. Determine how many corners and edges it has. Show the equations used.

An icosahedron has twenty triangular faces.

Twenty separate triangles

Chapter 5: SYMMETRY AND REGULAR FIGURES

17. Determine how many corners and edges it has. Show the equations used.

18. Copy and complete the following table. Use your results for exercises 14 through 17 to fill in the missing numbers.

Regular polyhedron	Number of faces	Number of corners	Number of edges
Tetrahedron	4	4	6
Cube	▧	▧	▧
Octahedron	▧	▧	▧
Dodecahedron	▧	▧	▧
Icosahedron	▧	▧	▧

19. How is the number of edges of each polyhedron related to its numbers of faces and corners?

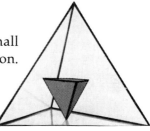

In the model at the right, each of the four corners of a small tetrahedron touches one of the four faces of a larger tetrahedron.

20. What does the model below show?

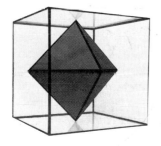

21. What does the model at the right show?

Look again at your table for exercise 18.

22. Could any other models like these be built? If so, what would they look like?

Set III

A clever puzzle called Back in the Box consists of a cubical box that is packed with seventeen tetrahedral pieces.* The idea is to dump the pieces out of the box and then figure out how to fit all of them back into it again.

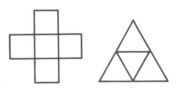

To figure out how to put just the largest piece, a regular tetrahedron, back into the box is something of a puzzle in itself. Make a model of it by constructing an open box from five squares measuring 5 centimeters on each side and a tetrahedron from four equilateral triangles measuring 7 centimeters on each side. Convenient arrangements for doing this are shown at the left.

When you have figured out how to fit the tetrahedron completely inside the box, make a drawing to illustrate the solution.

---

* Back in the Box is manufactured by Classic Games Company, Inc., Deer Park, New York 11729.

Lesson **5**

Semiregular Polyhedra

Leonardo da Vinci, the great Renaissance artist, scientist, and inventor, began one of his books with these words: "Let no one who is not a mathematician read my works." Da Vinci was so obsessed with mathematics for a time that he neglected his painting. He built a set of wooden models of geometric solids and from them created a series of beautiful illustrations for a book titled *The Divine Proportion* by Luca Pacioli, published in 1509. Three of these drawings appear at the top of this page. Each drawing is of a *semiregular polyhedron*.

▶ A **semiregular polyhedron** is a solid that has faces in the shape of more than one kind of regular polygon, yet every corner is surrounded by the same kinds of polygons in the same order.

Among the semiregular polyhedra are thirteen solids called the Archimedean solids.* Named for Archimedes, who wrote a book

---

*Recall that there are only five *regular polyhedra,* or *Platonic solids.*

Courtesy of AMF Voit, Inc.

about them, the Archimedean solids have interested mathematicians and designers ever since.

Compare the polyhedron illustrated in the third of da Vinci's drawings with the pattern on the soccerball shown in this photograph. Both consist of regular pentagons and regular hexagons with one pentagon and two hexagons meeting at each corner. The symbol for this pattern is 5-6-6.

The polyhedron, called a *truncated icosahedron,** is almost as round as the ball. It is round enough, in fact, that it could be put inside a sphere so that each corner would touch the surface.

The other solids pictured in da Vinci's drawings at the beginning of this lesson are a *truncated octahedron,* in which each corner is surrounded by a square and two hexagons, and a *cuboctahedron,* in which each corner is alternately surrounded by two equilateral triangles and two squares. The symbols for these solids are 4-6-6 and 3-4-3-4, respectively.

The cuboctahedron is an especially popular shape with designers. It has been used, for example, in the design of lamps, paperweights, and soap. The semiregular polyhedra, like so many other creations of mathematics, have been found to be useful in a wide variety of applications.

---

* This name, like those of the other Archimedean solids, is rather complicated and not worth remembering.

Exercises

Set I

Photographs of models of the Archimedean solids are shown below. Look at the regular polygons that surround each corner and write the appropriate symbol for each solid.

1.

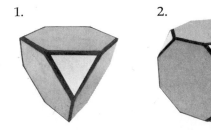

2.

3.

4.

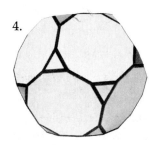

Chapter 5: SYMMETRY AND REGULAR FIGURES

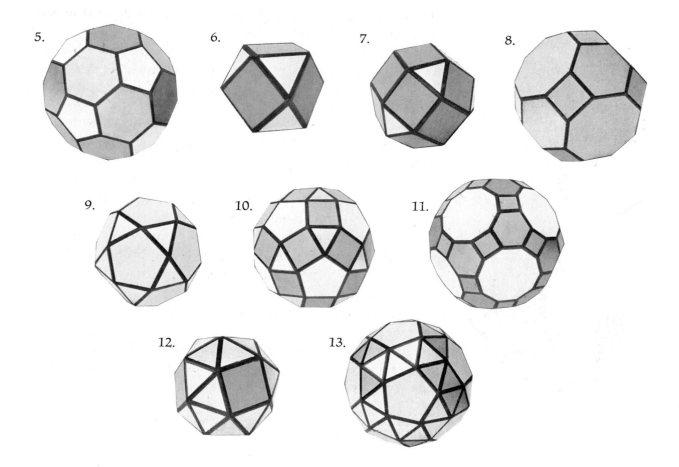

5.   6.   7.   8.

9.   10.   11.

12.   13.

The *truncated cube* is a cube whose corners have been cut off.

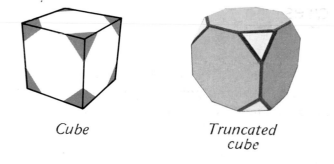

Cube *Truncated cube*

14. How many square faces meet at each corner of a cube?

15. What regular polygon is formed in place of each corner of the cube when the corners are cut off?

16. What regular polygon does each square face of the cube become when the corners are cut off?

The corners of the other regular polyhedra also can be cut off to form Archimedean solids. The *truncated dodecahedron* is shown here.

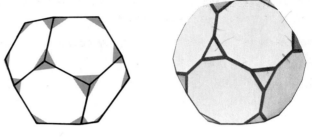

Dodecahedron   *Truncated dodecahedron*

17. How many pentagonal faces meet at each corner of a dodecahedron?

18. What regular polygon is formed in place of each corner of the dodecahedron when the corners are cut off?

19. What regular polygon does each pentagonal face of the dodecahedron become when the corners are cut off?

The other three truncated polyhedra are shown here.

Tetrahedron   *Truncated tetrahedron*

Octahedron   *Truncated octahedron*   *Icosahedron*   *Truncated icosahedron*

20. What do the tetrahedron, octahedron, and icosahedron have in common?

21. What regular polygon does each face of these polyhedra become when the corners are cut off?

22. How many faces meet at each corner of the tetrahedron, octahedron, and icosahedron?

23. What regular polygons are formed in place of each corner of the tetrahedron, octahedron, and icosahedron when the corners are cut off?

Another Archimedean solid, the *cuboctahedron*, can be obtained from either of two regular polyhedra as shown below.

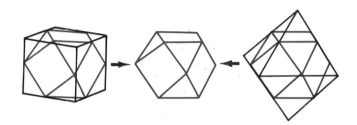

24. What are the two regular polyhedra?

25. How many square faces does a cuboctahedron have?

26. How many triangular faces does it have?

The *icosidodecahedron* also can be obtained from either of two regular polyhedra.

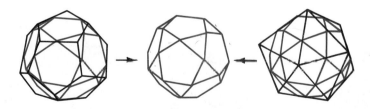

27. What are the two regular polyhedra?

28. How many pentagonal faces does an icosidodecahedron have?

29. How many triangular faces does it have?

This photograph shows a *rhombicuboctahedron* and its reflection in a mirror. This solid is symmetrical in the sense that it and its reflection are identical.

Two of the Archimedean solids are not identical with their reflections; they are reversed in a mirror instead.

30. Which two do you think they are?

Set II

This drawing is from a book by Albrecht Dürer published in 1525. It is a pattern for one of the Archimedean solids.

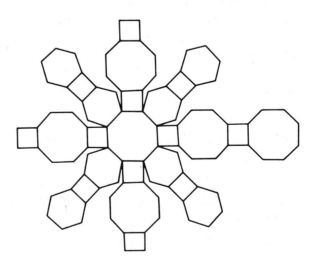

1. What is the name of the solid?

2. What regular polygons are its faces?

3. How many faces of each type does it have?

4. How many faces does it have altogether?

5. Use the method explained on page 271 to determine how many corners and edges it has. Show the equations used.

Use the following information and the same method of reasoning to find the number of corners and the number of edges in each of these solids. (Refer to the figures in Set I.) In each case, show the equations used.

6. A *truncated cube* contains 14 faces, of which 8 are triangles and 6 are octagons. (Figure 2)

7. A *truncated icosahedron* contains 32 faces, of which 12 are pentagons and 20 are hexagons. (Figure 5)

8. A *great rhombicosidodecahedron* contains 62 faces, of which 30 are squares, 20 are hexagons, and 12 are decagons. (Figure 11)

9. A *snub dodecahedron* contains 92 faces, of which 80 are triangles and 12 are pentagons. (Figure 13)

Here is a table for the other Archimedean solids.

Polyhedron	Number of faces	Number of corners	Number of edges
Truncated tetrahedron	8	12	18
Truncated octahedron	14	24	36
Truncated dodecahedron	32	60	90
Cuboctahedron	14	12	24
Rhombicuboctahedron	26	24	48
Icosidodecahedron	32	30	60
Rhombicosidodecahedron	62	60	120
Snub cube	38	24	60

10. How is the number of edges of each polyhedron in this table related to its numbers of faces and corners?

11. Does the same relation hold for the solid whose pattern by Dürer is shown on the preceding page? (Look at your answers to exercises 4 and 5.)

12. Does it hold for each of the solids in exercises 6 through 9?

Set III

It is easy for a child playing with blocks to discover that the cube can be used to fill space. Is it possible to fill space with any other regular or semiregular polyhedron?

The answer is yes: truncated octahedra can be fitted together to fill space. The photograph below shows a set of them.

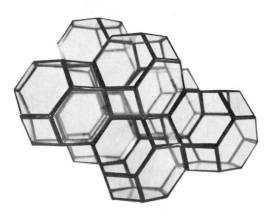

On tracing paper, make a copy of the pattern on the next page. (All that you need to do is to mark the corners of the hexagons.) Use your tracing to make five copies of the pattern on stiff paper. Cut each pattern out and fold along each of the lines. Use tape to make each pattern into a truncated octahedron as shown in this

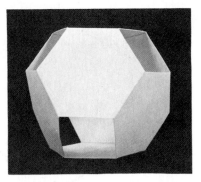

photograph. For simplicity, the pattern is designed so that each of the square faces of the solid is open.

After you have made the five truncated octahedra, see if you can fit them together. Use tape to hold them in place. (Even though you can refer to the photograph above, you will probably find that fitting them together is not as easy to do as it might seem.)

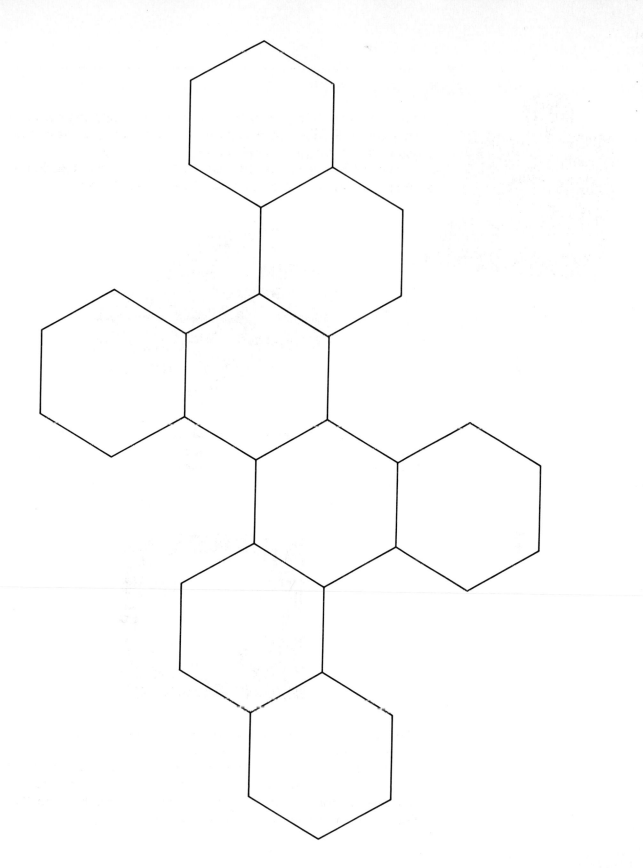

Photograph from the Moody Institute of Science

Lesson **6**

Pyramids and Prisms

The largest of all man-made geometric solids was built in about 2600 B.C. It is the Great Pyramid in Egypt, the only one of the "seven wonders of the world" still in existence. This pyramid, one of about eighty such structures built by the ancient Egyptians, was put together from more than two million stone blocks, weighing between 2 and 150 tons each.

The Great Pyramid is one member of an unlimited set of geometric solids called *regular pyramids*.

▶ A **regular pyramid** is a solid that has a regular polygon for its base and triangles that are identical in size and shape for the rest of its faces.

Although the Egyptians consistently chose the square for the shape of the bases of their pyramids, other polygons also can be used. Examples of pyramids are shown at the top of the next page.

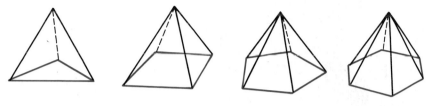

Triangular
pyramid

Square
pyramid

Pentagonal
pyramid

Hexagonal
pyramid

These illustrations show that a pyramid is named according to the shape of its base. Because all of the faces of a regular tetrahedron are equilateral triangles identical in size and shape, the tetrahedron is also a triangular pyramid.

There is another unlimited set of geometric solids called the *regular prisms*.

▶ A **right regular prism** is a solid that has a pair of regular polygons for its bases and rectangles that are identical in size and shape for the rest of its faces.

The bases of a prism are always parallel to each other, and a prism, like a pyramid, is named according to the shape of its bases. Examples of right regular prisms are shown below. Because all of

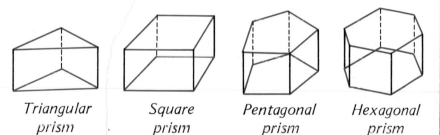

Triangular
prism

Square
prism

Pentagonal
prism

Hexagonal
prism

the faces of a cube are squares identical in size and shape, the cube is also a square prism.

Prisms are frequently found in nature. An awesome display of them is the Devil's Postpile in the Sierra Nevada mountains of California. The Devil's Postpile is a group of tall columns of rock, some of which are 60 feet high. Many of these columns have the shape of right regular prisms, most of them hexagonal and others pentagonal. So even geologists find mathematics in their study of the earth.

Devil's Postpile

Exercises

Set I

The word "prism" is often used to mean a block of glass by which light can be broken up into a spectrum of colors.

1. What shape do the bases of the glass prism shown in the photograph at the left have?

2. What shape do the rest of the faces of this prism have?

3. What is the name of a prism of this shape?

One of the oldest games known was played in the biblical city of Ur of the Chaldees in 2500 B.C. Dice like those shown in the photograph below were used in the game.

4. What type of geometric solid do the dice appear to be?

5. What shape are their bases?

6. How many faces does each die have?

Snow crystals grow around atmospheric dusts having particles of several different shapes, two of which are shown here.

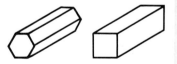

Hematite   Vermiculite

What shapes do the faces of particles

7. of hematite have?

8. of vermiculite have?

What type of geometric solid do particles

9. of hematite appear to be?

10. of vermiculite appear to be?

Chapter 5: SYMMETRY AND REGULAR FIGURES

If the pattern shown at the right were cut out, it could be folded to form a square pyramid. What solids could be formed from the following patterns?

11. 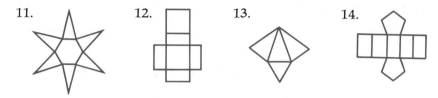 12. 13. 14.

The figures below show a series of regular polygons, in which each successive polygon has twice as many sides as the one preceding it.

15. What geometric figure does a regular polygon become more and more like as the number of its sides becomes larger and larger?

The figures below show a series of regular pyramids having the polygons above as their bases.

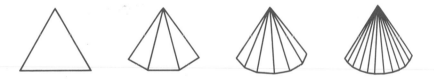

16. What geometric solid does a regular pyramid become more and more like as the number of sides in its base becomes larger and larger? (The answer contains four letters and starts with "c.")

The figures below show a series of regular prisms having for their bases the same polygons as those of the pyramids in exercise 16.

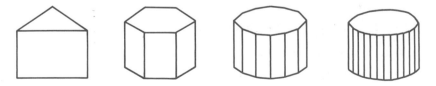

17. What geometric solid does a regular prism become more and more like as the number of sides of its bases becomes larger and larger? (The answer contains eight letters and starts with "c.")

The drawing at the left illustrates a prism all of whose faces are regular polygons.

18. Do the same number and type of faces meet at each of its corners?

19. Write a symbol to represent them.

20. Write a symbol to represent a prism whose faces are regular octagons and squares.

21. Do you think that there are an unlimited number of prisms all of whose faces are regular polygons?

The drawing below illustrates a pyramid all of whose faces are regular polygons.

22. Do the same number and type of faces meet at each of its corners?

23. Explain.

24. Do you think that there are an unlimited number of such pyramids?

25. Could a pyramid be made by folding together the pattern of regular polygons shown at the left?

Set II

In 1752, the Swiss mathematician Leonard Euler stated a formula relating the numbers of faces, corners, and edges of certain geometric solids. The formula is

$$F + C = E + 2$$

in which F represents the number of faces, C represents the number of corners, and E represents the number of edges. You have already found that this formula applies to both the regular polyhedra and the Archimedean solids.

Chapter 5: SYMMETRY AND REGULAR FIGURES

In the exercises that follow, we will see if Euler's formula also applies to pyramids and prisms.

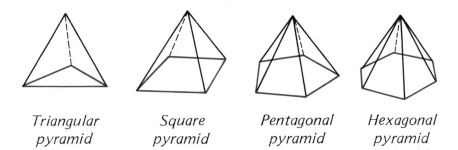

Triangular pyramid   Square pyramid   Pentagonal pyramid   Hexagonal pyramid

1. Refer to the figures above to copy and complete the following table.

Pyramid	Number of sides in base	Number of faces	Number of corners	Number of edges
Triangular	3	4	4	6
Square				
Pentagonal				
Hexagonal				

2. Does Euler's formula fit every pyramid in your table?

Suppose that a pyramid has a base with 12 sides.

3. How many faces would it have?

4. How many corners would it have?

5. How many edges would it have?

If a pyramid has a base with n sides, its number of faces is $n + 1$.

6. What is its number of corners in terms of n?

7. What is its number of edges in terms of n?

Substituting these expressions into Euler's formula, we get

$$F + C = E + 2$$
$$n + 1 + n + 1 = 2n + 2, \text{ or}$$
$$2n + 2 = 2n + 2.$$

This result shows that the formula is true regardless of what number n represents. So Euler's formula is true for all pyramids.

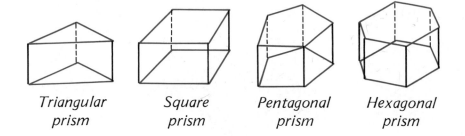

| Triangular prism | Square prism | Pentagonal prism | Hexagonal prism |

8. Refer to the figures above to copy and complete the following table.

Prism	Number of sides in bases	Number of faces	Number of corners	Number of edges
Triangular	3	5	6	9
Square				
Pentagonal				
Hexagonal				

9. Does Euler's formula fit every prism in your table?

Suppose that a prism has a base with 10 sides.

10. How many faces would it have?

11. How many corners would it have?

12. How many edges would it have?

Suppose that a prism has a base with *n* sides.

13. What is its number of faces in terms of *n*?

14. What is its number of corners in terms of *n*?

15. What is its number of edges in terms of *n*?

16. Substitute these expressions into Euler's formula to show that it is true for all pyramids.

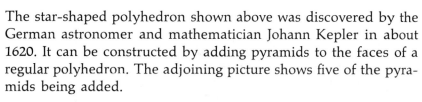

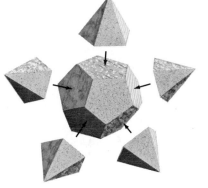

The star-shaped polyhedron shown above was discovered by the German astronomer and mathematician Johann Kepler in about 1620. It can be constructed by adding pyramids to the faces of a regular polyhedron. The adjoining picture shows five of the pyramids being added.

1. What kind of pyramids are they?

2. To what regular polyhedron are they added?

3. How many points does the star have altogether?

Kepler discovered another star-shaped polyhedron, shown

above, at the same time. It also can be formed by adding pyramids to the faces of a regular polyhedron.

4. What kind of pyramids are they?

5. To what regular polyhedron are they added?

6. How many points does this star have altogether?

Icosapirale, a sculpture by Charles Perry at the Alcoa Golden Gate Plaza in San Francisco

Chapter 5 / Summary and Review

In this chapter, we have studied symmetry, regular polygons, and their relation on a flat surface and in space.

Symmetry (*Lesson 1*) A figure has *line symmetry* if there is a line along which a mirror can be placed to reflect either half of the figure so that it reproduces the other half.

A figure has *rotational symmetry* if it can be rotated through an angle of less than 360° so that it coincides with its original position.

Regular polygons (*Lesson 2*) A regular polygon is a figure of which all sides are the same length and all angles are equal.

Regular polygons are named for their numbers of sides. The names most frequently used are: *equilateral triangle* (3 sides), *square* (4 sides), *pentagon* (5 sides), *hexagon* (6 sides), *heptagon* (7 sides), *octagon* (8 sides), *nonagon* (9 sides), *decagon* (10 sides), and *dodecagon* (12 sides).

Geometric constructions are drawings made with only two tools: a straightedge and a compass.

Mathematical mosaics (*Lesson 3*) A *mathematical mosaic* is a set of regular polygons arranged so that the polygons share their sides and the polygons at each corner point are alike in number, kind, and order.

Regular polyhedra: the Platonic solids (*Lesson 4*) A *regular polyhedron* is a solid having faces in the shape of a regular polygon. All of its faces, edges, and corners are identical.

There are only five regular polyhedra: the *tetrahedron* (4 triangular faces), *cube* (6 square faces), *octahedron* (8 triangular faces), *dodecahedron* (12 pentagonal faces), and *icosahedron* (20 triangular faces).

Semiregular polyhedra (*Lesson 5*) A *semiregular polyhedron* is a solid that has faces in the shape of more than one kind of regular polygon, yet every corner is surrounded by the same kinds of polygons in the same order.

Among the semiregular polyhedra are the thirteen *Archimedean solids.*

Pyramids and prisms (*Lesson 6*) A *regular pyramid* is a solid that has a regular polygon for its base and triangles that are identical in size and shape for the rest of its faces.

A *right regular prism* is a solid that has a pair of regular polygons for its bases and rectangles that are identical in size and shape for the rest of its faces.

Pyramids and prisms are named according to the shapes of their bases.

Exercises

Set I

The four symbols used on playing cards are symmetrical.

Heart   Spade   Diamond   Club

1. Which symbols have exactly one line of symmetry?

2. Which symbol has two lines of symmetry?

3. Which symbol has rotational symmetry?

If a knot is tied in a strip of paper and then pressed flat, a regular polygon is formed.

4. Which one is it?

It is fairly easy to construct a square with a straightedge and compass.

5. Name two regular polygons that are very easy to construct.

6. Name a regular polygon that is fairly difficult to construct.

7. Name a regular polygon that is impossible to construct.

The photographs below are of some stained glass windows created by Sheryl Cotleur of San Rafael, California. Write the symbol for the mosaic illustrated by each.

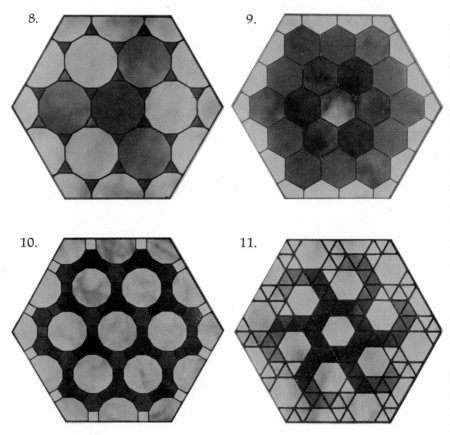

8.

9.

10.

11.

Designed by Sheryl Cotleur. Stained Glass Tesselations Posters, © 1977, Creative Publications, P.O. Box 10328, Palo Alto, California 94303.

12. What is the sum of the measures of the angles that surround each corner point of a mosaic?

Chapter 5: SYMMETRY AND REGULAR FIGURES

These diagrams are parts of floor plans created by the Swiss architect Le Corbusier.

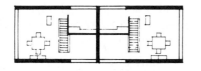

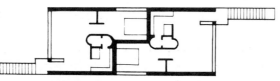

From *Handbook of Regular Patterns* by Peter S. Stevens, M.I.T. Press, 1980

13. What type of symmetry does the first plan illustrate?

14. What is a simple way to prove it?

15. What type of symmetry does the second plan illustrate?

16. What is a simple way to prove it?

The photograph at the right is of a child's ball. Although it is spherical in shape, it looks very much like a regular polyhedron.

17. Which polyhedron does it look like?

18. How many faces does this polyhedron have?

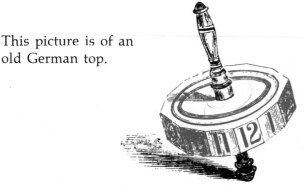

This picture is of an old German top.

19. What type of geometric solid does the part of the top shown in color appear to be?

20. What shape are its bases?

21. What shape do the rest of its faces have?

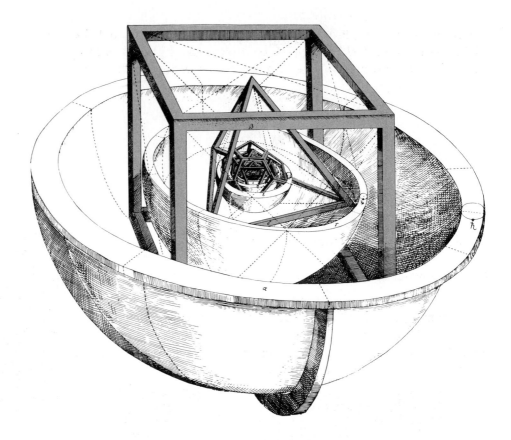

The astronomer Johann Kepler once tried to explain the spacing of the planets in the solar system by means of the regular polyhedra. He imagined a series of six spheres, one for each planet known at the time, with the sun at their center. These spheres were separated by the five regular polyhedra, as the drawing above by Kepler shows.

22. Name the polyhedra as they appear in the figure in order from largest to smallest. (The smallest one is barely visible.)

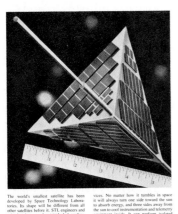
Set II

Small space satellites have been made in the shape shown in the advertisement at the left.

1. Give two different names for its shape.

2. How many faces, corners, and edges does it have?

3. Do these numbers fit Euler's formula: $F + C = E + 2$?

There are three simple ways of planting an orchard so that the distances between a tree and each of its closest neighbors are the same. The three ways are based on regular polygon mosaics. One of them is shown in this figure.

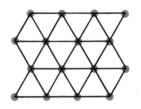

4. Make sketches showing the other two.

Certain viruses have the shape of one of the regular polyhedra.

5. From the microphotograph of the tipula iridescent virus shown at the right, which polyhedron do you think it is?

The picture below, created by means of a computer program, gives a strong impression of being three-dimensional.

Courtesy of Kenneth M. Smith and
Robley C. Williams

From Computer Graphics, Computer Art by
Herbert W. Franke, Verlag F. Bruckmann KG

6. On the assumption that we can see all of its faces except for its base, what type of geometric solid does it appear to be?

7. Do the same number and type of faces meet at each of its corners?

8. Explain.

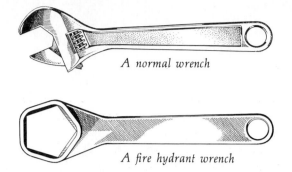

A normal wrench

A fire hydrant wrench

The nut used to turn on a fire hydrant is frequently made in the shape of a regular pentagon. It is much easier to put a wrench on this shape than on a shape that is not regular.

9. Why?

10. Which shape would be harder to turn using a normal wrench: a regular pentagon or a regular hexagon?

11. Why?

The geometric solid shown at the left consists of two intersecting regular polyhedra, one of which is white in this picture and the other brown.

12. What are the two regular polyhedra?

13. What kinds of pyramids appear in the solid?

In addition to the thirteen Archimedean solids, which have been known for more than two thousand years, a fourteenth solid was discovered in 1930 by J. C. P. Miller. Called a *pseudorhombicuboctahedron*, it is shown here beside a rhombicuboctahedron.

The pseudorhombicuboctahedron (the solid on the right) may have been overlooked for such a long time because of its symbol.

14. Explain why.

15. How do you think the numbers of faces, edges, and corners of the two solids compare?

Chapter 5: SYMMETRY AND REGULAR FIGURES

Set III

Most drums make sounds with no definite pitch. The kettledrum, however, can be tuned to play notes of different pitches by adjusting the tension on the skin that is its head.

Imagine that we have a set of four special kettledrums whose heads have the shape of regular polygons and an ordinary drum whose head is circular in shape. If all the heads of these drums have the same area and the same tension, then their pitches (or frequencies) will vary according to their shape.

Triangular drum

Square drum

Hexagonal drum

Octagonal drum

Circular drum

The table below shows how the pitches of the drums might compare under these conditions.

Shape of head	Pitch (in vibrations per second)
Equilateral triangle	146
Square	136
Regular hexagon	132
Regular octagon	131
Circle	130

1. What happens to the pitch of a "regular polygon" drum as the number of sides of its head increases?

Do you think that "regular polygon" drums could be built having the same area and tension as do the drums listed in the table but having approximately the following pitches? Explain your answers.

2. 140 vibrations per second.

3. 134 vibrations per second.

4. 125 vibrations per second.

Chapter 5 / Problems for Further Exploration

Lesson 1

1. Here is a game for two players that has a winning strategy based on symmetry. A set of pennies is arranged in a circle so that the pennies touch each other, as shown here.

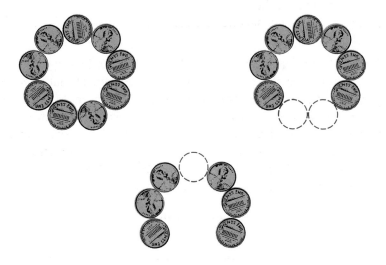

The two players take turns removing either one penny or two pennies that touch each other. The player who picks up the last penny wins.

The game is not fair, because the player who starts second can always win. Suppose that the first player picks up two pennies, leaving the board as shown in the second figure. The second player should then take the penny exactly opposite, so that the board looks like the third diagram.

a) What should the second player do from this point on to be sure of winning the game?

b) What does this strategy have to do with symmetry?

c) If the first player picks up only one penny at the start, what should the second player's original move be?

Suppose that the game is being played with eleven pennies. The first player takes a penny as shown in the first figure below.

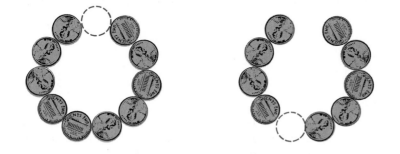

The second player then takes a penny nearly opposite as shown in the second figure.

d) Show that the second player's move will allow the first player to win.

2. Several years ago, a bill was submitted to Congress proposing that Federal Reserve notes be printed in a manner that enables a person who is blind to determine the denomination of each note. The idea was to trim one or more corners of each note so that the values of the notes could be told from their shapes.

There are seven denominations of bills currently in circulation: $1, $2, $5, $10, $20, $50, and $100. One way in which corners of a bill might be cut is shown here.

a) Make drawings like this to show all of the other ways in which the corners of bills can be cut (or not cut) so that blind people can tell the bills apart.

b) How many different ways are there altogether?

c) Do you have any ideas about how each particular denomination of bill should be cut? If so, explain your reasoning.

Lesson 2

1. The Swiss artist and designer Max Bill created a series of sixteen lithographs between 1934 and 1938 titled *Fifteen Variations
on a Single Theme*. Three of the lithographs, "The Theme," "Variation 2," and "Variation 5," are shown below.

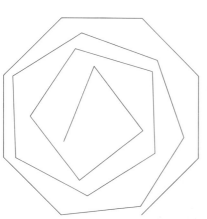

The Theme
Courtesy of Max Bill

Variation 2
Courtesy of Max Bill

Variation 5
Courtesy of Max Bill

Describe, as specifically as you can, what you see in
a) "The Theme."   b) "Variation 2."   c) "Variation 5."

2. The following puzzle is from *Amusements in Mathematics* by
Henry Ernest Dudeney, originally published in London in 1917
and still in print.*

Drawing by Henry Ernest Dudeney

The man at the table is showing his friends an equilateral
triangle that he has cut from paper. The puzzle is to figure out a
way to cut it into five pieces that can be put together to form
either two or three smaller equilateral triangles, using all five
pieces in each case. In other words, it should be possible to put
the five pieces together to form either the original triangle or to
form two triangles or to form three triangles, all equilateral.

As a hint, one of the cuts is shown in the figure below. Trace
the figure on your paper and cut it out.

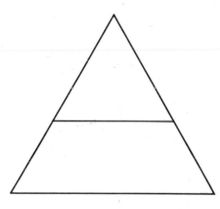

Can you solve the puzzle? If so, make drawings to show the
arrangements of the pieces in each of the three cases.

---

* The book is now available from Dover Publications, Inc., New York.

Lesson 3

1. Mosaics can be made from triangles that are not equilateral. Two examples are shown below.

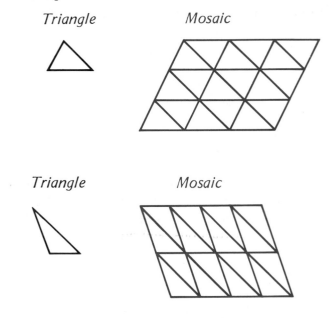

Mosaics can also be made from quadrilaterals that are not squares.

a) Make a copy on a file card of the quadrilateral shown below.

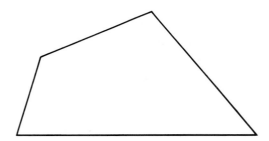

Cut your figure out, place it on a sheet of paper, and make a copy of it by tracing its border. Now see if you can figure out how to make additional copies of the figure on the paper in such a way that the space around each corner of the figure is filled with additional figures. (As in the triangle mosaics illustrated above, it will be necessary to turn the quadrilateral 180° before making some of the copies. Also, as in the triangle mosaics, pairs of equal sides should be touching.)

The examples of the triangle mosaics above suggest that a mosaic can be made from *any* triangle, regardless of its shape. It

Chapter 5: SYMMETRY AND REGULAR FIGURES

is rather surprising to realize that a mosaic can also be made from *any* quadrilateral, regardless of its shape.

b) Make a copy on a file card of the quadrilateral shown below and see if you can figure out how to use it to make a mosaic.

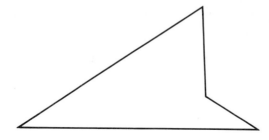

2. The pattern shown in the figure at the right can be extended to form a mathematical mosaic having the symbol 3-4-6-4.

If we take the four numbers in this symbol, make them the denominators of unit fractions, and add the fractions, we get

$$\frac{1}{3} + \frac{1}{4} + \frac{1}{6} + \frac{1}{4} =$$

$$\frac{4}{12} + \frac{3}{12} + \frac{2}{12} + \frac{3}{12} = \frac{12}{12} = 1.$$

The result turns out to be a very simple number: 1.
There are ten other mathematical mosaics:

3-3-3-3-3-3	3-12-12
3-3-3-3-6	4-4-4-4
3-3-3-4-4	4-6-12
3-3-4-3-4	4-8-8
3-6-3-6	6-6-6

a) Take the numbers in the symbol of each mosaic, make them the denominators of unit fractions as in the example above, and add the fractions.

b) What do you notice about the results?

Lesson 4

1. Jean J. Pedersen, a mathematics teacher at the University of Santa Clara, has discovered ways to weave the regular polyhedra from strips of paper. A nice model of a cube can be made from three strips of paper identical with the one shown here.

    Make the strips from heavyweight construction paper, each of a different color. Fold the strips along the brown lines.

The strips can now be woven together to make a rigid cube in which each pair of opposite faces has the same color. No glue or tape is needed. See if you can figure out how to do it.

2. Alexander Graham Bell, the inventor of the telephone, was fascinated by the regular tetrahedron. He invented and flew a number of large kites made from networks of tetrahedrons several years before the Wright brothers built their first airplane. The architecture of a museum of Bell's inventions in Nova Scotia is based on the tetrahedron shape.

   The following tetrahedron puzzle is available in some stores, but you can easily make it yourself. Make two copies of the pattern shown below on 4-inch-by-6-inch cards and cut them out. Fold each one along the brown lines and tape the edges of each together to form a pair of identical solids.

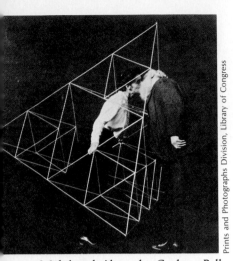

Mabel and Alexander Graham Bell sharing a kiss inside the frame of his tetrahedral kite, October 16, 1903

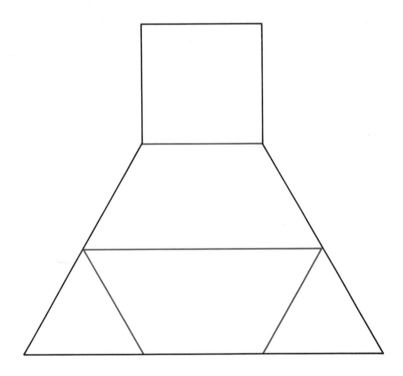

The puzzle is to put the two solids together to form a tetrahedron. Can you do it? If so, make a drawing to show the solution.

Lesson 5

1. In 1640, Descartes made an interesting discovery about the angles in geometric solids that you may be able to discover, too.

   The sum of the angles at each corner of a geometric solid is

Chapter 5: SYMMETRY AND REGULAR FIGURES

always less than 360°. Exactly how much less varies from one solid to another.

Look at the figures below and at the truncated tetrahedron pictured on page 278. An equilateral triangle and two hexagons

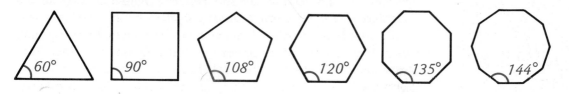

meet at each corner of the solid. The sum of the three angles surrounding each corner is

$$60° + 120° + 120° = 300°$$

which falls 60° short of 360°. We will call this difference, 60°, the *angular deficiency* of the truncated tetrahedron.

a) Use the same reasoning with each of the other Archimedean solids pictured in exercises 1 through 13 on pages 276-277 to copy and complete the following table.

Polyhedron	Number of corners	Sum of angles around each corner	Angular deficiency
1	12	300°	60°
2	24	▨	▨
3	24	▨	▨
4	60	▨	▨
5	60	▨	▨
6	12	▨	▨
7	24	▨	▨
8	48	▨	▨
9	30	▨	▨
10	60	▨	▨
11	120	▨	▨
12	24	▨	▨
13	60	▨	▨

b) How does the angular deficiency of a geometric solid vary with the number of corners that the solid has? Can you write a formula for the angular deficiency, a, in terms of the number of corners, n?

A cuboctahedron

2. Doris Schattschneider and Wallace Walker have created many interesting three-dimensional models based on the art of M. C. Escher.* One of them is a cuboctahedron whose surface is covered by one of Escher's animal mosaics.

   To make this model, you need to make eight copies of pattern A and six copies of pattern B on tracing paper. Use a black felt-point pen to make the tracings so that they are easy to see.

Pattern A

Pattern B

After making the copies, glue the tracing paper to a stiff sheet of paper and cut them out.

Tape the triangles and squares together to make a cuboctahedron, putting the tape on the inside of the model. After the model has been put together, look at one fish and the five others around the solid that are in its path. Use a felt-tip marker to color these fish the same color. Then use three other colors to color the fish in the other three paths.

Lesson 6

1. In Lesson 5, a semiregular polyhedron was defined as a solid having faces in the shape of more than one kind of regular polygon, yet every corner is the same. Among the semiregular polyhedra are the Archimedean solids, of which three have faces that are triangles and squares. They are shown at the top of the next page with their symbols.

---

*They appear in *M. C. Escher Kaleidocycles* (Ballantine Books, 1977).

Chapter 5: SYMMETRY AND REGULAR FIGURES

3-4-3-4   3-4-4-4   3-3-3-3-4

There is another semiregular polyhedron, called a *square antiprism*, whose symbol is 3-3-3-4.

a) Cut out eight triangles and two squares from stiff paper. (The patterns on page 260 are of a convenient size.) Tape them together to make a model of a square antiprism.

b) In what way is a square antiprism like a square prism?

c) How many faces, edges, and corners does a square antiprism have?

In general, antiprisms are semiregular polyhedra having the symbol 3-3-3-*n*, in which *n*, the number of sides in each of the bases, is a number larger than 3.

d) If an antiprism has a base with *n* sides, how could its numbers of faces, corners, and edges be expressed in terms of *n*?

e) Is Euler's formula true for all antiprisms? Show why or why not.

2. This picture is a drawing of the radiolarian *Aulonia hexagona*.* From its name and appearance, you might assume that all of the cells covering its surface are hexagonal in shape. If you look at it carefully, however, you will notice some cells of other shapes.

a) What shapes are they?

A biologist once claimed that he had seen a radiolarian that was covered with only hexagonal cells. It is possible to prove, using mathematics, that he must have been mistaken.

The proof is based on the fact that mathematicians have proved that the formula

$$F + C = E + 2$$

is true for every "spherical" polyhedron, even though not all of its faces are regular.

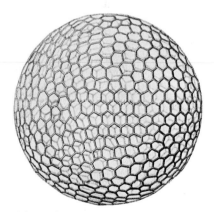

The radiolarian Aulonia hexagona

From Ernst Haeckel's *Monograph of the Challenger Radiolaria*, 1887; Science and Technology Research Center, The New York Public Library, Astor, Lenox, and Tilden Foundations

---

* Other radiolarians are pictured on page 265.

Suppose that a radiolarian has n hexagonal cells. Then, if the cells did not share their corners, there would be $6n$ separate corners. However, three corners come together at each corner of the solid, and so it must have $\frac{6n}{3} = 2n$ corners.

b) Use the same type of reasoning to show how many edges the solid must have.

c) Substitute these results into the formula

$$F + C = E + 2$$

and show that the result is false. If you can do this, you have proved that no radiolarian in the past, present, or future can ever be covered with only hexagonal cells.

Chapter **6**

MATHEMATICAL CURVES

Lowell Observatory photograph

Lesson 1

The Circle and the Ellipse

Recorded observations of Halley's comet are known to have been made as long ago as 467 B.C. Since then, it has traveled around the sun thirty-one times, taking about 76 years for each trip. When the comet is near the sun, it becomes a brilliant object in the sky, as can be seen in the photograph above, showing it with the planet Venus in 1910.

The orbit of Halley's comet is illustrated in the drawing below by the brown curve. Some of the comet's positions between 1910

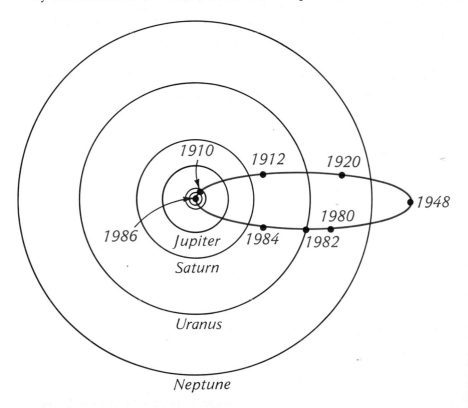

and 1986 are shown on this curve. The point at the center of the diagram represents the sun and the black curves represent the orbits of some of the planets, starting with the Earth and Mars.

The curve in which Halley's comet orbits the sun is called an *ellipse*. Although the shapes of the paths of comets were not known before the seventeenth century, the ellipse was given its name by the Greek mathematician Apollonius in the third century B.C. Apollonius made a thorough study of the curve and discovered its properties, even though there was no use for such knowledge at the time.

In 1609, Johann Kepler established the importance of the ellipse when he discovered that the orbits of the planets around the sun are not circular. Although the circle is the simplest mathematical curve and the orbits of the planets are very close to being circles, they, like the orbit of Halley's comet, are ellipses.

▶ A **circle** is the set of all points in a plane that are at the same distance from a fixed point in the plane. The fixed point is called the *center* of the circle and the distance its *radius*.

The figure below shows two points, A and B, that are 2 centimeters from point C. These two points, together with the rest of the

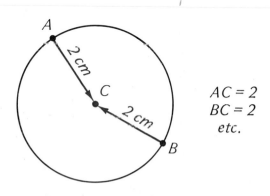

$AC = 2$
$BC = 2$
etc.

points on the page that are 2 centimeters from point C, make up the circle shown.

▶ An **ellipse** is the set of all points in a plane such that the sum of the two distances from each point to two fixed points is the same. The two fixed points are called the *foci** of the ellipse.

The figure below shows two points, A and B, the sums of whose distances from points F_1 and F_2 are 5 centimeters. These two points, together with the rest of the points on the page having the same property, make up the ellipse shown.

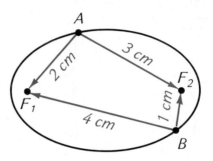

$$AF_1 + AF_2 = 2 + 3 = 5$$
$$BF_1 + BF_2 = 4 + 1 = 5$$
$$etc.$$

All circles, regardless of their size, have the same shape. Ellipses, however, have many different shapes, ranging from nearly round to long and narrow. For this reason, ellipses are measured with two numbers rather than one. The two numbers are the lengths of its *major* (longer) *axis* and *minor* (shorter) *axis*. These axes are illustrated in the figures below.

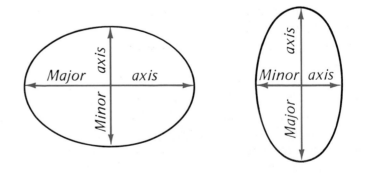

---

* *Foci* (pronounced "fō-sī") is the plural of *focus*.

Chapter 6: MATHEMATICAL CURVES

Exercises

Set I

Experiment: DRAWING ELLIPSES

Part 1. Turn a sheet of graph paper (4 units per inch or 2 units per centimeter) sideways and draw a pair of perpendicular axes with their origin at the center of the paper. Label the long axis x, the

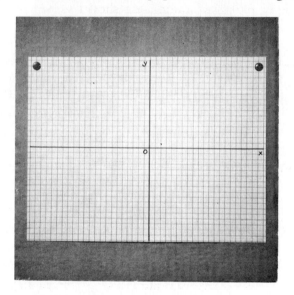

short axis y, and the origin O. Tack the graph paper on a square piece of corrugated cardboard as shown in the figure above.

Mark the two points on the x-axis that are 6 units to the left and 6 units to the right of the origin and label them A_1 and A_2. Put a tack in each point, but do not press them all the way into the cardboard.

Take a piece of string about 10 inches long (20 centimeters if you are using metric graph paper) and, as accurately as you can, tie a knot in it to make a loop exactly 32 units around. (If you are using graph paper ruled 4 units per inch, 32 units would be 8 inches. If you are using graph paper ruled 2 units per centimeter, 32 units would be 16 centimeters.) Put the loop on the board around the tacks and, using your pencil, pull it taut to form a triangle, as shown on the next page. Now move the pencil around the paper, keeping the string taut, to draw an ellipse. (While drawing the ellipse, you may have to hold the tacks in place with your

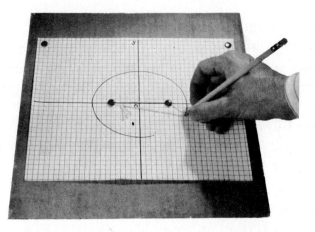

other hand to prevent them from popping out of the board.) Label the ellipse by writing the letter A on it at some point.

Mark the two points on the x-axis that are 4 units to the left and right of the origin and label them B_1 and B_2. Move the tacks to these points and use the same loop of string to draw another ellipse. Label it B.

Mark the two points on the x-axis that are 2 units to the left and right of the origin and label them C_1 and C_2. Move the tacks to these points and draw a third ellipse. Label it C.

Part 2. Remove the sheet of graph paper from the cardboard and tack a new sheet in place. Draw and label a pair of axes as before. Mark the two points on the x-axis that are 9 units to the left and right of the origin, label them D_1 and D_2, and put tacks in these points.

Take a longer piece of string and make, as accurately as you can, a loop that is 48 units (12 inches or 24 centimeters) around. Put the loop around the tacks, draw an ellipse, and label it D.

Mark the two points on the x-axis that are 11 units to the left and right of the origin and label them E_1 and E_2. Move the tacks to these points, draw an ellipse, and label it E.

Part 3. Use your drawings of the five ellipses as a guide in answering the following questions.

1. The roundest of the five ellipses is comparable in shape to the orbits of some of the planets around the sun. Which ellipse is the roundest?

2. The most elongated ellipse is comparable in shape to the orbits of some comets around the sun. Which ellipse is the most elongated?

Chapter 6: MATHEMATICAL CURVES

3. Two of the ellipses that you have drawn have the same shape. Which two are they?

4. The two points at which you put the tacks to draw each ellipse are its foci. On which axis, the major or the minor, are the foci of an ellipse always located?

5. How many units apart are the foci of ellipse A?

6. How many units around was the loop that you used to draw ellipse A?

7. What is the sum of the distances of each point on ellipse A from its foci? (See the figure at the right.)

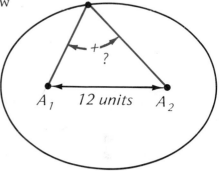

8. Copy and complete the following table.

Ellipse	Length of loop used	Distance between foci	Sum of distances of each point on ellipse from foci
A	32	12	20
B			
C			
D			
E			

9. Look at ellipses A, B, and C. What happens to the *size* of these ellipses as the sum of the distances from each point on the ellipse to the foci increases?

10. What happens to the *shape*?

11. Look at ellipses D and E. What happens to the size of these ellipses as the sum of the distances from each point on the ellipse to the foci decreases?

12. What happens to the shape?

Set II

When a circle is seen in perspective, it usually looks like an ellipse. This is illustrated by this drawing of a girl holding a circular hoop.

Every mathematical curve can be associated with an equation.*

---

* You have already studied examples of this in Chapter 3. Look at page 166.

Because circles and ellipses are closely related, their equations are similar. In fact, every circle and ellipse can be associated with an equation of the form $\frac{x^2}{a^2} + \frac{y^2}{b^2} = 1$ in which a and b represent positive numbers.

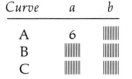

Curve	a	b
A	6	▦
B	▦	▦
C	▦	▦

1. Compare the three curves shown below with their equations. Then copy and complete the table at the left.

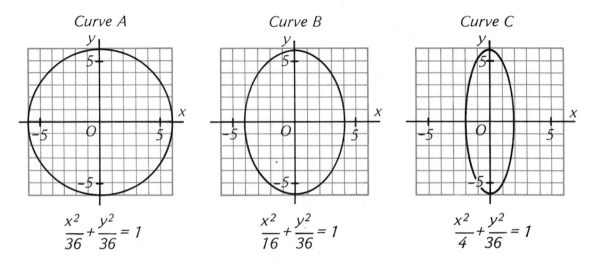

Curve A

$$\frac{x^2}{36} + \frac{y^2}{36} = 1$$

Curve B

$$\frac{x^2}{16} + \frac{y^2}{36} = 1$$

Curve C

$$\frac{x^2}{4} + \frac{y^2}{36} = 1$$

2. For what kind of curve are a and b equal?

3. For what kind of curve are a and b unequal?

4. Look at the three curves below. Which curve has the equation
   $$\frac{x^2}{25} + \frac{y^2}{4} = 1?$$

Curve D

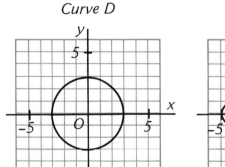

Curve E

Curve F

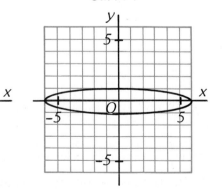

5. Write equations for the other two curves.

These photographs are of an ostrich egg as seen from the top and from the side.

6. Graph the shape of the egg as seen from the top by plotting the points in the table below and connecting them with a smoothly curved line.

x	0	1	2	3	4	3	2	1	0	−1	−2	−3	−4	−3	−2	−1
y	4	3.9	3.5	2.6	0	−2.6	−3.5	−3.9	−4	−3.9	−3.5	−2.6	0	2.6	3.5	3.9

7. Use the table below to graph the shape of the egg as seen from the side.

x	0	1	2	3	4	3	2	1	0	−1	−2	−3	−4	−3	−2	−1
y	5	4.8	4.3	3.3	0	−3.3	−4.3	−4.8	−5	−4.8	−4.3	−3.3	0	3.3	4.3	4.8

8. Write an equation for the graph of the egg as seen from the top.

9. Write an equation for the graph of the egg as seen from the side.

Elliptipool is a game played on a pool table whose shape is an ellipse. The table has only one pocket, which is located at one focus. A ball at the other focus can be hit in *any* direction and, if there are no other balls in the way, it will end up in the pocket! This drawing shows some of the paths that the ball might take.

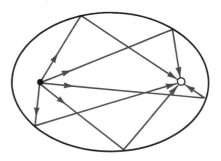

1. Suppose that you are playing Elliptipool and that you want to hit the ball at A, in the figure at the top of the next page, so that it goes into the pocket without hitting ball B. What will you do?

Chapter 6: MATHEMATICAL CURVES

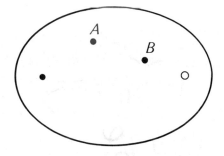

2. Trace the figure and show the path of the ball.

3. Suppose that a ball is at C on an Elliptipool table that has no pocket. You hit the ball toward the focus on the right with enough force so that it rebounds from the cushion several times before coming to a stop. Trace the figure and show the path of the ball.

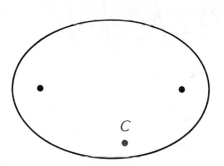

4. If the ball were to keep on going without stopping, where would its path eventually be?

Lesson **2**

The Parabola

When a ball is hit into the air, it follows a path in the shape of a curve called a *parabola*. If the ball is hit at a 45° angle with the ground, its path looks like the one in the cartoon above.

Galileo proved that the path of an object thrown through space is this curve. You probably recognize the parabola because you have drawn several graphs of functions with its shape.*

▶ A **parabola** is the set of all points in a plane such that the distance of each point from a fixed point is the same as its distance from a fixed line. The fixed point is called the *focus* of the parabola.

The figure below shows three points, A, B, and C, whose distances from point F, the focus, are equal to their distances from line ℓ. These three points, together with all other points having the same property, make up the parabola shown.

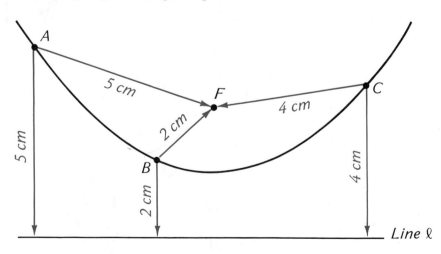

---

* See Chapter 3, Lesson 4.

322

By permission of Johnny Hart and Field Enterprises, Inc.

Because we can extend line ℓ indefinitely in both directions, the parabola can be extended indefinitely as well. This means that, unlike an ellipse, a parabola cannot be drawn in its entirety.

Cables that are hung between towers and used to suspend bridges form curves that are very close to the shape of a parabola. The Golden Gate Bridge in San Francisco is a well-known example of this curve.

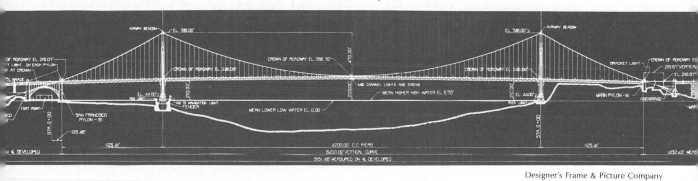

Designer's Frame & Picture Company

The parabola can be rotated about this line to form a curved surface that has a very useful property. If a mirror is made in this shape and a light placed at the focus of the mirror, the light is reflected in rays parallel to the axis. This forms a straight beam of light.

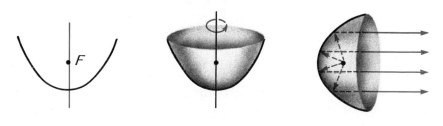

Parabolic mirrors are used in automobile headlights and, on a larger scale, in searchlights. They are also used in reflecting telescopes and in antennas to collect light and radio waves from outer space. In the latter uses, the beam comes *toward* the parabolic surface and is brought together at the focus.

Exercises

Set I

This photograph, taken with a strobe light, shows a bouncing golf ball. A formula for the path of the outlined bounce is $y = 12x - x^2$.

1. Copy and complete the following table for this formula.

x	0	1	2	3	4	5	6	7	8	9	10	11	12
y	0	11	20										

2. Draw and label a pair of axes as shown here. Then plot the

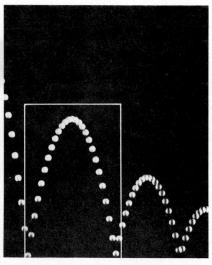

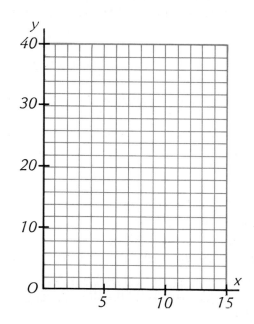

points corresponding to the numbers in the table and connect them with a smooth curve.

Chapter 6: MATHEMATICAL CURVES

3. A formula for the path of another bounce of the golf ball is $y = 10x - x^2$.

Copy and complete the following table for this formula.

x	0	1	2	3	4	5	6	7	8	9	10
y											

4. Plot the points corresponding to the numbers in the table on the axes that you used for the first bounce. Connect them with a smooth curve.

5. On which bounce does the ball reach the greater height?

A formula that approximates the curve of the cables of the Golden Gate Bridge is $y = 700 + 5(x^2) - 100x$.

6. Copy and complete the following table for this formula.

x	0	2	4	6	8	10	12	14	16	18	20
y	700	520									

7. Draw and label a pair of axes as shown here. Also draw the three heavy segments shown. The x-axis represents the water

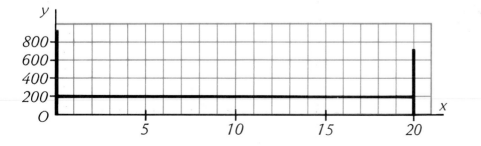

level and the three heavy segments represent the towers and roadway of the bridge.

Plot the points corresponding to the numbers in the table and connect them with a smooth curve to represent the cables.

8. Although the two axes are scaled differently, one unit on each axis represents 200 feet. How far is it between the two towers of the bridge?

9. How far above the water is the roadway?

10. How far above the roadway do the cables reach?

Set II

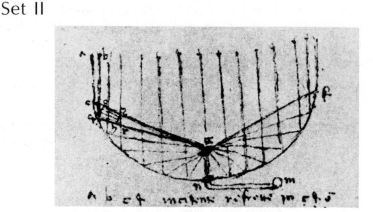

This drawing, by Leonardo da Vinci, shows how a parabolic mirror can be used to produce a beam of light. In the following exercises, we will consider the reflective property of the parabola that makes such a mirror possible.

1. Copy and complete the following table for the formula $y = x^2$.

x	−8	−7	−6	−5	−4	−3	−2	−1	0	1	2	3	4	5	6	7	8
y	64																

2. Draw and label a pair of axes as shown here. Plot the points

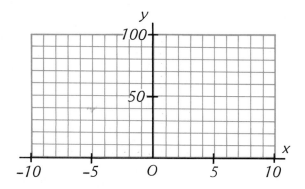

corresponding to the numbers in the table and connect them with a smooth curve. The focus of the parabola is at (0, 25). Label it F.

Draw lines from the focus to each plotted point on the parabola. Then draw lines upward from each of these points so that each line is parallel to the y-axis. When you have done this, your drawing should look similar to the drawing of da Vinci.

Chapter 6: MATHEMATICAL CURVES

3. If the parabola was a mirror and there was a light at the focus, the lines that you drew from the focus to the parabola and from the parabola upward would represent rays of light. How does the angle at which each light ray hits the parabola seem to compare in size with the angle at which it is reflected?

4. How does the angle at which each light ray hits the parabola change in size the farther the ray is from the y-axis?

5. The light rays leave the focus in many different directions. What happens to them when they are reflected from the parabola?

High beam

The lightbulb in an automobile headlight has two filaments: one for high beam and one for low beam. The filaments are located in different positions with respect to the mirror.

6. Where is the filament for the high beam located?

7. Where is the filament for the low beam located?

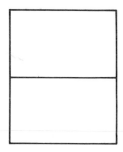

Low beam

Set III

Experiment: PARABOLAS BY FOLDING PAPER

Part 1. Fold a sheet of unlined paper in half as illustrated in the adjoining figure, and tear it along the fold into two equal pieces. Mark a point about one inch above the center of the lower edge of one of the pieces. Fold the paper, as shown here, so that the lower edge touches the point. Make a sharp crease. Open the paper flat and fold again in a different direction, again making sure that the lower edge touches the point. Repeat this about twenty times, folding the paper in a different direction each time, and a parabola should appear. Trace the curve with your pencil and label it A.

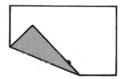

Part 2. Mark a point in the center of the other piece of paper and do the same as before. Trace the curve that results and label it B.

Part 3.

1. What are the two points called with respect to the parabolas?

2. Although the two curves seem to have different shapes, they are actually the same. Which parabola looks like an enlargement of part of the other?

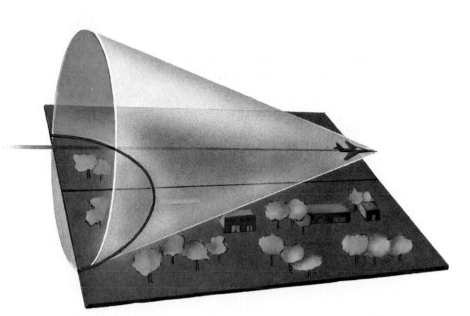

The Hyperbola

In 1953, a pilot flying faster than the speed of sound over a western air base created a shock wave that damaged almost every building on the base. This was the first evidence of the destructive power of the shock wave that is heard as a "sonic boom."

The sonic boom shock wave has the shape of a cone, and it intersects the ground in part of a curve called a *hyperbola*. It hits every point on this curve at the same time, so that people in different places along the curve on the ground hear it at the same time. Because the airplane is moving forward, the hyperbolic curve moves forward and eventually the boom can be heard by everyone in its path.

The definition of a hyperbola is very much like the definition of an ellipse.

▶ A **hyperbola** is the set of all points in a plane such that the difference between the two distances from each point to two fixed points is the same. The two fixed points are called the *foci* of the hyperbola.

329

The figure below shows three points: A, B, and C. The differ-

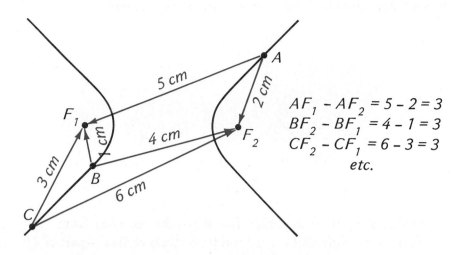

$$AF_1 - AF_2 = 5 - 2 = 3$$
$$BF_2 - BF_1 = 4 - 1 = 3$$
$$CF_2 - CF_1 = 6 - 3 = 3$$
$$etc.$$

ence between the two distances of each of these points from points F_1 and F_2 is 3 centimeters. Points A, B, and C, together with the rest of the points on the page having the same property, make up the hyperbola shown. Notice that the hyperbola consists of two separate parts, called its branches. The sonic-boom curve described earlier is only one branch.

The early Greek mathematicians knew all about the hyperbola. In fact, it was Apollonius* who gave the curve its name. The Greeks were equally familiar with the parabola, as well as the ellipse. How did they come to be acquainted with these curves so long ago? Their knowledge was recorded in eight books written by Apollonius on the curves that could be produced by slicing a cone in different directions. If the slice is in the same direction as the side of the cone, the curve that results is a parabola.

Side view

Perspective view

Parabola

---

* See page 313.

Chapter 6: MATHEMATICAL CURVES

If the slice is tilted from the direction of the side of the cone toward the horizontal, the curve is an ellipse instead.

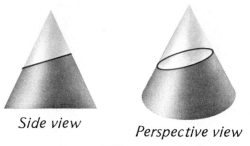

Side view   *Perspective view*

Ellipse

If the slice is tilted in the other direction—that is, away from the direction of the side of the cone and toward the vertical—part of a hyperbola is formed.

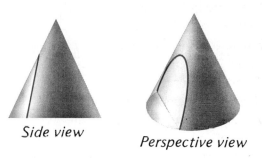

Side view

Perspective view

Hyperbola

To see the rest of the hyperbola, imagine a second cone balanced upside down on the point of the first one. If the slice is continued through the second cone, the other branch of the hyperbola is formed.

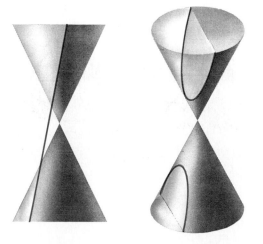

Because the parabola, ellipse, and hyperbola are formed when a cone is sliced into sections, they are called *conic sections*. Apollonius titled his work *The Conics*. Although it seemed to have no practical use at the time, many important applications of the conic sections have been discovered since.

Exercises

Set I

Experiment: A HYPERBOLA BY FOLDING PAPER

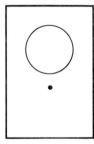

Draw a circle having a radius of about 4 centimeters on a sheet of tracing paper. Mark a point about 3 centimeters below the circle.

Now fold the paper so that the point falls on the circle and make a sharp crease. Open the paper flat and fold in a different direction so that the point again falls on the circle. Continue in this way until the point has been folded into different positions *all the way around* the circle.

You should be able to find both branches of a hyperbola when you are finished. Trace them in on the paper.

1. Where are the two foci of the hyperbola located?

2. How many lines of symmetry does the hyperbola have?

Set II

Part of the two branches of a hyperbola can be seen in this photograph of the McDonnell Planetarium in St. Louis. The designer, Gyo Obata, was inspired to use the hyperbola by the fact that certain comets travel in hyperbolic orbits.

Photograph by Kiku Obata

Chapter 6: MATHEMATICAL CURVES

Just as ellipses are associated with equations of the form

$$\frac{x^2}{a^2} + \frac{y^2}{b^2} = 1,$$

hyperbolas are associated with equations of the form

$$\frac{x^2}{a^2} - \frac{y^2}{b^2} = 1.$$

1. Compare the ellipse and hyperbola shown below with their equations.

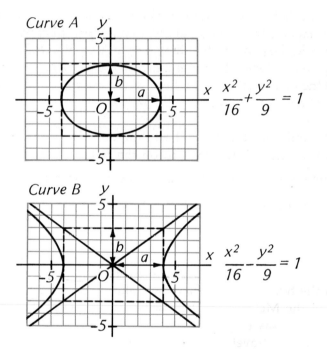

Curve A

$$\frac{x^2}{16} + \frac{y^2}{9} = 1$$

Curve B

$$\frac{x^2}{16} - \frac{y^2}{9} = 1$$

Then copy and complete the following table.

Curve	a	b
A	4	▓
B	▓	▓

The lines through the corners of the dashed rectangle in the drawing of the hyperbola are called its *asymptotes*. The branches of the hyperbola get closer and closer to these lines as the curve is continued to the left and right.

2. Compare the three hyperbolas shown below with their equations.

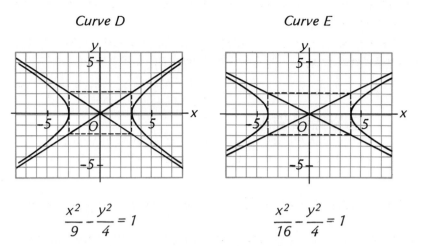

Curve C

$$\frac{x^2}{4} - \frac{y^2}{4} = 1$$

Curve D

$$\frac{x^2}{9} - \frac{y^2}{4} = 1$$

Curve E

$$\frac{x^2}{16} - \frac{y^2}{4} = 1$$

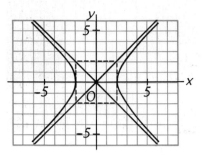

Curve	a	b
C	2	▓▓▓
D	▓▓▓	▓▓▓
E	▓▓▓	▓▓▓

Then copy and complete the table at the left.

3. What happens to the distance between the branches of these hyperbolas as a increases?

4. What happens to the shape of the branches as a increases?

5. What are the lines through the corners of the dashed rectangles in the figures called?

6. What happens to the distance between the branches of the hyperbolas and these lines as the curves are continued to the left and right?

7. Look at the two hyperbolas below. Which one has the equation $\dfrac{x^2}{25} - \dfrac{y^2}{9} = 1$?

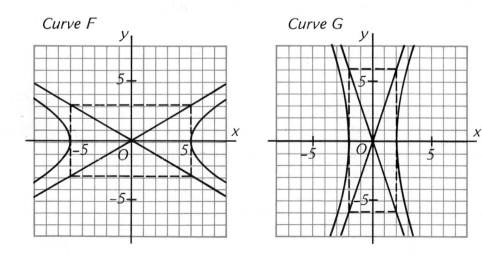

Curve F

Curve G

334

8. Write an equation for the other hyperbola.

The shadows cast on a wall by a lamp with a circular shade form the branches of a hyperbola.

9. Graph the hyperbola by plotting the points in the tables below and connecting each set with a smooth curve.

x	−4	−3	−2	−1	0	1	2	3	4
y	4.2	3.8	3.4	3.1	3	3.1	3.4	3.8	4.2

x	−4	−3	−2	−1	0	1	2	3	4
y	−4.2	−3.8	−3.4	−3.1	−3	−3.1	−3.4	−3.8	−4.2

The equation of the hyperbola that you graphed is $\dfrac{y^2}{9} - \dfrac{x^2}{16} = 1$.

10. How does the graph differ from those in the preceding exercises?

11. How does its equation differ from those of the hyperbolas in the preceding exercises?

Set III

If two sheets of glass are taped together along one pair of edges, separated slightly along the opposite pair, and placed in a pan of water to which some dye has been added, one branch of a hyperbola will be formed by capillary action.

A formula for a hyperbola produced in this way is $y = \dfrac{4}{x} + 1$.

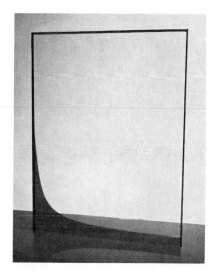

1. Copy and complete the following table for this formula.

x	$\dfrac{1}{3}$	$\dfrac{1}{2}$	1	2	3	4	6	8	10	12
y	13	9								

2. Draw an appropriate pair of axes, plot the ten points in the table, and connect them with a smooth curve.

3. You have drawn the branch of the hyperbola formed by the water between the two sheets of glass. If the other branch of the hyperbola were added to the graph, where would it be?

4. Where are the asymptotes of the hyperbola?

Lesson 4

The Sine Curve

When a musical sound wave is changed into a visual image by an oscilloscope, it has a regular pattern that repeats itself many times each second. Such a repeating pattern, produced by a note blown by Louis Armstrong on his trumpet, is shown in the diagram below.

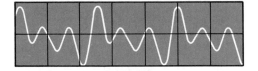

The French mathematician Joseph Fourier, who lived during the time of Napoleon, showed that sound waves that repeat themselves like this one are related to a curve called the *sine curve*.

The sine curve can be best understood by considering a point moving around a circle. To locate the various positions of the point, the circle is numbered in degrees from 0 to 360. Some of the positions of the point, together with the corresponding numbers, are shown in the left-hand figure below. Next a pair of coordinate axes is centered on the circle as shown in the right-hand figure.

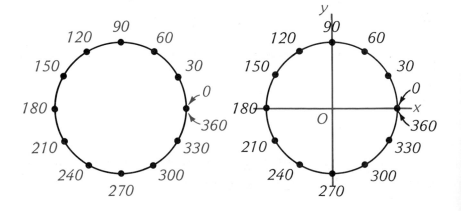

Suppose that a point starts at 0° on the circle and moves counterclockwise around it at a steady rate. As the point moves around the upper half of the circle, its distance from the x-axis first increases until the point gets to 90° and then decreases as the point continues to 180°. As the point moves around the lower half of the circle, its distance from the x-axis increases until the point reaches 270° and then decreases as the point continues to 360°. (This is illustrated in the figure below.)

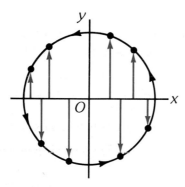

The distances can be measured in terms of the radius of the circle, which is one unit long. Those measured upward from the x-axis give positive y values, and those measured downward from the x-axis give negative y values.

Measure y values   *Plot y values*

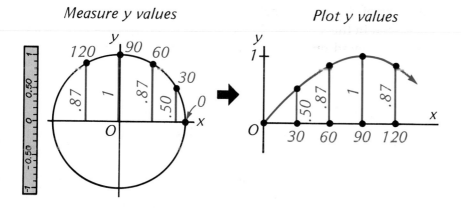

The distances are called the *sines* of the positions of the point on the circle. A graph of these sines on another pair of axes produces the *sine curve*.

One trip of the point around the circle results in one "wave" of the sine curve, shown below.

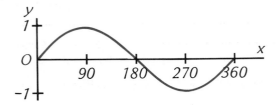

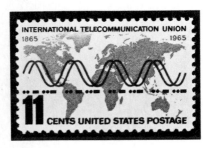

If the graph is continued, the curve is repeated over and over. Four waves appear in the pictures of the sine curves on the postage stamp shown here.

Although the sine curve is especially important in the study of waves, it has many other applications as well. We shall consider some of them in the exercises of this lesson.

Exercises

Set I

This diagram can be used to estimate some sines to the nearest hundredth.

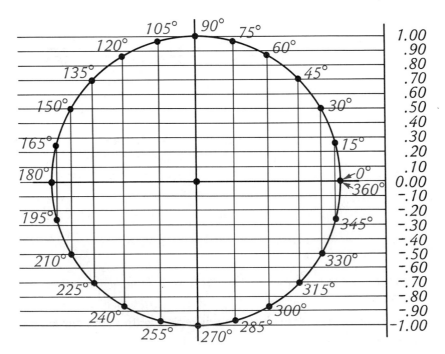

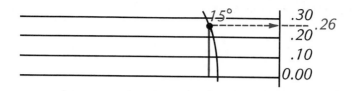

For example, the sine of 15° is approximately 0.26.

1. Copy the table of sines shown below.

x	sine of x
0°	0
15°	0.26
30°	0.50

Continue the table to 360°, using the diagram to estimate the sines.

2. What are the largest and smallest sines in your table?

3. Draw and label a pair of axes as shown in the figure below.

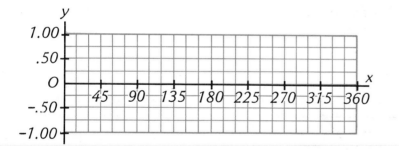

Note that 1 unit on the x-axis represents 15° and 1 unit on the y-axis represents 0.25. Use your table to plot points corresponding to each sine from 0° through 360° and connect the points to form a smooth sine curve.

4. Near what values of x do the sines undergo the least change?

5. What do you notice about the direction of the sine curve near these values?

6. Near what values of x do the sines undergo the greatest change?

7. What do you notice about the direction of the sine curve near these values?

The part of the sine curve that you have graphed starts at 0° and stops at 360°. The curve can be continued endlessly in both directions.

8. If it were continued to the right, what would the sine of 450° be?

9. What would the sine of 540° be?

10. What would the sine of –90° be?

Set II

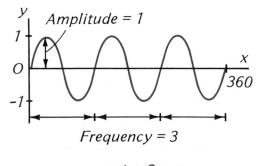

The drawing above shows the sound waves produced by a tuning fork being displayed as a sine curve on an oscilloscope. A typical equation for such a curve is $y = $ sine $3x$,* the graph of which is shown below.

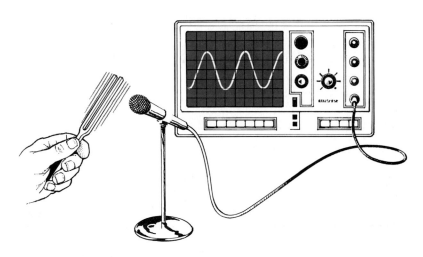

$y = $ sine $3x$

---

* This equation says that y is the sine of the number that is 3 times x.

Chapter 6: MATHEMATICAL CURVES

The *amplitude* of a sine curve is its maximum distance from the x-axis. The *frequency* of a sine curve is the number of complete waves that occur when x increases by 360°. The distance along the x-axis required for one complete wave is its *wavelength*.

1. Copy and complete the following table by obtaining the information from the graphs below.

Equation	Amplitude	Frequency
$y = \text{sine } 3x$	1	3
$y = 4 \text{ sine } x$	▓▓▓	▓▓▓
$y = 2 \text{ sine } 5x$	▓▓▓	▓▓▓
$y = 3 \text{ sine } \frac{1}{2}x$	▓▓▓	▓▓▓

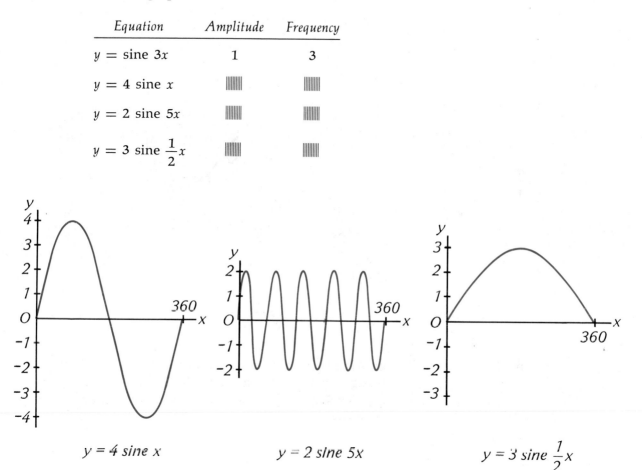

$y = 4\ sine\ x$ $y = 2\ sine\ 5x$ $y = 3\ sine\ \dfrac{1}{2}x$

2. The loudness of a sound depends on the amplitude of its wave. Which equation in the table in exercise 1 corresponds to the loudest sound?

3. Which one corresponds to the softest sound?

4. The pitch of a sound depends on its frequency. Which equation in the table corresponds to the sound having the highest pitch?

5. Which one corresponds to the sound having the lowest pitch?

6. What happens to the wavelength of a sound as its pitch increases?

7. Compare the equations of the sound waves in exercise 1 with their amplitudes and frequencies. Then write equations for the following three curves.

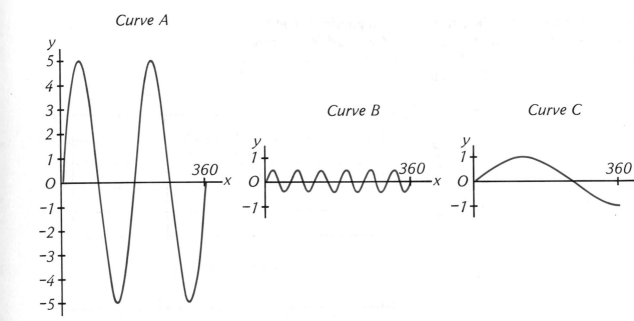

Curve A

Curve B

Curve C

Radio stations broadcast in either AM or FM. Examples of waves sent in each form are shown below.

AM radio wave

FM radio wave

8. What is continually changing in the sine curve associated with the AM wave?

9. What stays constant?

Chapter 6: MATHEMATICAL CURVES

10. What is continually changing in the sine curve associated with the FM wave?

11. What stays constant?

Set III

The giraffe is a remarkable animal. When a giraffe 18 feet tall is standing up, its heart has to pump blood upward for 10 feet to reach its brain. As a result, the blood pressure of the giraffe is higher than that of any other animal.

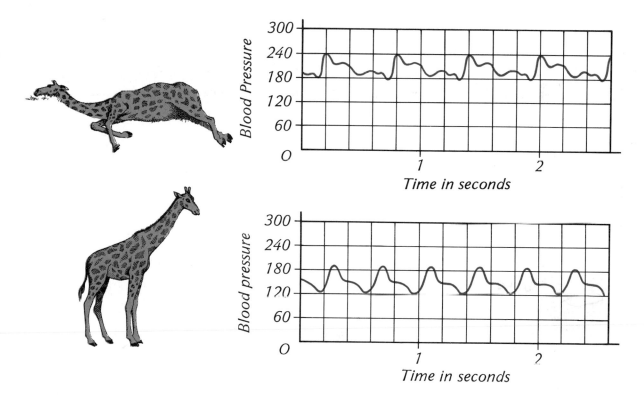

The graphs above are of the blood pressure near a giraffe's head when the animal is lying down and when it is standing up.* The changes in pressure with time can be used to measure the animal's heart rate.

1. What happens to the blood pressure near the giraffe's head when the animal stands up?

---

* After "The Physiology of the Giraffe" by James V. Warren. Copyright © 1974 by Scientific American, Inc. All rights reserved.

2. What is the wavelength of the graph for the giraffe when it is lying down? (Estimate it as a fraction of a second.)

3. What is the wavelength of the graph for the giraffe when it is standing up?

4. Estimate the number of times that the giraffe's heart beats each minute when it is lying down.

5. Make a comparable estimate for the giraffe when it is standing up.

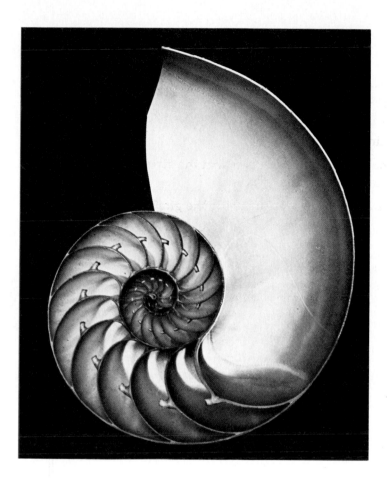

Lesson 5

Spirals

One of the most beautiful designs in nature is the shell of the chambered nautilus, a sea creature that lives in the South Pacific. As the nautilus grows, it builds and moves through a series of ever-larger chambers. Each chamber has the same shape as the one before it. The photograph above shows the shell of a chambered nautilus that has been cut in half to reveal these chambers. It has the shape of a curve called a *spiral*.

▶ A **spiral** is a curve traced by a point that moves around a fixed point from which it moves farther and farther away.

There are several kinds of spirals. The groove of a phonograph record is an *Archimedean spiral*. Named after Archimedes, who wrote a book on spirals, the Archimedean spiral is one whose loops are spaced at equal intervals. Because of this equal spacing, the successive distances of the loops from the center of the spiral

345

form an *arithmetic sequence*. In the figure below, these distances are measured in millimeters.

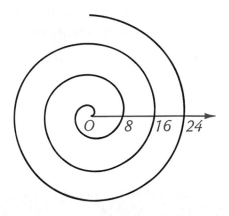

An Archimedean spiral

The curve of the shell of the chambered nautilus is a spiral of a different kind. Called a *logarithmic spiral*, it was discovered by Descartes, the inventor of coordinate geometry. The successive

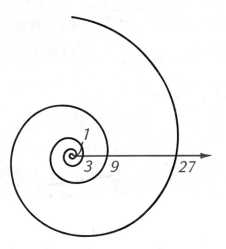

A logarithmic spiral

distances of the loops from the center of a logarithmic spiral form a *geometric sequence*. As a result, the loops of a logarithmic spiral are spaced farther and farther apart as it winds outward.

Spirals exist in nature in many different ways. They appear in the heads of daisies, in elephant tusks, and in the webs of certain spiders. The internal parts of the ear that sense sound are in a

Chapter 6: MATHEMATICAL CURVES

spiral arrangement. The most spectacular spirals in nature can be seen only through a telescope—they are the galaxies and nebulas of the universe. Most galaxies have spiral shapes, including the Milky Way, the one in which our sun is a star. Our solar system is about three-fourths of the way from the center of the spiral to the edge.

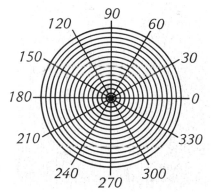

Palomar Observatory, California Institute of Technology

Exercises

Set I

Courtesy of Santa Cruz Seaside Company

A person walking from the center of a merry-go-round at a steady speed along a radius of the floor travels, with respect to the ground, along an Archimedean spiral. To draw a picture of such a path, it is convenient to use polar graph paper.*

---

* Polar graph paper can be made by using a compass to draw a set of 15 circles all having the same center and with radii of 7.5, 7, 6.5, 6, 5.5, 5, 4.5, 4, 3.5, 3, 2.5, 2, 1.5, 1, and 0.5 cm. Use a protractor to draw a series of lines through this center, called the *pole* of the graph, at 30° angles to one another. Number the lines 0°, 30°, 60°, 90°, and so forth, to 330°.

1. Plot a point on each of the angle lines at a distance from the pole as given in the table below. (We will consider 0.5 centimeter to be 1 unit, and so the radius of the largest circle is 15 units.) The points corresponding to the angles included in the table have been plotted on the diagram below as an example.

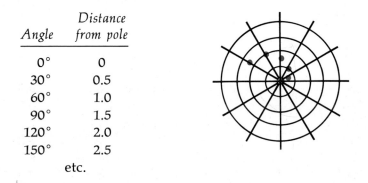

Angle	Distance from pole
0°	0
30°	0.5
60°	1.0
90°	1.5
120°	2.0
150°	2.5
etc.	

Plot points following this pattern to the outermost circle. Connect the points in order with a smooth curve. The result is an Archimedean spiral.

2. The spiral that you have drawn represents the person's path on the merry-go-round with respect to the ground. Through how many degrees will the merry-go-round have turned by the time the person is halfway from the center to the edge?

3. Through how many degrees in all will the merry-go-round have turned by the time the person reaches its edge?

4. Copy and complete the following table for the curve that you have drawn.

Number of revolutions, x	0	1	2
Distance from pole, y	0	6	▓▓▓

5. What kind of number sequence do the distances form?

6. Write a formula for y, the distance from the pole, in terms of x, the number of revolutions.

7. What would the distance from the pole be after three revolutions if the curve were continued?

The spiral of the chambered nautilus is a logarithmic spiral.

8. To draw a logarithmic spiral, plot the points represented in the table below on another sheet of polar graph paper. Connect them with a smooth curve.

Angle	Distance from pole	Angle	Distance from pole
0°	1	1 rev., 90°	3.9
30°	1.1	1 rev., 120°	4.3
60°	1.2	1 rev., 150°	4.7
90°	1.3	1 rev., 180°	5.2
120°	1.4	1 rev., 210°	5.7
150°	1.6	1 rev., 240°	6.2
180°	1.7	1 rev., 270°	6.8
210°	1.9	1 rev., 300°	7.5
240°	2.1	1 rev., 330°	8.2
270°	2.3	2 revolutions	9
300°	2.5	2 rev., 30°	9.9
330°	2.7	2 rev., 60°	10.8
1 revolution	3	2 rev., 90°	11.8
1 rev., 30°	3.3	2 rev., 120°	13.0
1 rev., 60°	3.6	2 rev., 150°	14.2

From *Medical Radiography and Photography*, published by Radiography Markets Division, Eastman Kodak Company

9. Copy and complete the following table for this curve.

Number of revolutions, x	0	1	2
Distance from pole, y	1	▨	▨

10. What kind of number sequence do the distances form?

11. Write a formula for y, the distance from the pole, in terms of x, the number of revolutions.

12. What would the distance from the pole be after three revolutions if the curve were continued?

Set II

The leaves of the sago palm curl in spirals, as this photograph illustrates. If the successive distances of the loops of a spiral from its center form an *arithmetic* sequence, the spiral is *Archimedean*. If they form a *geometric* sequence, the spiral is *logarithmic*.

Write a sequence for the distances of the loops from the center of each of the following spirals. Measure the distances in millimeters along the lines indicated.

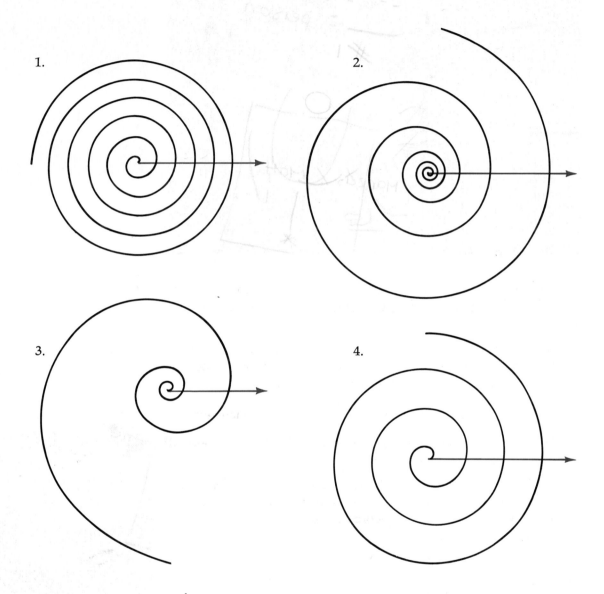

1.

2.

3.

4.

5. Which of these spirals are Archimedean?

6. Which are logarithmic?

7. Look again at the photograph of the sago palm. What type of spirals does it seem to have?

8. What type of spirals do the arms of the galaxy shown in this photograph seem to be?

Lick Observatory photograph

The groove in a phonograph record is an Archimedean spiral. In each revolution of the record, the needle travels one loop of the spiral.

9. Before 1948, the standard phonograph record made 78 revolutions each minute. How many loops did a groove of one of these records have if it played for 3 minutes?

10. A long-playing record makes $33\frac{1}{3}$ revolutions each minute. How many loops does a groove of one of these records have if it plays for 21 minutes?

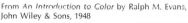
From *An Introduction to Color* by Ralph M. Evans, John Wiley & Sons, 1948

11. An MCA videodisc makes 1,800 revolutions each minute and one of its grooves plays for 60 minutes. How many loops does a groove of one of these records have?

Set III

The figure at the right, invented by the British psychologist J. Frazer, seems to be something other than what it really is.

1. What does the figure seem to be?

2. What does it actually consist of?

Photograph by J. H. Lartigue; courtesy of the Association des Amis de Jacques-Henri Lartigue

Lesson **6**

The Cycloid

As a car moves forward, the tops of its wheels move faster with respect to the road than do the bottoms. This is illustrated in the picture above, taken many years ago by the French photographer Jacques-Henri Lartigue. The spokes closest to the ground can be seen distinctly, whereas those farthest from the ground are blurred.

As a wheel rolls along a straight line, its center travels at the same rate along another straight line. The first diagram on the next page shows five positions of such a wheel as it makes one revolution. The distances from A to B, from B to C, from C to D, and from D to E are equal because, if the wheel rolls forward at a steady speed, its center does also.

Other points on the wheel travel along more complicated paths. A point on the rim—a pebble caught in the tread, for example—follows a path called a *cycloid*.

▶ A **cycloid** is the curve traced by a point on the rim of a wheel as the wheel rolls along a straight line.

The second diagram on the next page shows part of this curve. Again, five positions of a wheel as it makes one revolution are

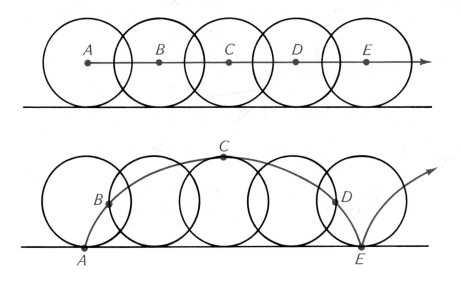

illustrated. In each successive position, the wheel has made a quarter turn. The first and last show the wheel with the point on the rim touching the ground; the other three show the wheel after turning 90°, 180°, and 270°.

Notice that, as the wheel rolls equal distances along the ground, the point on the rim does not travel equal distances along its cycloid path. The distance from B to C is longer than the distance from A to B, yet each distance represents a quarter turn of the wheel. This shows that the point on the rim must move faster as it approaches C. Its greatest speed is at the moment that it has reached the top. As the point moves downward to E, it slows down until it stops for an instant at the bottom of the curve before beginning another arc. Points A and E of the cylcoid, where the curve suddenly changes directions, are called *cusps*.

Although the early Greek mathematicians knew a lot about the conic sections (the circle, ellipse, parabola, and hyperbola) and about the spiral of Archimedes, they apparently never thought of the cycloid. In fact, no one knows who discovered this remarkable curve. Galileo, in the seventeenth century, suggested that it would be a good shape for an arch of a bridge. The great French mathematician Pascal later studied the cycloid as a result of suffering from a toothache! He decided that he needed something interesting to think about to take his mind off the pain and, as a result, made many discoveries about this "curve of a rolling wheel."

Exercises

Set I

Experiment: DRAWING CYCLOIDS

Part 1. The Cycloid The easiest way to draw a cycloid is to roll a wheel along a straight line and mark some of the positions taken by a point on its rim. Our procedure is based on this method.

Using a compass, draw a circle with a radius of 1 inch (2 centimeters) on a 3-inch-by-5-inch card.* Cut the circle out, center it on the first figure below and mark 12 points on the rim as indicated by the lines. Take your circle away from the page and, using a

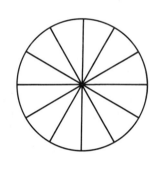

ruler, draw diameters to connect the points that you have marked as shown in the second figure. You now have a wheel with "spokes." Number the spokes 0 through 11 as shown in the third figure, and cut a small notch in the rim of the wheel at the spoke numbered 0.

Turn a sheet of graph paper ruled either 4 units per inch or 2 units per centimeter sideways. Draw a line across it about 3 inches (7 centimeters) from the top. Mark the line with a series of 20 points that are 2 units ($\frac{1}{2}$ inch or 1 centimeter) apart. Number the points from the left 0 through 11 and 0 through 7 as shown in the diagram at the top of the next page.

---

*The radius should be 1 inch if you are using graph paper ruled 4 units per inch. It should be 2 centimeters if you are using graph paper ruled 2 units per centimeter.

Chapter 6: MATHEMATICAL CURVES

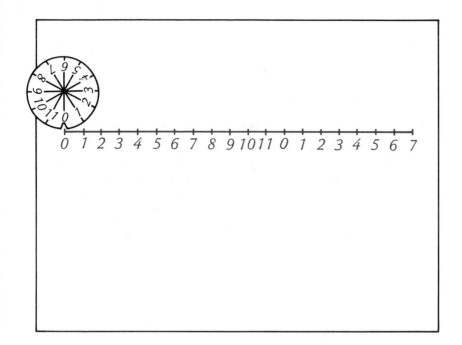

Place your wheel on the line so that the spoke numbered 0 points downward and the notch is at the point on the line numbered 0. Mark a heavy dot on the graph paper at the notch. Then roll the wheel to the right so that the spoke numbered 1 points downward and just touches the point on the line numbered 1. (You will have to adjust the position of the wheel slightly to make these numbers line up.) Mark another heavy dot on the graph paper through the notch in the wheel.

Repeat this procedure all the way across the paper so that each spoke points downward to the point on the line with its corresponding number; mark a dot in the notch each time.

Connect the dots with a smooth curve to form one full arc of a cycloid and part of a second one.

Part 2. A Prolate Cycloid The path of a point on the *inside* of a wheel that rolls along a straight line (other than the center of the wheel) is a *prolate cycloid*. To draw this curve, we will use the method used in drawing the cycloid.

Use a paper punch to punch a hole in the center of the spoke numbered 0 on your wheel. Place the wheel back on the line that you drew for Part 1 and roll it along the line, marking a dot on the paper through the center of the punched hole each time. When the dots are connected with a smooth curve, a prolate cycloid is the result.

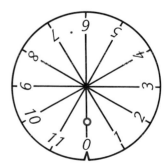

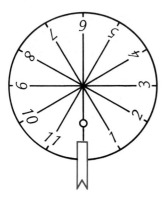

Part 3. A Curtate Cycloid The wheels of a train have flanges that extend beyond the part of the wheel that rides on the track. Points on the flange are *outside* of the wheel and follow a *curtate cycloid*.

First, cut a small strip from the card out of which you cut the circle, cut a notch in the end of this strip, and tape it along the spoke numbered 0, as shown in the diagram at the left.

Roll the wheel along the line again, marking dots in the notch of the strip. A curtate cycloid results when the dots are connected with a smooth curve.

Part 4. Use your drawings as a guide in answering the following questions.

1. Where is each point on the rim of a rolling wheel when it is moving the fastest?

2. Where is each point on the rim of a rolling wheel when it comes to a stop for an instant?

3. Does each point inside a rolling wheel also come to a stop at certain instants?

4. Where is each point inside a rolling wheel when it is moving the slowest?

5. Is there any point inside a rolling wheel that moves forward at a steady rate?

6. What is strange about the motion of a point outside a rolling wheel?

Set II

The gears in an automatic transmission include wheels that roll inside and outside of other wheels. The paths of the points on the rims of such wheels are epicycloids and hypocycloids.

Experiment: DRAWING EPICYCLOIDS AND HYPOCYCLOIDS

Part 1. An Epicycloid The path of a point on the rim of a wheel that rolls around the outside of a circle is an *epicycloid*.

Remove the small strip that was taped to the wheel made for the Set I experiment. Place the wheel on the center of a sheet of paper and trace its circumference. Mark points on the circle corresponding to the 12 points on the wheel, remove the wheel from the paper, and draw diameters connecting the points. Number the "spokes" that you have just drawn clockwise, starting with 0 at the top. The result should look like the bottom wheel in the adjoining figure.

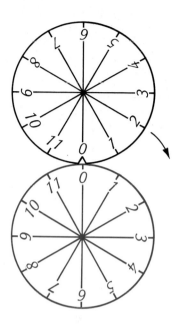

Put the wheel at the top of the circle on your paper so that the spoke marked 0 on the wheel touches the spoke marked 0 on the circle and is in line with it. This is illustrated in the figure. Mark a heavy dot at the notch of the wheel.

Now roll the wheel to the right so that the spokes numbered 1 are in line with each other and mark another heavy dot through the notch of the wheel. Continue around the circle in the same way, being careful that the corresponding spokes are lined up each time. Finally, connect the dots with a smooth curve.

Part 2. A Hypocycloid The path of a point on the rim of a wheel that rolls around the inside of a circle is a *hypocycloid.*

Use a compass to draw a circle with a radius of 3 inches (6 centimeters) on a sheet of paper and a protractor to mark points around the circle at 10° intervals. Number the points starting from the bottom as shown in the figure below.

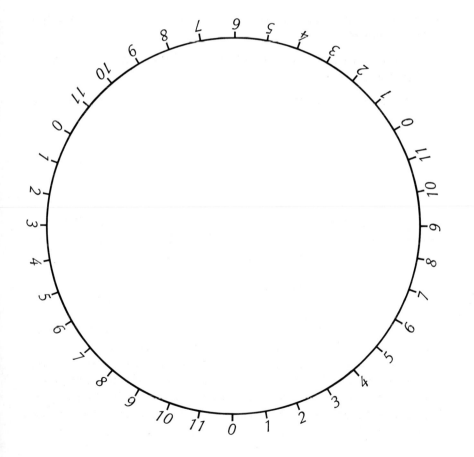

Place your wheel inside the circle so that the spoke marked 0 on the wheel touches the point marked 0 at the bottom of the circle. This is illustrated in the figure shown here. Mark a heavy dot at the notch of the wheel.

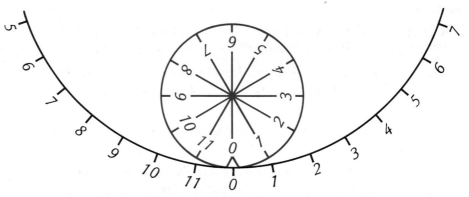

Roll the wheel counterclockwise around the circle, marking dots in the notch of the wheel as you do so. Connect the dots with a smooth curve.

Part 3. Another Hypocycloid Use a compass to draw a circle with a radius of 2 inches (4 centimeters) on a sheet of paper and a protractor to mark points around the circle at 15° intervals. Number the points starting from the bottom as shown in this figure.

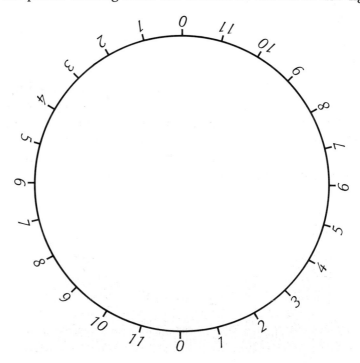

Chapter 6: MATHEMATICAL CURVES

Follow the rest of the procedure described in Part 2 to finish the drawing.

Part 4. Use your drawings as a guide in answering the following questions.

1. The epicycloid that you drew in Part 1 is also called a *cardioid*. Why? (*Hint:* The term "cardiac" has a related meaning.)

2. The radius of the circle that you drew in Part 2 is three times the radius of the wheel. How many cusps does the hypocycloid that you drew have?

3. If you drew the figure for Part 3 accurately, you discovered that the resulting hypocycloid is not a curve. What is it?

4. The radius of the circle that you drew in Part 3 is two times the radius of the wheel. How many cusps does the hypocycloid that you drew have?

5. How does the radius of the circle in the upper adjoining figure compare with the radius of the wheel that produced the hypocycloid?

6. How does the radius of the circle in the lower adjoining figure compare with the radius of the wheel that produced the hypocycloid? (*Hint:* How many arcs of the curve were produced by one trip around the circle?)

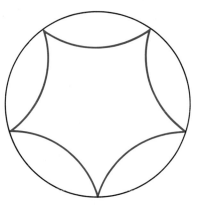

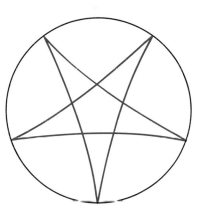

Set III

The cycloid has an interesting property when it is turned upside down. If a curved ramp is built in the shape of a cycloid and a couple of marbles are released from any two points on the ramp, they will always reach the bottom at the same time!

One marble will usually have farther to roll than the other, so how can this be possible?

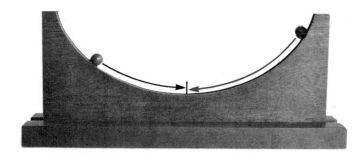

Circle
(a is constant)

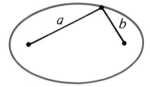

Ellipse
(a + b is constant)

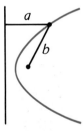

Parabola
(a = b)

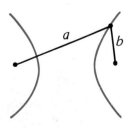

Hyperbola
(a − b is constant)

Chapter 6 / Summary and Review

In this chapter we have become acquainted with:

The conic sections (*Lessons 1, 2, and 3*) The conic sections are curves that are formed when a cone is sliced into sections. They include:

The *circle:* the set of all points in a plane that are at the same distance from a fixed point in the plane called the center.

The *ellipse:* the set of all points in a plane such that the sum of the two distances from each point to two fixed points, called the foci, is the same.

The *parabola:* the set of all points in a plane such that the distance of each point from a fixed point, called the focus, is the same as its distance from a fixed line.

The *hyperbola:* the set of all points in a plane such that the difference between the two distances from each point to two fixed points, called the foci, is the same.

The sine curve (*Lesson 4*) The *amplitude* of a sine curve is its maximum distance from the x-axis. The *frequency* of a sine curve is the number of complete waves when x increases by 360°.

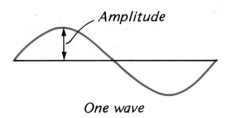

One wave

Spirals (*Lesson 5*)   A spiral is a curve traced by a point that moves around a fixed point from which the point moves farther and farther away.

The successive distances of the loops of an *Archimedean spiral* from its center form an arithmetic sequence.

The successive distances of the loops of a *logarithmic spiral* from its center form a geometric sequence.

The Cycloid (*Lesson 6*)   A cycloid is the curve traced by a point on the rim of a wheel as the wheel rolls along a straight line.

Exercises

Set I

These photographs show ice cream cones that have been cut in different directions.

1. What is the set of curves formed by the cuts called?

2. Name each curve.

1

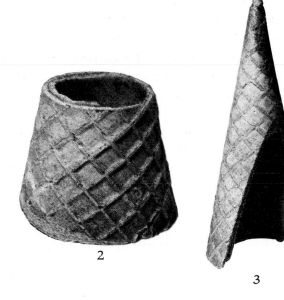

2

3

4

There is a whispering gallery in the Capitol building in Washington, D.C. The ceiling of this room is a curved surface, formed by rotating an ellipse about one of its axes. If you stand at one spot in this large room and whisper very softly, someone standing at a certain spot many meters away can hear you clearly.

3. What are these two spots called with respect to the ellipse?

Here is a photograph of a fountain in which the jets of water are pointed in many directions.

4. What curve is the shape of the path of each jet?

This map shows part of the last orbit of *Skylab* before it broke apart and fell to earth over the Indian Ocean and Australia.

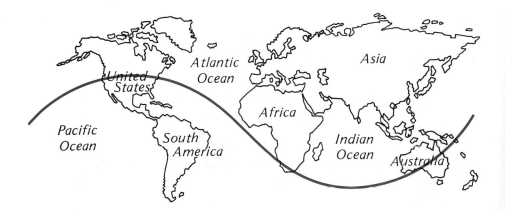

Chapter 6: MATHEMATICAL CURVES

5. What kind of curve does its orbit seem to be with respect to the map?

6. How much of the curve is shown?

When an alpha particle is shot toward the nucleus of an atom, the particle is repelled and changes direction. The great British scientist Ernest Rutherford showed that the particle's path is along one of the two branches of a curve that we have studied. It is illustrated on this postage stamp.

7. Which curve is it?

This photograph shows a round placemat made of straw.

8. What curve does it illustrate?

9. Does the curve seem to be Archimedean or logarithmic?

10. Explain the reason for your answer.

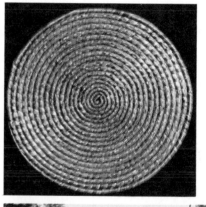

Galileo found that the curve traced by a point on the rim of a rolling wheel makes the strongest possible arch for a bridge.

11. What curve is it?

Courtesy of New York State Department of Transportation

Set II

The graph of an alternating electric current is a sine curve.

1. How many waves of this curve are shown?

2. The graph of the alternating current used in the United States has 60 waves per second. What fraction of a second is shown above?

Points have been marked at equal intervals on each side of the angle shown below.

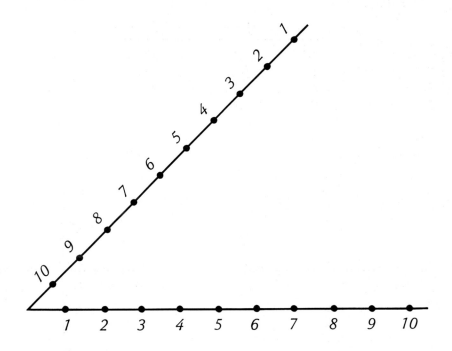

3. Trace the figure and connect each pair of points having the same number with a straight line segment.

4. What kind of curve is the result?

This drawing was made by Sir Isaac Newton to illustrate the gravitational forces on a planet in its orbit around the sun.

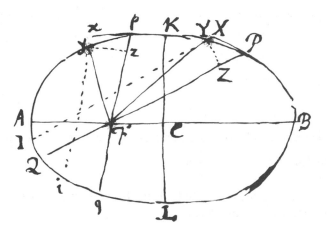

5. What kind of curve is it?

6. Copy and complete the following table for the four curves below.

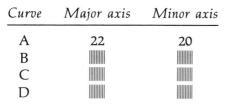

Curve	Major axis	Minor axis
A	22	20
B		
C		
D		

Curve A

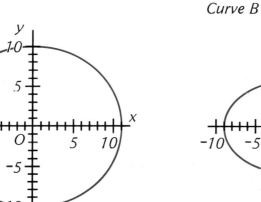

Curve B

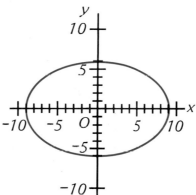

Curve C

Curve D

7. Find the ratio of the major and minor axes for each curve. The ratio for curve A is $\frac{22}{20} = 1.1$.

8. Two of the curves have the same shape as the curve drawn by Newton. Which are they?

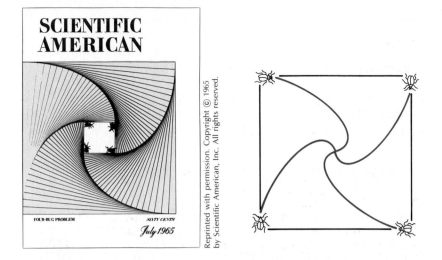

In the "four-bug problem," illustrated on the cover of *Scientific American* shown at the left above, four bugs are at the corners of a square. At the same time as the others do, each bug begins crawling directly toward the bug at the next corner. If all the bugs move at the same speed, they will follow the paths shown in the diagram at the right and will meet in the center of the square.

9. What kind of curve is each bug's path?

Sir Christopher Wren, a great English architect, discovered that an arc of a cycloid is a certain number of times as long as the diameter of the wheel that produces the cycloid. Look at this drawing of a cycloid and its wheel.

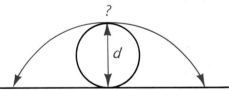

10. What do you think the number is?

Ships at sea use the LORAN system to determine their locations by radio. On the map at the left, points A, B, and C show the positions of radio transmitters. The two curves represent possible locations of a ship receiving certain signals from the stations.

11. What kind of curves are they?

12. In how many different points do the curves intersect?

13. From the signals received, the captain of a ship knows that the ship is at one of these points. Which one?

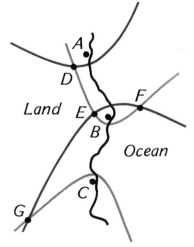

Chapter 6: MATHEMATICAL CURVES

Set III

Experiment: CURVES FROM CUTTING A CANDLE

Cut out a strip of paper about 10 centimeters long and 3 centimeters wide. Wind it tightly around a small "birthday cake" candle and, with a razor blade, carefully cut through the paper and candle

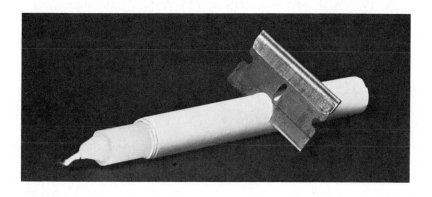

at a slant, as shown in this photograph. Look at the cross section of the candle that you have cut.

1. What curve do you see?

2. The length of one axis of this curve was determined by the diameter of the candle. Which one?

3. What determined the length of the other axis of the curve?

   Unwind the paper.

4. What curve results?

5. What dimension of the curve was determined by the size of the candle?

6. What dimension of the curve was determined by the slant of the cut?

Chapter 6 / Problems for Further Exploration

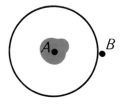

Lesson 1

1. An island is located at the center of a deep circular lake. Trees are located at the points labeled A and B. A man who cannot swim has a rope a few meters longer than 300 meters, the diameter of the lake. How can he use the rope to get to the island?*

2. Experiment: *Curves by Folding Paper*

   Use a compass to draw a circle with a radius of about 8 centimeters on a sheet of tracing paper. Mark a point approximately 3 centimeters inside the circle and label it A. Cut the circle out and fold it so that its edge falls on the point that you marked; make a sharp crease.

   Unfold the paper and fold it again in a different direction so that the edge again falls on the point. Repeat this many times so that points all around the edge have been folded onto the point.

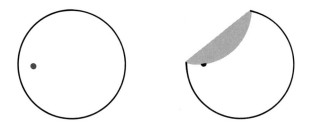

   a) What kind of curve is the result?

   Draw the curve on the paper. Also mark the center of the original circle and label it B.

   b) Where are the foci of the curve?

   _____

   *From "Mathematical Games" by Martin Gardner, *Scientific American,* July 1971.

368

Draw a radius of the circle as shown in this figure. Fold the paper so that the outer endpoint of the radius falls on point A. Notice that the fold touches the curve at one point.

c) How is the sum of the distances from this point to points A and B related to the radius of the circle?

Open the paper flat again, draw another radius of the circle, and repeat the directions in the preceding paragraph.

d) What do you notice?

e) What do these results have to do with the definition of an ellipse?

Do the entire experiment two more times: once with a point near the center of the circle and once with a point close to the edge.

f) How does the location of the point determine the shape of the curve formed?

Lesson 2

Experiment: *The Parabola and the Cone*

The parabola is related to the cone in an interesting way that you can discover by doing the following experiment.*

First, draw and label a pair of axes as shown below on a sheet of graph paper ruled either 4 units per inch or 2 units per centimeter.

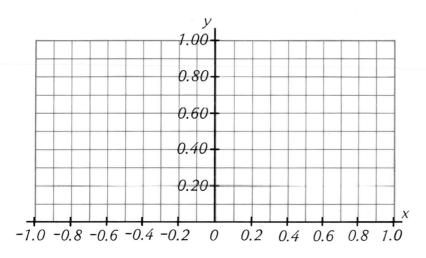

* After an exercise by Donald W. Stover in *The Conics: A Geometric Approach* (Houghton Mifflin, 1971).

a) Use the equation $y = 0.75(x^2)$ to copy and complete the following table.

x	−1.0	−0.8	−0.6	−0.4	−0.2	0	0.2	0.4	0.6	0.8	1.0
y	0.75	0.48									

b) Plot the points corresponding to the numbers in the table and connect them with a smooth curve. The result is part of a parabola.

Next, use a compass to draw a circle with a radius of 3.75 inches (7.5 centimeters) on a sheet of unlined paper. Keeping the compass adjusted to the radius of the circle, mark six equally spaced points around the circle and label four of them as shown in the first figure below. Label the center of the circle O. Use a ruler to draw the three line segments shown in the second figure.

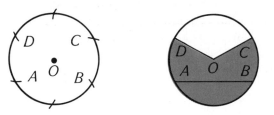

Cut out the shaded part of the figure and tape the edges OC and OD together to make a cone. Cut the cone along line AB and throw away the smaller piece.

c) Place the cut cone on the graph paper and describe what you discover.

Lesson 3

A point lies on a curve only if its coordinates fit the equation of the curve. The equation of the ellipse below is $\dfrac{x^2}{9} + \dfrac{y^2}{4} = 1.$

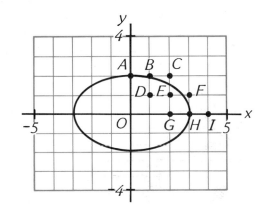

Chapter 6: MATHEMATICAL CURVES

a) Copy and complete the table at the right for the nine lettered points.

Use a calculator to find the values of $\dfrac{x^2}{9} + \dfrac{y^2}{4}$ and record them to the nearest hundredth in the table.

b) How is the value of $\dfrac{x^2}{9} + \dfrac{y^2}{4}$ for a given point related to where that point is located with respect to the ellipse?

The equation of the hyperbola shown below is $\dfrac{x^2}{9} - \dfrac{y^2}{4} = 1$.

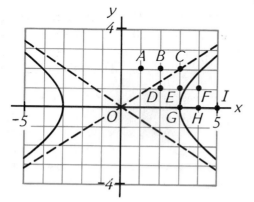

Point	x	y	$\dfrac{x^2}{9} + \dfrac{y^2}{4}$
A	0	2	1
B	1	2	1.11
C			
D			
E			
F			
G			
H			
I			

c) Copy and complete the table at the right for the nine lettered points, again recording the values in the last column to the nearest hundredth.

d) How is the value of $\dfrac{x^2}{9} - \dfrac{y^2}{4}$ for a given point related to where that point is located with respect to the hyperbola and its asymptotes?

Point	x	y	$\dfrac{x^2}{9} - \dfrac{y^2}{4}$
A	1	2	−0.89
B			
C			
D			
E			
F			
G			
H			
I			

Lesson 4

If sine x represents a musical note, sine x, sine $1.25x$, and sine $1.5x$ represent the notes of a major chord. For example, if sine x represents C, sine $1.25x$ represents E and sine $1.5x$ represents G. From the following exercise, you will discover what these notes would look like on the screen of an oscilloscope.

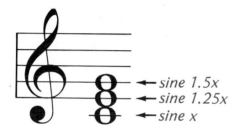

←— sine 1.5x
←— sine 1.25x
←— sine x

Draw and label two pairs of axes as shown in the figure below.

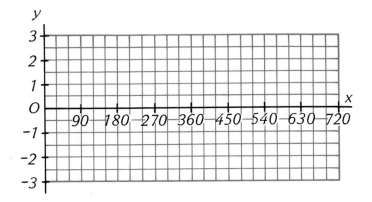

Note that 1 unit on the x-axis represents 30° and 1 unit on the y-axis represents 0.5.

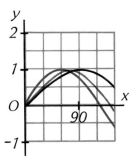

a) Use the values of sine x in the table on the next page to draw a sine curve for C. Part of the curve is shown in black in the adjoining figure. Next, use the values of sine 1.25x in the table to draw a sine curve for E on the same pair of axes. Part of this curve is shown in gray in the figure. Then use the values of sine 1.5x in the table to draw a sine curve for G also on the same pair of axes. Part of this curve is shown in brown in the figure. The three curves that you have drawn show the three notes of the chord as they would appear individually on the oscilloscope screen.

By turning a dial on the oscilloscope, these three simple curves can be combined into one complicated curve that represents the major chord itself. The oscilloscope does this by adding the values of the three curves. This is illustrated in the table below.

x	sine x		sine 1.25x		sine 1.5x		chord
0°	0	+	0	+	0	=	0
30°	0.5	+	0.6	+	0.7	=	1.8
60°	0.9	+	1.0	+	1.0	=	2.9
90°	1.0	+	0.9	+	0.7	=	2.6
120°	0.9	+	0.5	+	0	=	1.4
150°	0.5	+	−0.1	+	−0.7	=	−0.3

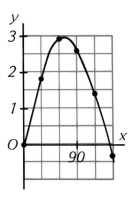

b) Find the values of the chord for the remaining values of x through 720°.
c) Then plot the corresponding points on the second pair of axes and connect them with a smooth curve. Part of the curve is shown in the adjoining figure.

Chapter 6: MATHEMATICAL CURVES

x	sine x	sine $1.25x$	sine $1.5x$
0°	0	0	0
30°	0.5	0.6	0.7
60°	0.9	1.0	1.0
90°	1.0	0.9	0.7
120°	0.9	0.5	0
150°	0.5	−0.1	−0.7
180°	0	−0.7	−1.0
210°	−0.5	−1.0	−0.7
240°	−0.9	−0.9	0
270°	−1.0	−0.4	0.7
300°	−0.9	0.3	1.0
330°	−0.5	0.8	0.7
360°	0	1.0	0
390°	0.5	0.8	−0.7
420°	0.9	0.3	−1.0
450°	1.0	−0.4	−0.7
480°	0.9	−0.9	0
510°	0.5	−1.0	0.7
540°	0	−0.7	1.0
570°	−0.5	−0.1	0.7
600°	−0.9	0.5	0
630°	−1.0	0.9	−0.7
660°	−0.9	1.0	−1.0
690°	−0.5	0.6	−0.7
720°	0	0	0

Lesson 5

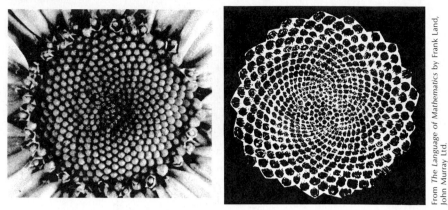

From *The Language of Mathematics* by Frank Land, John Murray Ltd.

Two opposite sets of spirals appear in the head of the chrysanthemum shown in the photograph above.

a) What kind of spirals does the head of the chrysanthemum seem to have?

b) How many spirals are in the set that winds clockwise from the center?

c) How many spirals are in the set that winds counterclockwise from the center?

Your answers to parts b and c are part of a famous number sequence.

d) What sequence is it?

Lesson 6

Experiment: *More Hypocycloids*

With a compass, draw circles with radii of 1 centimeter and 3 centimeters on a 4-inch-by-6-inch card. Cut the circles out, place

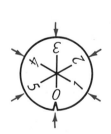

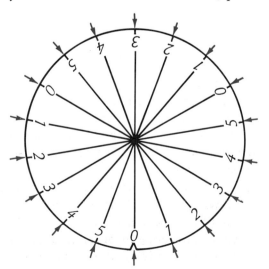

Chapter 6: MATHEMATICAL CURVES

them on the circles on the preceding page, and mark points on their rims as indicated by the arrows. Take your circles away from the page and, with a ruler, draw diameters to connect the points that you have marked to form two wheels with "spokes." Number the spokes as shown, and cut a small notch in the rim of each wheel at the spoke at the bottom numbered 0.

a) Use a compass to draw a circle with a radius of 4 centimeters on a sheet of paper and a protractor to mark points around the circle at 15° intervals. Number the points starting from the bottom as shown in this figure.

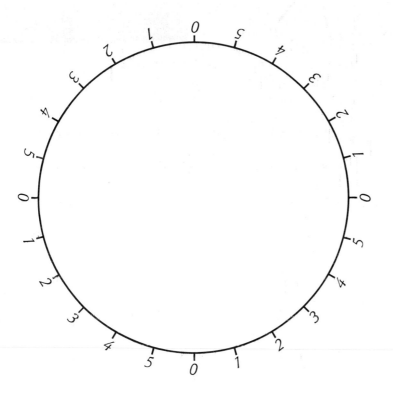

Place the small wheel inside the circle so that the spoke marked 0 on the wheel touches the point marked 0 at the bottom of the circle. Mark a heavy dot at the notch of the wheel.

Roll the wheel counterclockwise around the circle, marking dots in the notch of the wheel as you do so. Connect the dots with a smooth curve.

b) Now repeat the directions in the preceding three paragraphs, drawing a new circle and rolling the larger wheel around it. This time the wheel will have to be rolled around the circle more than once in order to produce the entire curve.

c) What is surprising about the results?

Chapter 7

METHODS OF
COUNTING

Lesson 1

The Fundamental Counting Principle

© 1968 United Feature Syndicate, Inc.

Suppose that you are taking a true-or-false test and that, like Linus, you are guessing the answers. If there are 10 questions, how many different patterns of answers are possible?

Before trying to answer this question, we will consider a simpler one. Suppose that a true-or-false test has just 3 questions. In how many different ways can they be answered? The ways are shown in the diagram below. Called a *tree diagram* because of its branches,

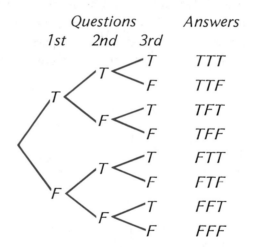

it reveals that there are 8 ways in which 3 questions can be answered.

Notice that this number can be obtained by multiplying the number of ways of answering the first question, 2, by the number of ways of answering the second question, 2, by the number of ways of answering the third question, 2:

$$2 \times 2 \times 2 = 8$$

Instead of drawing a tree diagram for a true-or-false test of 10 questions, we can count the number of ways of answering it by following the same procedure:

$$2 \times 2 \times 2 \times 2 \times 2 \times 2 \times 2 \times 2 \times 2 \times 2 = 1{,}024$$

378

There are 1,024 ways.

We will call the method used in figuring this out the **fundamental counting principle.**

▶ To find the number of ways of making several decisions in succession, multiply the numbers of choices that can be made in each decision.

Here is another example of how this principle works. Suppose that you are taking a matching test in which you are supposed to match 3 answers with 3 questions. In how many different ways can the questions be answered? The tree diagram below, in which the answers are represented by A, B, and C, shows that there are 6 ways in which the 3 questions can be answered.

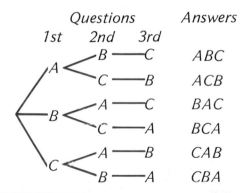

Questions   *Answers*

1st   2nd   3rd

A ⟨ B —— C   ABC

      C —— B   ACB

B ⟨ A —— C   BAC

      C —— A   BCA

C ⟨ A —— B   CAB

      B —— A   CBA

This example differs from those about the true-or-false tests because the number of ways of answering each successive question is not the same. When an answer has been chosen for a question, it is "used up," and so the number of answers from which to choose for the next question is 1 less. You have 3 choices for the first answer, 2 choices for the second, and just 1 choice for the third:

$$3 \times 2 \times 1 = 6$$

In true-or-false tests, the choices made in answering the questions are *independent* of each other. In matching tests, the choices are *dependent*.

Exercises

Set I

This diagram shows the different ways in which a true-or-false test having 4 questions can be answered.

1. How many ways are there?

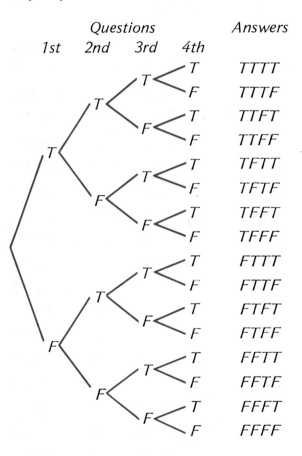

	Questions			Answers
1st	2nd	3rd	4th	

2. What numbers would you multiply to find this answer by using the fundamental counting principle?

3. Copy and complete the following table showing the numbers of ways in which true-or-false tests having different numbers of questions can be answered.

Number of questions	1	2	3	4	5	6
Number of ways to answer	2	4	8	‖‖‖‖	‖‖‖‖	‖‖‖‖

4. What kind of number sequence do the numbers on the second line of this table form?

5. Are the choices made in answering the questions of a true-or-false test *independent* or *dependent*?

This diagram shows the different ways in which a matching test having 4 questions can be answered if each of the 4 answers is used exactly once.

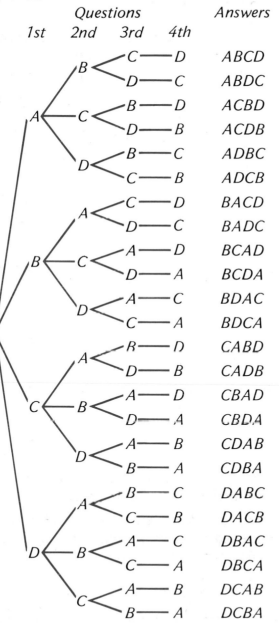

Questions				Answers
1st	2nd	3rd	4th	
A	B	C — D		ABCD
		D — C		ABDC
	C	B — D		ACBD
		D — B		ACDB
	D	B — C		ADBC
		C — B		ADCB
B	A	C — D		BACD
		D — C		BADC
	C	A — D		BCAD
		D — A		BCDA
	D	A — C		BDAC
		C — A		BDCA
C	A	B — D		CABD
		D — B		CADB
	B	A — D		CBAD
		D — A		CBDA
	D	A — B		CDAB
		B — A		CDBA
D	A	B — C		DABC
		C — B		DACB
	B	A — C		DBAC
		C — A		DBCA
	C	A — B		DCAB
		B — A		DCBA

6. How many ways are there?

7. What numbers would you multiply to find this answer by using the fundamental counting principle?

8. Copy and complete the following table showing the numbers of ways in which matching tests having different numbers of questions can be answered.

Number of questions	1	2	3	4	5	6
Number of ways to answer	1	2	6	▥	▥	▥

9. Are the choices made in answering the questions of a matching test in which each of the answers is used exactly once *independent* or *dependent*?

The discovery of the genetic code—the way in which the genes of a living organism determine the structure of the organism's proteins—is one of the most remarkable achievements of molecular biology.

A gene contains a long molecule that is made up of 4 kinds of smaller molecules. These smaller molecules are often referred to by the letters A, C, G, and T.* Groups of these small molecules along the large molecule make up the code.

The diagram at the left shows the different sequences of 2 molecules that are possible.

10. How many sequences of 2 molecules are there?

11. What numbers would you multiply to find this answer by using the fundamental counting principle?

12. Copy and complete the following table showing the numbers of sequences possible for different numbers of molecules.

Number of molecules	1	2	3	4
Number of sequences	4	16	▥	▥

On a tour of the Hawaiian Islands, Captain and Mrs. Cook plan to travel from the island of Oahu to the island of Kauai, and then to Lanai, Maui, and Hawaii in that order, before returning to Oahu. The number of ways in which they can travel between each island (including both planes and boats) is shown on the map. For exam-

Molecules

1st	2nd	Sequences

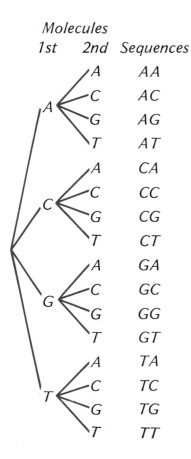

* The molecules are nucleotides, and they contain the bases *adenine, cytosine, guanine,* and *thymine.*

Chapter 7: METHODS OF COUNTING

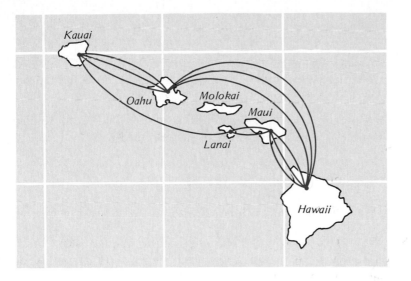

ple, they can travel from Lanai to Hawaii in $2 \times 3 = 6$ different ways.

13. In how many different ways can they travel from Maui to Oahu?

14. In how many different ways can the Cooks plan their entire tour?

A member of the U.S. Public Health Service once came up with a list of words from which impressive-sounding 3-word phrases can be chosen at random. Here is his list.*

1	2	3
integrated	management	options
total	organizational	flexibility
systematized	monitored	capability
parallel	reciprocal	mobility
functional	digital	programming
responsive	logistical	concept
optional	transitional	time-phase
synchronized	incremental	projection
compatible	third-generation	hardware
balanced	policy	contingency

The phrases are made up by choosing a word from each column. Examples are: "integrated management programming" and "total digital capability."

---

* *Time,* September 13, 1968, page 22.

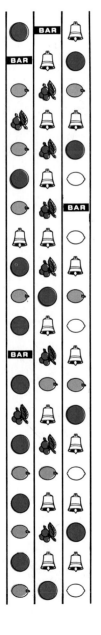

15. How many different 3-word phrases can be made up from this list in the same way?

16. How many different 3-word phrases would be possible if each column of the list contained only 5 words?

17. How many different 3-word phrases would be possible if each column of the list contained 15 words?

Suppose that you are at Bob's Restaurant and see the following items on the menu.

Dinners		Beverages		Desserts	
Big Boy Hamburger	1.70	Coffee	.45	Hot Fudge Sundae	1.20
Fried Chicken	3.95	Coca-Cola	.45	Apple Pie	1.00
Shrimp	3.95	Root Beer	.45	Banana Cream Pie	1.00
Fish and Fries	3.75	Milk	.50	Strawberry Pie	1.05
Bob's Special Steak	5.50	Iced Tea	.45	Cheese Cake	.90
				Fudge Brownies	.55

18. If you order a dinner and a beverage, without dessert, how many different combinations can you choose from?

19. If you also order a dessert, how many combinations are possible?

20. How many choices do you have if you order a dinner, beverage, and dessert, but decide that you do not want seafood?

Set II

Most slot machines have 3 dials with 20 symbols on each dial. The symbols on each dial of a typical machine are shown in the figure at the left and summarized in the table below.

Symbol	Dial 1	Dial 2	Dial 3
Bar	2	1	1
Bell	1	8	7
Plum	7	2	3
Orange	8	2	4
Cherry	2	7	0
Lemon	0	0	5
	20	20	20

Each dial can stop on any one of its 20 symbols.

1. In how many different positions can the three dials of the machine stop?

2. The biggest payoff is for 3 bars. How many ways are there of getting 3 bars?

3. Copy and complete the following table of ways in which the 3 dials can stop on the same symbol.

Combination	Number of ways
3 bars	2
3 bells	
3 plums	
3 oranges	
3 cherries	
3 lemons	

4. What arrangement of 3 symbols can turn up in only 1 way?

5. What arrangement of 3 symbols can turn up in the most ways? In how many ways can it turn up?

6. How many arrangements that *look* different can turn up on the machine? (*Hint:* 5 symbols that look different can turn up on the first dial.)

Keys of different shapes are designed by choosing from several patterns for each of their parts. The keys of General Motors cars have 6 parts.

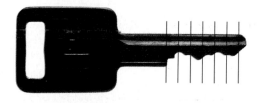

7. Before 1967, there were 2 patterns for each part. How many different key designs were possible?

8. Now General Motors uses 3 patterns for each part. How many different key designs do they have now?

9. If the number of patterns for each part were increased to 4, how many key designs would be possible?

The call letters of radio and television stations in the United States begin with either K or W. Those west of the Mississippi River start with K and those east of it with W.

10. Some stations, such as KID in Idaho Falls, Idaho, and WOW in Omaha, Nebraska, have 3 call letters. How many sets of call letters having 3 letters are possible?

There are 11,160 radio and television stations in the United States.

11. Could every station be named with a different set of 3 letters beginning with K or W?

12. Other stations, such as KUZZ in Bakersfield, California, and WARF in Jasper, Alabama, have 4 call letters. How many sets of call letters having 4 letters are possible?

Telephone area codes in the United States and Canada consist of 3 digits, in which the first is a digit from 2 through 9, the second is either 0 or 1, and the third can be any digit except 0.

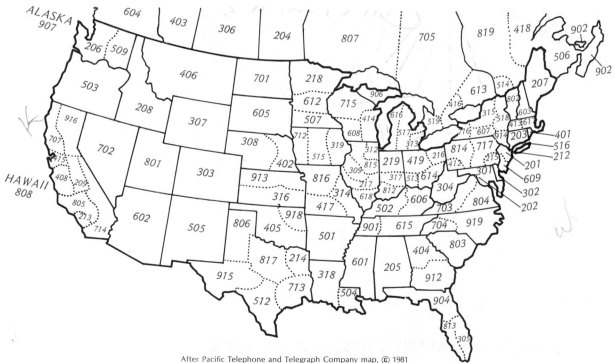

After Pacific Telephone and Telegraph Company map, © 1981

Chapter 7: METHODS OF COUNTING

13. According to these rules, how many different area codes can *begin* with the digit 2?

14. How many different area codes can *end* with the digit 2?

15. How many different area codes are possible altogether?

Set III

A book titled *Cent mille milliards de poèmes* by Raymond Queneau contains a hundred thousand billion poems. The poems are sonnets, each one consisting of 14 lines. They are printed on the right-hand pages of the book only, the left-hand pages being blank. Each page is sliced into horizontal strips, with 1 line of the sonnet on a strip. The strips can be turned individually so that the lines can be selected in many different ways. Regardless of how they are chosen, however, every sonnet formed is structurally correct and makes sense.

How many pages must the book contain to produce a hundred thousand billion poems? Explain your answer.

Cent mille milliards de poèmes by Raymond Queneau; © Editions Gallimard

Lesson 2

Permutations

Drawing by Bill Hoest; © 1967 *Look* Magazine

"Next time, let me handle the seating arrangements."

BILL HOEST

The host of this dinner party has not done a very good job of planning the seating arrangements. It looks as though everyone would rather be somewhere else. Suppose that the 14 people at this table had decided to exchange places before they sat down. In how many different ways could they arrange themselves?

The person taking the first chair could be any one of the 14 people. After that person was seated, the person taking the second chair could be any one of the 13 people remaining. By the fundamental counting principle, there are

$$14 \times 13 = 182$$

ways in which 2 people can take the first two chairs.

Continuing to reason in this fashion, we find that there are

$$14 \times 13 \times 12 = 2,184$$

ways in which 3 people can be seated at the first three places,

$$14 \times 13 \times 12 \times 11 = 24,024$$

ways in which 4 people can be seated at the first four places, and

$$14 \times 13 \times 12 \times 11 \times 10 \times 9 \times 8 \times 7 \times 6 \times 5 \times 4 \times 3 \times 2 \times 1 = 87{,}178{,}291{,}200$$

ways in which all 14 people can be seated at the table.

This number is so large that if they were able to take a different seating arrangement every second without stopping, it would take more than 2,700 *years* for the 14 people to arrange themselves in every possible order.

Mathematicians call these arrangements *permutations*.

▶ A **permutation** is an arrangement of things in a definite order.

We have found that the number of permutations of 14 people is

$$14 \times 13 \times 12 \times 11 \times 10 \times 9 \times 8 \times 7 \times 6 \times 5 \times 4 \times 3 \times 2 \times 1.$$

Mathematicians use an exclamation mark to indicate this product by writing 14!. The exclamation mark is also called a **factorial** symbol and 14! is read as "14 factorial."

▶ In general, $n!$ means to multiply the consecutive numbers from n all the way down to 1.

▶ The number of permutations of n different things is $n!$.

Although the symbol ! is not used in mathematics to express surprise, the numbers that it represents are, like 14! above, often surprisingly large. For example, consider 5!, the number of different ways in which just 5 people can be seated in a row. They are illustrated in the tree diagram shown on the next page, in which A, B, C, D, and E represent the 5 people. Because

$$5! = 5 \times 4 \times 3 \times 2 \times 1 = 120$$

there are 120 different ways.

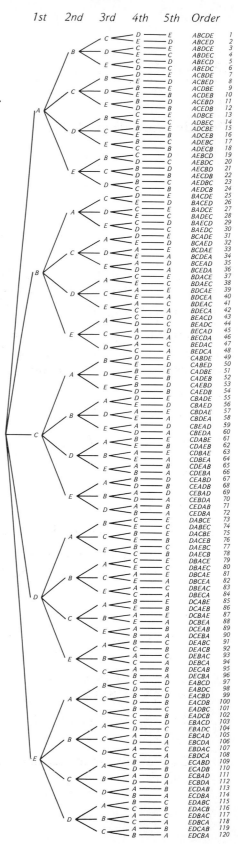

Exercises

Set I

The values of 1! through 4! are shown below.

$$1! = 1$$
$$2! = 2 \times 1 = 2$$
$$3! = 3 \times 2 \times 1 = 6$$
$$4! = 4 \times 3 \times 2 \times 1 = 24$$

1. Continue this list by finding the values of 5! through 13!. If you do not make any mistakes, you should find that $13! = 6,227,020,800$.

Although the equation $2! + 2! = 4!$ "looks" true, it is false because $2! = 2$, $4! = 24$, and $2 + 2 \neq 24$. Use your list to decide which of the following equations are true and which are false.

2. $3! + 3! = 6!$

3. $10! = 10 \times 9!$

4. $1! + 4! + 5! = 145$

5. $\dfrac{8!}{4!} = 2!$

6. $6! \times 7! = 10!$

Use your list to answer each of the following questions.

7. In how many orders can the names of 5 candidates for the same office be listed on a ballot?

8. In how many orders can 6 runners finish a race if there are no ties?

9. In how many orders can 7 different digits of a telephone number be arranged?

10. How many ways are there of scrambling the letters of the word SCRAMBLE?

11. In how many ways can 10 people line up at a theater box office?

12. In how many orders can the 13 cards in a bridge hand be arranged?

Set II

The 1958 CBS television schedule for Sunday evening between 9 and 11 o'clock was

> 9:00 "G.E. Theater"
>
> 9:30 "Alfred Hitchcock Presents"
>
> 10:00 "$64,000 Question"
>
> 10:30 "What's My Line?"

The tree diagram at the right shows the different ways in which the network could have scheduled the 9:00 and 9:30 programs from these four.

9:00 p.m. 9:30 p.m.

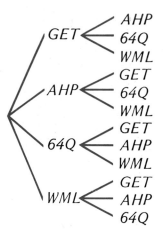

1. How many ways are there?

2. What numbers would you multiply to find this answer by using the fundamental counting principle?

3. How many ways are there for scheduling all four programs?

4. Are the choices made in selecting the programs *independent* or *dependent*?

The number of floats in Pasadena's Rose Parade has varied from just a few at the beginning to 100 in 1921. In recent years, there have been about 60 floats in each parade. In how many different ways can the first and second floats be chosen in the parade if there are

5. 30 floats in all?

6. 60 floats in all?

Chapter 7: METHODS OF COUNTING

Eight horses are entered in a race in which bets are placed on which horses will win, place, and show (that is, finish first, second and third). Suppose that the race is run and that there are no ties.

"Blue Boy seems to be holding back a bit."

Courtesy of Joseph Zeis

7. In how many ways can the win and place be taken?

8. In how many ways can the win, place, and show be taken?

9. In how many orders can all eight horses finish the race?

A club has four officers: president, vice-president, secretary, and treasurer. If a member cannot hold more than one office, in how many ways can the officers be elected if the club has

10. 12 members?

11. 15 members?

A newspaper is running a contest with four prizes of $500, $300, $200, and $100. No one is allowed to win more than one prize. If 1,003 people send in entries, in how many different ways can

12. the top two prizes be won?

13. the top three prizes be won?

14. all four prizes be won?

Set III

There was a table set out under a tree in front of the house, and the March Hare and the Hatter were having tea at it: a Dormouse was sitting between them, fast asleep, and the other two were using it as a cushion, resting their elbows on it, and talking over its head. . . .

The table was a large one, but the three were all crowded together at one corner of it. "No room! No room!" they cried out when they saw Alice coming. "There's plenty of room!" said Alice indignantly, and she sat down in a large arm-chair at one end of the table.

—*Alice's Adventures in Wonderland* by LEWIS CARROLL

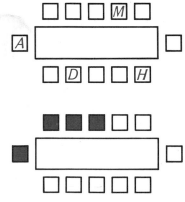

1. If the table had 12 places around it, in how many different ways could Alice, the March Hare, the Hatter, and the Dormouse be seated at it? (One way is shown in the first diagram at the left.)

2. How many different sets of 4 consecutive places are there at the table? (One set is shown in the second diagram at the left.)

3. In how many ways can Alice, the March Hare, the Hatter, and the Dormouse be seated next to each other with no empty places in between?

Lesson 3

More on Permutations

The name of the magazine published by the San Diego Zoo is *ZOONOOZ*. This name is unusual because spelled backwards it is the same word. In other words, reversing the order of the letters ZOONOOZ leaves their order unchanged.

In how many *different* orders can the letters be arranged? This question is about the number of permutations of 7 things, and it might seem that the answer would be 7!, or 5,040. This is clearly incorrect, however, because not all the letters of ZOONOOZ are different. We have just seen, for example, that reversing the order of the letters does not result in a new arrangement.

Even though not all the letters are different, we can find the number of orders by first labeling the letters with numbers to make them different. If the 2 Z's are labeled Z_1 and Z_2 and the 4 O's, O_1, O_2, O_3, and O_4, the 7 letters of

$$Z_1 O_1 O_2 N O_3 O_4 Z_2$$

can be arranged in 7! different orders. Among these 7! orders are such arrangements as

$$Z_1 Z_2 N O_1 O_2 O_3 O_4 \quad \text{and} \quad Z_2 Z_1 N O_1 O_2 O_3 O_4.$$

Notice that, without the numbers, these arrangements are the same: ZZNOOOO. Some other orders that lead to the same pattern are

$$Z_1Z_2NO_1O_2O_4O_3, \quad Z_1Z_2NO_1O_3O_2O_4, \quad \text{and} \quad Z_1Z_2NO_1O_3O_4O_2.$$

How many such orders leading to the same pattern are there altogether? There are $2! \times 4!$ such orders because the 2 Z's can be arranged in $2!$ ways and the 4 O's can be arranged in $4!$ ways.

By the same reasoning, it follows that not only are there $2! \times 4!$ numbered orders equivalent to the unnumbered order ZZNOOOO, but there are also $2! \times 4!$ numbered orders equivalent to *every* unnumbered order. This means that the number of different permutations of the 7 letters of ZOONOOZ is not $7!$, but $\dfrac{7!}{2! \times 4!}$. This fraction can be simplified by first writing out each factorial and then reducing:

$$\frac{7 \times 6 \times 5 \times 4 \times 3 \times 2 \times 1}{2 \times 1 \times 4 \times 3 \times 2 \times 1} = \frac{7 \times \overset{3}{\cancel{6}} \times 5 \times \cancel{4} \times \cancel{3} \times \cancel{2} \times \cancel{1}}{\cancel{2} \times 1 \times \cancel{4} \times \cancel{3} \times \cancel{2} \times \cancel{1}} = \frac{105}{1} = 105$$

There are 105 ways in which the letters of ZOONOOZ can be arranged. The arrangements are shown, in alphabetical order, on the facing page. Merely counting them, as we have done, is much easier than making the list.

►In general, the number of permutations of n things, of which a things are alike, another b things are alike, another c things are alike, and so forth, is

$$\frac{n!}{a! \times b! \times c! \cdots}.$$

Chapter 7: METHODS OF COUNTING

1. NOOOOZZ
2. NOOOZOZ
3. NOOOZZO
4. NOOZOOZ
5. NOOZOZO
6. NOOZZOO
7. NOZOOOZ
8. NOZOOZO
9. NOZOZOO
10. NOZZOOO
11. NZOOOOZ
12. NZOOOZO
13. NZOOZOO
14. NZOZOOO
15. NZZOOOO
16. ONOOOZZ
17. ONOOZOZ
18. ONOOZZO
19. ONOZOOZ
20. ONOZOZO
21. ONOZZOO
22. ONZOOOZ
23. ONZOOZO
24. ONZOZOO
25. ONZZOOO
26. OONOOZZ
27. OONOZOZ
28. OONOZZO
29. OONZOOZ
30. OONZOZO
31. OONZZOO
32. OOONZZOZ
33. OOONZZO
34. OODNZZO
35. OOOONZZ

36. OOOOZNZ
37. OOOOZZN
38. OOOZNOZ
39. OOOZNZO
40. OOOZONZ
41. OOOZOZN
42. OOOZZNO
43. OOOZZON
44. OOZNOOZ
45. OOZNOZO
46. OOZNZOO
47. OOZONOZ
48. OOZONZO
49. OOZOONZ
50. OOZOOZN
51. OOZOZNO
52. OOZOZON
53. OOZZNOO
54. OOZZONO
55. OOZZOON
56. OZNOOOZ
57. OZNOOZO
58. OZNOZOO
59. OZNZOOO
60. OZONOOZ
61. OZONOZO
62. OZONZOO
63. OZOONOZ
64. OZOONZO
65. OZOOONZ
66. OZOOOZN
67. OZOOZNO
68. OZOOZON
69. OZOZNOO
70. OZOZONO

71. OZOZOON
72. OZZNOOD
73. OZZONOO
74. OZZOONO
75. OZZOOON
76. ZNOOOOZ
77. ZNOOOZO
78. ZNOOZOO
79. ZNOZOOO
80. ZNZOOOO
81. ZONOOOZ
82. ZONOOZO
83. ZONOZOO
84. ZONZOOO
85. ZOONOOZ
86. ZOONOZO
87. ZOONZOO
88. ZOOONOZ
89. ZOOONZO
90. ZOOOOZN
91. ZOOOZNO
92. ZOOOZON
93. ZOOZNOO
94. ZOOZONO
95. ZOOZOON
96. ZOZNOOO
97. ZOZONOO
98. ZOZOONO
99. ZOZOOON
100. ZZNOOOO
101. ZZNOOOO
102. ZZONOOO
103. ZZOONOO
104. ZZOOONO
105. ZZOOOON

Exercises

Set I

The ways in which the letters of the name JACK can be arranged are shown below.

JACK	AJCK	CJAK	KJAC
JAKC	AJKC	CJKA	KJCA
JCAK	ACJK	CAJK	KAJC
JCKA	ACKJ	CAKJ	KACJ
JKAC	AKJC	CKJA	KCJA
JKCA	AKCJ	CKAJ	KCAJ

1. Show how the number of ways can be found without making a list.

2. Make a list of all the ways in which the letters of the name JILL can be arranged.

3. Why is the number of ways of arranging the letters of JILL less than the number of ways of arranging the letters of JACK?

4. Show how the number of ways of arranging the letters of JILL can be found without making a list.

5. Which of the following names have the same number of arrangements as JILL?

ANNE   DOUG   EMMA   JANE   MARK   PETE

6. Make a list of all the ways in which the letters of the name LULU can be arranged.

7. Show how the number of ways can be found without making a list.

8. The word ANGERED is remarkable in that its letters can be rearranged to form another word with the same meaning, ENRAGED. In how many different ways can the seven letters of these words be arranged?

9. There was once a British race horse whose name was POTOOOOOOOO. This strange name was pronounced like the name of a common vegetable. Do you see why? In how many different ways can the letters of POTOOOOOOOO be arranged?

10. In how many different ways can the letters of the word ANTEATEREATER be arranged?

Set II

Riddle: A customer walks into a hardware store to buy something and asks the clerk how much 1 would cost. The clerk tells him 50¢. He then asks how much 10 would cost. The clerk says $1.00. The customer says, "I'll buy 15155," and pays the clerk $2.50. What was the customer buying?

Answer: Digits for his house number.

1. How many different 5-digit house numbers can be made from the digits 15155?

2. Show that your answer is correct by making a list of the house numbers.

Mrs. Olson, who drinks a lot of coffee and likes it with cream, claims that by tasting a cup she can tell whether the cream was poured into the cup before or after the coffee was. To find out whether in fact she can tell which has been poured in first, we prepare several cups of coffee. We pour the cream into some of them first and the coffee into the others first. The cups are given to her one at a time for her to tell how they were made. One possible order of the cups is illustrated here.

How many different orders are possible if there are

3. 4 cups, with 2 cups of each type?

4. 6 cups, with 3 cups of each type?

5. 8 cups, with 4 cups of each type?

The stack of poker chips at the left contains 12 chips, of which 3 are light brown, 4 are white, and 5 are dark brown. Into how many different looking stacks can

6. the light-brown and the white chips be piled?

7. the light-brown and the dark-brown chips be piled?

8. the white and the dark-brown chips be piled?

9. all 12 chips be piled?

In studying the genetic code,* a biologist wonders about the number of possible arrangements of 4 kinds of molecules in a chain.

10. If the chain contains just 4 molecules with 1 of each kind, how many different arrangements are possible?

11. If the chain contains 8 molecules with 2 of each kind, how many different arrangements are possible?

12. If the chain contains 12 molecules with 3 of each kind, how many different arrangements are possible?

_____

* See page 382.

_____

Set III

One of the most frequently sung songs in the world is "Happy Birthday to You," written by Mildred and Patty Hill in 1893. If only the pitches and not the time values are considered, the number of different tunes that can be made by rearranging its notes is

HAPPY BIRTHDAY TO YOU!

Mildred J. Hill
Patty S. Hill

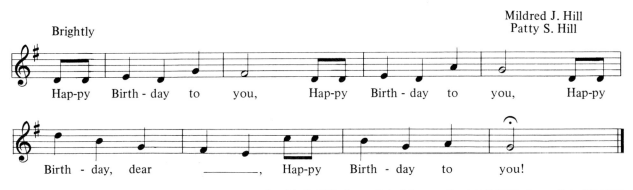

$$\frac{25!}{8! \times 5! \times 3! \times 2! \times 2! \times 2! \times 2!} = 33,394,325,524,560,000.$$

1. Where do the factorials in this expression come from?

2. Write an expression for finding the number of ways in which the first 6 notes of the song can be arranged.

3. How many are there?

4. Write an expression for finding the number of ways in which the first 12 notes of the song can be arranged.

5. How many are there?

Lesson 4

Combinations

The best hand in the game of poker is a royal flush, which consists of an ace, king, queen, jack, and 10 of the same suit. Because there are four suits of cards, there are four hands that are royal flushes. How many poker hands are possible in all?

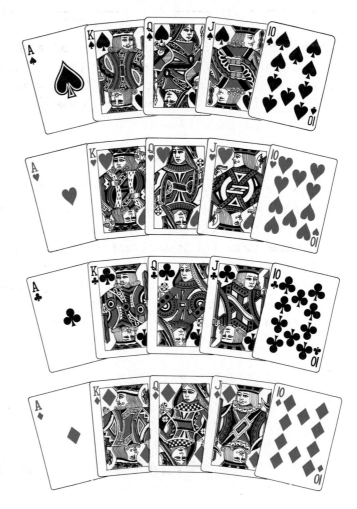

"Then again, he could be bluffing."

Drawing by Herb Green

A poker hand consists of 5 cards from a deck of 52 cards, and so it would seem that we could find the answer to this question by using the fundamental counting principle. With 52 choices for the first card, 51 for the second card, and so on, it would seem that there are

$$52 \times 51 \times 50 \times 49 \times 48$$

different poker hands possible.

If we look at two of the hands counted in this way, however, we find that they are not really different. As shown below, they contain the same cards but in different orders. The product $52 \times 51 \times 50 \times 49 \times 48$, then, counts the poker hands as if the order of the cards in them was important. To make up for the fact that the order is *not* important, we need to know the number of ways in which 5 cards of a given hand can be arranged. It is 5!, or 120.

This means that the product $52 \times 51 \times 50 \times 49 \times 48$ is 120 times as large as the actual number of different hands. Dividing, we get

$$\frac{52 \times 51 \times 50 \times 49 \times 48}{5 \times 4 \times 3 \times 2 \times 1} = 2,598,960.$$

The number of poker hands that are really different is 2,598,960.

The problem that we have just solved is called a *combination* problem.

▶ A **combination** is a selection of things in which the *order does not matter.*

▶ To find the number of combinations possible, *first use the fundamental counting principle and then divide by the number of ways in which the things can be arranged.*

Here is another example of how this works. How many different pairs of cards can be chosen from the 5 cards in a royal flush? By

the fundamental counting principle, there are 5×4 ways to choose the 2 cards in order. The number of ways in which 2 cards can be arranged is $2! = 2 \times 1$. Therefore, the number of pairs of cards that can be chosen is

$$\frac{5 \times 4}{2 \times 1} = \frac{20}{2} = 10.$$

The ten pairs are illustrated in the figure below.

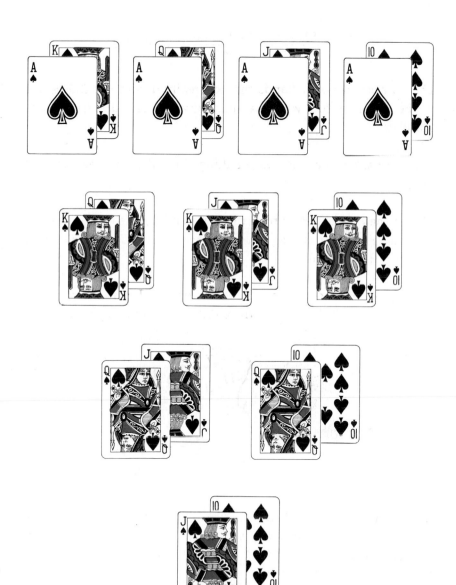

Exercises

Set I

The ways in which 2 people can be chosen from a set of 4 is shown in the tree diagram below.

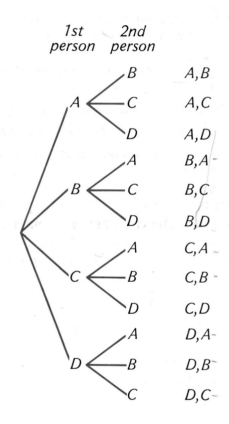

1. How many ways are shown in the diagram if the order is important (A, B, for example, is different from B, A)?

2. What numbers would you multiply to find this answer by using the fundamental counting principle?

3. How many ways are there if the order is not important (A, B, for example, is the same as B, A)?

4. How can you find this answer without looking at the diagram?

Chapter 7: METHODS OF COUNTING

The first telephone exchange went into service in New Haven, Connecticut, in 1878. It linked 21 telephones. The first telephone directory, published in New Haven later the same year, listed 50 telephones.

5. How many different ways could the 21 telephones in the original exchange be paired for conversations?

6. How many different ways could the 50 telephones listed in the first directory be paired for conversations?

Ten standbys hope to get seats on a flight from New York to London.

7. If there are 3 seats available on the plane, in how many different orders can the names of 3 standbys be called?

8. How many different combinations of 3 standbys can be chosen for the flight?

The Braille system of writing used by blind people consists of a code of 63 characters. The characters are combinations of raised dots, ranging in number from 1 to 6. Examples of letters written in Braille are shown below. The solid circles represent raised dots

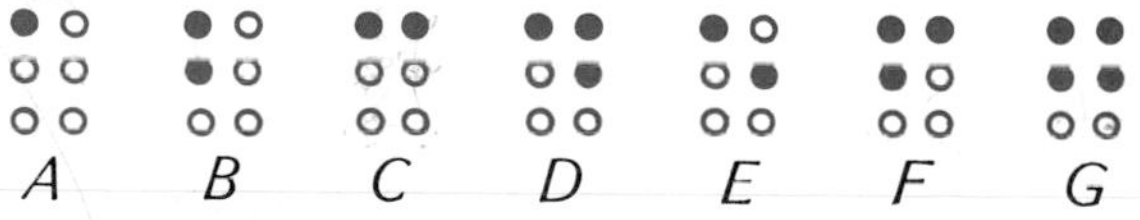

and the hollow circles show where the paper remains flat.
In how many different ways can

9. 1 of the 6 dots be raised?

10. 2 of the 6 dots be raised?

11. 3 of the 6 dots be raised?

12. 4 of the 6 dots be raised?

13. 5 of the 6 dots be raised?

14. all 6 dots be raised?

15. Does the Braille code use every possible way of combining from 1 to 6 dots?

Set II

When the National Baseball League was founded in 1876, it consisted of the following 8 teams:

> Boston Red Caps
> Chicago White Stockings
> Cincinnati Reds
> Hartford Blues
> Louisville Colonels
> New York Mutuals
> Philadelphia Athletics
> St. Louis Reds

1. How many games did the teams have to play in order for each team to play each of the others exactly once?

2. In how many different orders could one of the teams play each of the other teams exactly once?

A supermarket chain once held a contest in which each customer was given a card containing 16 spots covered with silver film. The customer could scratch the film off either 4 or 6 of the covered spots. Of the 16 spots, 6 were considered to be winning spots. A customer who chose those 6 spots would win $500. If someone chose any 4 of the 6 winning spots, the prize would be $100.

3. In how many different ways could a customer choose 4 spots?

4. In how many different ways could a customer choose 6 spots?

5. Which choice—4 or 6 spots—would give a customer a better chance of winning?

According to ancient Hindu astrologers, the world would end when the sun, the moon, and the 5 other planets then known (Mercury, Venus, Mars, Jupiter, and Saturn) seemed to come together in the same place in the sky. In fact, they thought, as do

Chapter 7: METHODS OF COUNTING

modern astrologers, that the coming together of just 2 of these heavenly bodies influenced events on the earth.

Abraham ben Ezra, a mathematician of the twelfth century, figured out the number of combinations of 2 or more heavenly bodies that might come together.

Find the number of combinations of the 7 heavenly bodies that contain:

6. 2 of them.

7. 3 of them.

8. 4 of them.

9. 5 of them.

10. 6 of them.

11. all 7 of them.

12. Add these numbers to find out what number ben Ezra got.

Set III

A 25¢ toll can be paid with as little as 1 coin (a quarter) or as many as 25 coins (25 pennies).

What other numbers of coins between 1 and 25 are possible in paying the toll?

Drawing by Jack Hadley; © 1966 Saturday Review, Inc.

Chapter 7 / Summary and Review

In this chapter we have become acquainted with:

The fundamental counting principle (*Lesson 1*)   To find the number of ways of making several decisions in succession, multiply the numbers of choices that can be made in each decision.

Permutations (*Lessons 2 and 3*)   A permutation is an arrangement of things in a definite order.

In general, $n!$ means to multiply the consecutive numbers from n all the way down to 1.

The number of permutations of n different things is $n!$.

The number of permutations of n things, of which a things are alike, another b things are alike, another c things are alike, and so forth, is

$$\frac{n!}{a! \times b! \times c! \cdots}.$$

Combinations (*Lesson 4*)   A combination is a selection of things in which the order doesn't matter.

To find the number of combinations possible, first use the fundamental counting principle and then divide by the number of ways in which the things can be arranged.

Exercises

Set I

The Morse code, invented about 1838 by Samuel Morse for use in telegraphy, uses patterns of dots and dashes to represent the letters of the alphabet. Part of the international version of the code is shown in the tree diagram below.

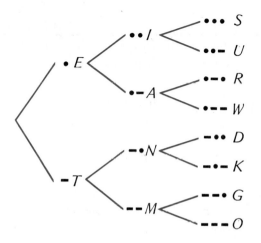

The diagram shows that 2 patterns are possible if just 1 symbol is used: •, which represents E, and —, which represents T.

1. How many patterns can be made by using 2 symbols?

2. How many patterns can be made by using 3 symbols?

3. How many patterns do you think can be made by using 4 symbols?

4. Can the entire alphabet be represented by patterns having 4 symbols or less? Explain.

A manufacturer of stationery is considering printing note paper personalized with initials.

5. How many different types of the note paper must be printed to include every possible set of 2 initials?

6. If the manufacturer decides not to print any note paper with the letters Q, X, or Z, how many types of note paper are eliminated?

"Well, finally! I thought this thing would never end."

Courtesy of Elmer Atkins

In a train made up of an engine followed by 100 cars followed by a caboose, the number of different orders in which the 100 cars can appear on the train is incredibly large. It is

93,326,215,443,944,152,681,699,238,856,266,700,490,715,968,264,381,621,468,592,963,895,217,599,

993,229,915,608,941,463,976,156,518,286,253,697,920,827,223,758,251,185,210,916,864,000,000,000,

000,000,000,000,000.

7. Write a shorter expression for this number.

8. Its value was determined with a computer. What did the computer have to do to find it?

The Pepsi-Cola Company had a "Matching Picture" contest in which the object was to match the pictures of four Miss Americas with their baby pictures. Contestants were allowed to enter as

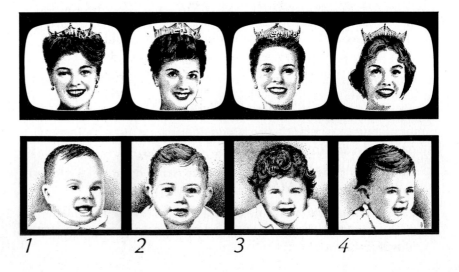

1   2   3   4

many times as they wished and the first prize was $10,000.

9. How many different entries would you have to send in to be sure of having all the pictures matched correctly?

10. Does it seem reasonable to think that you would win $10,000 if you sent in an entry with all of the pictures correctly matched?

Anagrams, the rearrangements of the letters of words to form different words, are the basis for many word puzzles and games. In how many different orders can the letters of the names of the following states be arranged? (*Note:* The order of the letters does not have to result in another word.)

11. ALABAMA

12. TENNESSEE

13. MISSISSIPPI

Set II

The flags of nine nations of the world have the pattern shown at the right with the three stripes in the colors listed below.

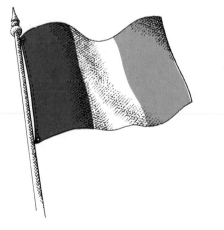

Nation	1st stripe	2nd stripe	3rd stripe
Belgium	black	yellow	red
Chad	blue	yellow	red
France	blue	white	red
Guinea	red	yellow	green
Ireland	green	white	gold
Italy	green	white	red
Ivory Coast	gold	white	green
Mali	green	yellow	red
Nigeria	green	white	green

1. How many different flags having this pattern are possible if the first and third stripes can be black, blue, gold, green, or red and the second stripe can be white or yellow?

2. How many different flags having this pattern are possible if each stripe may be any one of the seven colors but each stripe is a different color?

License plates in California used to consist of a digit, a letter, and four more digits.

3. How many license plates of this type were possible?

4. In 1956, the pattern was changed to three letters followed by three digits. How many license plates of this type were possible?

5. When this pattern was used up, the state began issuing license plates consisting of three digits followed by three letters. How many license plates of this type were possible?

6. In 1978, California began issuing plates consisting of a digit followed by three letters followed by three more digits. How many license plates of this type are possible?

Courtesy of Joseph Zeis

Ordinarily, all 12 members of a jury must agree before a case can be decided. There are 12 different ways in which a jury can be deadlocked when 1 person disagrees with the rest, because any one of the jurors can be that person.

7. How many different combinations of 2 members of a jury can cause the jury to be deadlocked?

8. How many different combinations of 3 members of the jury can disagree with the rest?

Chapter 7: METHODS OF COUNTING

Twenty players are entered in a tennis tournament in which each person is scheduled to play each of the others exactly once.

9. How many matches are necessary?

10. If one more person enters the tournament, how many more matches have to be added to the schedule?

The Baskin-Robbins Ice Cream Stores have 31 flavors of ice cream.

11. How many different 3-scoop ice cream cones are possible if each scoop is a different flavor and you want the scoops put on the cone in a particular order?

12. How many different 3-scoop cones are possible if each scoop is a different flavor and you don't care about their order on the cone?

Set III

The automobile industry offers a tremendous variety of models and features from which to choose. For example, one company manufactures a model that comes in seven colors, with a choice of either cloth or leatherette upholstery.

1. How many different choices of this model are available?

   The company offers the following options for each car:

   > automatic transmission
   >
   > clock
   >
   > radio
   >
   > air conditioner
   >
   > whitewall tires
   >
   > hinged rear side windows
   >
   > sliding steel sunroof

2. How many combinations of options are available for each car? (Notice that, in each case, you have two choices: you can take it or leave it.)

3. If a dealer wanted to have one car with each possible combination of color, upholstery, and options on his lot, how many cars would he have to have?

Chapter 7 / Problems for Further Exploration

Lesson 1

1. The ZIP code for routing mail was introduced by the United States Postal Service in 1963. It consists of 5 digits ranging from 00601 for Adjuntas, Puerto Rico, to 99950 for Ketchikan, Alaska.

   a) How many 5-digit ZIP codes are possible if each digit can be any one of the digits from 0 through 9?

   b) If the ZIP code were changed to 5 letters, how many codes would be possible?

   In 1980, the population of the United States was about 220,000,000.

   c) If the ZIP code consisted of digits only, how many digits would be necessary in order for every person in the United States to have his or her own personal code?

   d) If it consisted of letters only, how many letters would be necessary?

2. The Bell Telephone companies use a color code to keep track of the wires in their cables. The cables contain 50 wires, each of which is coded with two colors. The two colors appear as stripes on the wires: one color as a wide stripe and the other as

a narrow one. The list below shows the colors of the wires with the color of the wide stripe listed first.

1. blue-white	18. blue-yellow	35. gray-black
2. gray-yellow	19. black-gray	36. black-brown
3. black-blue	20. white-orange	37. white-green
4. orange-purple	21. red-green	38. green-purple
5. green-white	22. gray-white	39. purple-brown
6. yellow-orange	23. orange-red	40. yellow-green
7. brown-black	24. purple-green	41. brown-purple
8. orange-white	25. brown-yellow	42. gray-red
9. red-orange	26. green-red	43. orange-yellow
10. blue-black	27. yellow-brown	44. green-black
11. white-gray	28. red-brown	45. blue-purple
12. purple-orange	29. purple-blue	46. red-gray
13. brown-white	30. blue-red	47. purple-gray
14. white-blue	31. black-green	48. white-brown
15. black-orange	32. brown-red	49. gray-purple
16. yellow-blue	33. yellow-gray	50. orange-black
17. green-yellow	34. red-blue	

Can you figure out the system by which the colors are paired? If so, what is it and why are there exactly 50 possibilities according to the system?

Lesson 2

1. "Change ringing" is the practice of ringing a set of bells in every possible permutation. Each permutation is called a *change* and the maximum number of permutations possible for a given set of bells is called a *peal*.

A set of three bells—A, B, and C, for example—has a peal of 6 changes as shown in the tree diagram below. Bell ringers can

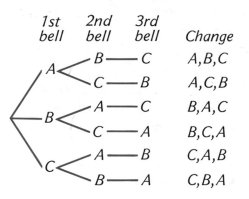

1st bell	2nd bell	3rd bell	Change
A	B — C		A,B,C
	C — B		A,C,B
B	A — C		B,A,C
	C — A		B,C,A
C	A — B		C,A,B
	B — A		C,B,A

ring a peal of three bells at the rate of about one change each second; a peal, then, takes about 6 seconds.

a) How many changes are there in a peal of 4 bells? About how long would such a peal take?

b) How many changes are there in a peal of 7 bells? About how long would such a peal take?

c) How many changes would there be in a peal of 12 bells? Could someone ring a peal of 12 bells? Explain.

2. The figure above illustrates a lip-reading puzzle invented many years ago by the American puzzle maker Sam Loyd. According to Loyd, the picture shows the 12 boys just as each one was beginning to say his name. The names are Alfred, Alden, Eastman, Oom, Arthur, Luke, Fletcher, Matthew, Theodore, Richard, Shirmer and Hisswald.

a) In how many different ways can the names of the boys be guessed?

b) If someone was certain of the names of half of the boys, in how many different ways could the names of the remaining boys be guessed?

c) Can you figure out which boy has which name?

Lesson 3

1. As every baseball fan knows, the World Series is held early in October each year between the winners of the American and National League pennants. The two teams play until one team has won 4 games.

   Suppose that the Yankees are playing the Dodgers. How many different orders of winners in the games played are possible?

   First, consider the number of orders possible in which the Yankees win the series. The number of games that it takes to win the series—that is, 4, 5, 6, or 7—depends on whether the Dodgers win 0, 1, 2, or 3 games.

   If it takes only 4 games, the Yankees win all 4, and there is only 1 possible order of winners: YYYY.

   If the series lasts 5 games, and the Yankees win it, there are several possible orders of winners: DYYYY, YDYYY, YYDYY, and YYYDY. Because the winner of the series always wins the last game, the last letter in the row is Y. Then the problem is one of finding the number of orders of the other 4 letters in the row: 3 Y's and 1 D.

   $$\frac{4!}{3!} = \frac{4 \times 3 \times 2 \times 1}{3 \times 2 \times 1} = 4$$

   a) How many orders of winners are possible if the series lasts 6 games? (The Yankees win the last game, and so the problem is: In how many ways can 5 letters, 3 Y's and 2 D's, be arranged?)

   b) How many orders of winners are possible if the Yankees win the series on the seventh game?

   c) Copy and complete the following table:

   Ways in which the Yankees win the series

Number of games in the series	4	5	6	7
Number of orders of winners	1	4	▦	▦

d) The sum of the numbers on the second line of the table is the number of orders of winners possible if the Yankees win. How many is that?

e) How many different orders of winners are possible if the Dodgers win the series?

f) How many different orders of winners in a World Series are possible altogether?

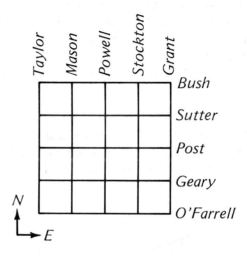

2. The map above shows part of downtown San Francisco. Someone who wants to go from the corner of Taylor and O'Farrell to the corner of Powell and Sutter has a choice of 10 direct routes. They are shown in the figures below.

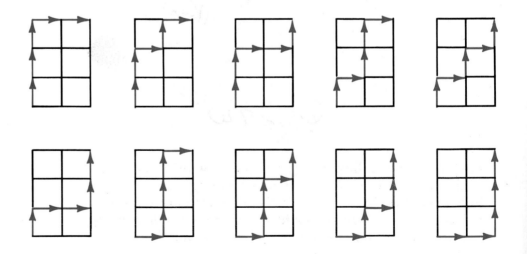

Chapter 7: METHODS OF COUNTING

The fact that there are 10 routes can be discovered without making a diagram. Going from Taylor and O'Farrell to Powell and Sutter requires traveling 3 blocks north and 2 blocks east. If each block north is represented by an N and each block east by an E, each route can be represented by an arrangement of the letters NNNEE.

a) Show, using the fundamental counting principle, that there are 10 such arrangements.

b) Now copy the map and figure out the number of direct routes from Taylor and O'Farrell to each of the other corners. Write the numbers on the corners as illustrated in the map at the right.

c) What patterns do you see in the numbers on the finished figure?

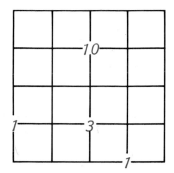

Lesson 4

1. Elaborate designs can be formed by winding string around a set of nails arranged in a circle. The figure below shows the pattern that would result from winding string around 21 nails in every possible way.

a) How many windings (lines) are in the figure? Explain how the answer can be found without counting them.

b) Suppose that the number of nails in the figure were doubled. How many windings would result from winding string around a set of 42 nails in every possible way?

2. A "double-six" set of dominoes contains the 7 "doubles" dominoes shown below.

The rest of the dominoes, called "singles," have different numbers of spots on each square. Examples of "singles" are shown below.

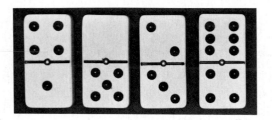

a) How many singles dominoes are there if every possible combination of squares is included?

b) How many dominoes are in a complete "double-six" set of dominoes?

The squares of a "double-nine" set of dominoes either are blank or have from 1 to 9 spots.

c) How many dominoes are in a complete "double-nine" set?

The squares of a "double-twelve" set of dominoes either are blank or have from 1 to 12 spots.

d) How many dominoes are in a complete "double-twelve" set?

Chapter 8

THE MATHEMATICS
OF CHANCE

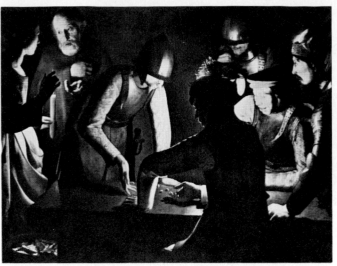

Georges de la Tour, *Le Reniement de Saint Pierre* [St. Peter's denial]; Musée des Beaux-Arts

Lesson 1

Probability: The Measure of Chance

The branch of mathematics called probability theory originated in the sixteenth century, when the Italian physician and mathematician Jerome Cardan wrote the first book on the subject, *The Book on Games of Chance*. Because games of chance have been played for thousands of years, it is not surprising that "the mathematics of chance" was at first used primarily to solve problems dealing with gambling. As the subject became better understood, scientists began to apply it to other fields of study. The applications have proved to be so significant that, according to one twentieth-century mathematician, probability theory has become "a cornerstone of all of the sciences."*

Probabilities are expressed as numbers on a scale ranging from 0 to 1 or as percentages on a scale ranging from 0% to 100%. Both scales are shown in the figure at the top of the next page. The larger the probability of a given event, the more likely it is that the event will occur.

The probabilities of some events can be calculated as the ratio of two numbers. Consider, for example, the probabilities associated with throwing a die. A die is a cube with six faces marked with

---

* Mark Kac, in his article titled "Probability," *Scientific American*, September 1964, pages 92–108.

```
     0   0.1  0.2  0.3  0.4  0.5  0.6  0.7  0.8  0.9  1.0
     ├────┼────┼────┼────┼────┼────┼────┼────┼────┼────┤
    0%  10%  20%  30%  40%  50%  60%  70%  80%  90% 100%
     ↑                      ↑                        ↑
 No chance              Even                    Absolute
   at all              chances                  certainty
```

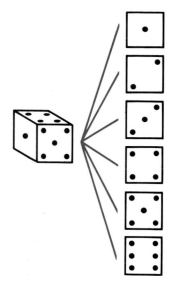

from one to six spots. If the die is a fair one, it is equally likely that any *one* of the *six* faces will turn up when it is thrown. The probability of a particular number turning up is $\frac{1}{6}$ or, as a percentage,

$$\frac{1}{6} \times 100 = \frac{100}{6} \approx 17\%.^*$$

In general, this method of finding a probability requires finding two numbers and dividing one by the other:

▶**Probability of an event** $= \dfrac{\text{number of ways in which the event can occur}}{\text{total number of ways}}$.

This formula works whenever chances are equally likely, as in throwing dice.

Probabilities used in betting are usually expressed as *odds*.

▶If the number of ways in which an event can occur is x and the number of ways in which it does not occur is y, then the **odds against the event** are said to be in the ratio of "y to x."

When a die is thrown, the odds against a particular number turning up are 5 to 1 because there are five ways in which some other number can turn up compared with one way in which the given number can turn up.

The methods of finding probabilities and odds that we have just considered are usually applied to games of chance. Probabilities are also determined by statistical methods and by educated guessing. A medical report that says that there is a 1% chance of a woman giving birth having twins is giving a probability based on statistical information. A weather report that says that there is a 60% chance of rain is giving a probability based on an educated guess that it is somewhat more likely to rain than not.

_____

* See pages 631–632 if you do not remember how to figure out percentages.

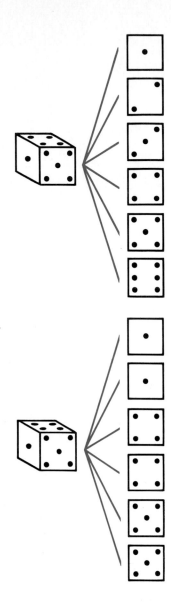

Exercises

Unless otherwise directed, express each probability as a fraction in simplest terms.

Set I

A die is thrown once.

1. If each face is equally likely to turn up, what is the probability of getting a 4?

2. What are the odds against getting a 4?

3. What is the probability of an odd number turning up?

4. What are the odds against an odd number turning up?

A die is misspotted with two faces marked with 1 spot, two faces marked with 4 spots, and two faces marked with 5 spots. It is thrown once.

5. What is the probability of getting a 4 with this die?

6. What are the odds against getting a 4?

7. What is the probability of an odd number turning up?

8. What are the odds against an odd number turning up?

Probabilities for various events occurring in major league baseball games are listed below.

The losing team doesn't score	0.15
Both games of a double-header are won by the same team	0.47
The winning run is scored in the last inning	0.19
No bases are stolen during a game	0.36
A game goes into extra innings	0.09
The home team wins the game	0.53
A game has at least one home run	0.64

9. Which one of these events is the most likely to occur?

10. Which one is the least likely to occur?

11. Which two are the most unpredictable?

*A nineteenth-century drawing
of a shell game by A.B. Frost*

In a shell game, the operator covers a pea with one of three shells and then moves the shells about. The player tries to keep track of the pea and then bets the operator that he knows where it is. Suppose that the operator moves the shells so rapidly that the player cannot follow them and must guess where the pea is.

12. If the game is an honest one, what is the probability that the player will guess the correct shell?*

13. If you play the game many times, how many games would you expect to win in the long run?

14. How many games would you expect to lose?

Lotteries are contests in which each entry has an equal chance of winning. Many such contests allow a person to enter more than once. Suppose that you have entries in two lotteries as shown here.

Lottery	Your number of entries	The rest of the entries
A	12	68
B	15	110

15. What is the probability that you will win lottery A?

16. What are the odds against your winning lottery A?

---

* The game is not usually played honestly.

17. What is the probability that you will win lottery B?

18. What are the odds against your winning lottery B?

19. Which lottery are you more likely to win?

Suppose that the eight cards shown here are shuffled and then one card is chosen at random. What is the probability that the card is

20. the king of diamonds?

21. not the king of diamonds?

22. a heart?

23. an ace?

24. a face card?

Set II

If someone in New York City is treated in an emergency room for a bite, the probability that the person was bitten by a dog is $\frac{9}{10}$.

1. Express this probability as a percentage.

The probability that the person was bitten by a cat is $\frac{1}{20}$.

2. Express this as a percentage.

The probability that the person was bitten by another person is $\frac{1}{25}$.

3. Express this as a percentage.

4. Do the three percentages that you have just determined add up to 100%?

5. Do you think they should? Explain.

The ways in which three questions on a true-or-false test can be answered correctly or incorrectly are shown in the tree diagram at the right. R stands for right and W stands for wrong.

If the answers are chosen at random, what is the probability that

6. all three questions are answered correctly?

7. none of the questions is answered correctly?

8. more questions are answered correctly than incorrectly?

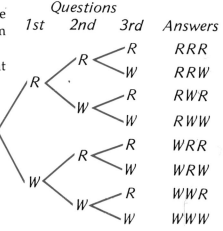

| Questions | | | |
1st	2nd	3rd	Answers
	R	R	RRR
		W	RRW
R		R	RWR
	W	W	RWW
	R	R	WRR
		W	WRW
W		R	WWR
	W	W	WWW

Courtesy of *Gambling Times*

An American roulette wheel has 38 compartments around its rim. Two of these are numbered 0 and 00 and are green; the others are numbered from 1 to 36, of which half are red and half are black.

When the wheel is spun in one direction, a small ivory ball is rolled in the opposite direction along its rim. If the wheel is a fair one, the chances of the ball falling into any one of the 38 compartments as it slows down are equally likely.

Express the probability of each of the following coming up both as a fraction and as a percentage:

9. The number 7.

10. A black number.

11. A red number.

12. A red number, if all 26 numbers that had come up previously were black.*

13. Any number from 1 through 12.

14. A green number.

15. A number that is not green.

The pricemaker at a racetrack estimates the probability of each horse winning as a percentage and then changes it into odds. For example, if the probability that a certain horse will win is 40%, the odds are found as follows:

$$40\% = \frac{40}{100} \quad \begin{array}{l} \leftarrow \text{ chances of winning} \\ \leftarrow \text{ total number of chances} \end{array}$$

$$100 - 40 = 60 \leftarrow \text{ chances of not winning}$$

$$\frac{\text{chances of not winning}}{\text{chances of winning}} = \frac{60}{40} = \frac{3}{2}$$

The odds on this horse are 3 to 2.

16. Use this method to complete the table at the left.

Probability that horse will win	Odds on horse
10%	9 to 1
20%	
30%	
40%	3 to 2
50%	
60%	
70%	
80%	
90%	

Set III

Triskaidekaphobia is fear of the number 13. Friday falls on the thirteenth day of the month 48 times every 28 years.

1. What is the probability that any given Friday falls on the thirteenth?

2. What are the odds against it?

---

* This actually happened on August 18, 1913, at the Casino in Monte Carlo.

Chapter 8: THE MATHEMATICS OF CHANCE

By permission of Johnny Hart and Field Enterprises, Inc.

Lesson 2

Binomial Probability

Which of the following is the better bet? If a coin is tossed four times, would you bet in favor of it coming up heads half the time and tails half the time or would you bet against it?

If a coin is tossed once, there are two equally likely outcomes: it can come up heads or tails. Each outcome has a probability of $\frac{1}{2}$.

If the coin is tossed twice, there are four equally likely outcomes. They are illustrated in the tree diagram at the right. The probability that the coin will come up heads once and tails once is 50% because there are 2 ways out of the 4 of this happening and

$$\frac{2}{4} = \frac{1}{2} = 50\%.$$

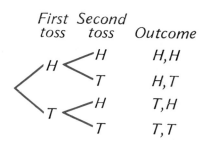

431

Continuing the tree diagram to illustrate four tosses, we find that there are 16 equally likely outcomes.

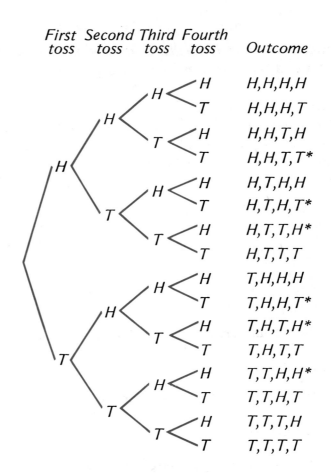

First toss	Second toss	Third toss	Fourth toss	Outcome
			H	H,H,H,H
		H	T	H,H,H,T
	H	T	H	H,H,T,H
			T	H,H,T,T*
H		H	H	H,T,H,H
			T	H,T,H,T*
	T	T	H	H,T,T,H*
			T	H,T,T,T
		H	H	T,H,H,H
	H		T	T,H,H,T*
		T	H	T,H,T,H*
T			T	T,H,T,T
		H	H	T,T,H,H*
	T		T	T,T,H,T
		T	H	T,T,T,H
			T	T,T,T,T

Of these 16 possible outcomes, there are exactly 6 (marked with asterisks in the diagram) in which the coin turns up heads half the time and tails half the time. This means that the probability of this occurring is

$$\frac{6}{16} = \frac{3}{8} = 37.5\%.$$

The probability that the coin will not turn up heads half the time and tails half the time is

$$\frac{10}{16} = \frac{5}{8} = 62.5\%.$$

Chapter 8: THE MATHEMATICS OF CHANCE

So betting *against* a coin turning up half heads and half tails when it is tossed four times is the better bet.

The chances of getting heads or tails when a coin is tossed is one example of *binomial* probability.

▶The probabilities in any situation in which there are *two possible outcomes* for each part are called **binomial.**

Examples of binomial probabilities vary widely. They include such problems as predicting the chances of a couple having boys or girls and figuring out how likely it is that a baseball player will get a certain number of hits when he comes to bat.

Exercises

Set I

The tree diagram at the right shows the possibilities of boys and girls in a family with two children. We will assume that the probability of a child being a boy is $\frac{1}{2}$, even though it is actually slightly

more than $\frac{1}{2}$.*

		Children
First	Second	in
child	child	family
B	B	B,B
	G	B,G
G	B	G,B
	G	G,G

1. Refer to the diagram to copy and complete the following table.

Number of boys	2	1	0
Number of possibilities	1	▨	▨
Probability	$\frac{1}{4}$	▨	▨
Percent probability	25%	▨	▨

What is the percent probability that, in a family with two children,

2. both children are of the same sex?

3. at least one child is a boy?

---

*The probability that a child born in the United States is a boy is 0.513.

This tree diagram shows the possibilities of hits for three successive times that a baseball player comes to bat. (H stands for hit and O for no hit.)

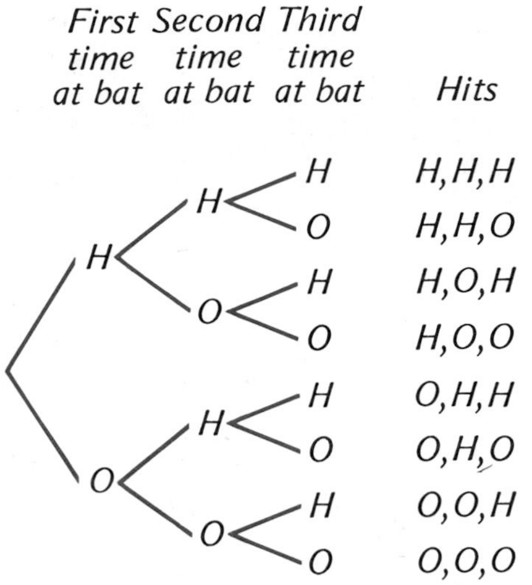

First time at bat	Second time at bat	Third time at bat	Hits
H	H	H	H,H,H
		O	H,H,O
	O	H	H,O,H
		O	H,O,O
O	H	H	O,H,H
		O	O,H,O
	O	H	O,O,H
		O	O,O,O

For simplicity, we will assume that the probability of a hit each time at bat is $\frac{1}{2}$.*

4. Refer to the diagram to copy and complete the following table.

Number of hits	3	2	1	0
Number of possibilities	1	▦	▦	▦
Probability	$\frac{1}{8}$	▦	▦	▦
Percent probability	12.5%	▦	▦	▦

What is the percent probability that, when this baseball player comes to bat three times in succession, he will get

5. at least 2 hits?

6. at least 1 hit?

The tree diagram at the top of the next page shows the possibilities of it snowing during four successive days at a ski resort. (S stands for snow and O for no snow.)

---

* We are assuming that the player's batting average is 0.500. As baseball fans know, this is considerably higher than that of the typical player. The world record average for a season is 0.438.

Chapter 8: THE MATHEMATICS OF CHANCE

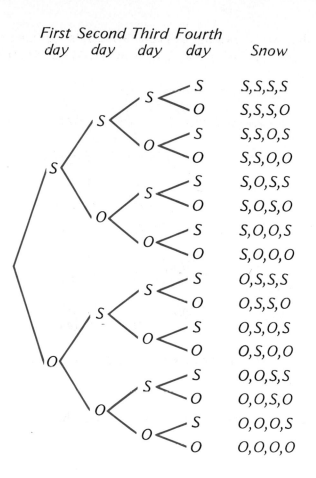

	First day	Second day	Third day	Fourth day	Snow
				S	S,S,S,S
			S	O	S,S,S,O
			O	S	S,S,O,S
		S		O	S,S,O,O
	S		S	S	S,O,S,S
		O		O	S,O,S,O
			O	S	S,O,O,S
				O	S,O,O,O
			S	S	O,S,S,S
		S		O	O,S,S,O
			O	S	O,S,O,S
	O			O	O,S,O,O
		O	S	S	O,O,S,S
				O	O,O,S,O
			O	S	O,O,O,S
				O	O,O,O,O

We will assume that the probability of snow on a given day is $\frac{1}{2}$.

7. Refer to the diagram to copy and complete the following table. Round each probability to the nearest percent.

Number of days that it snows	4	3	2	1	0
Number of possibilities	1				
Probability	$\frac{1}{16}$				
Percent probability	6%				

8. What is the percent probability that, in four successive days, it will snow on at least half of the days?

9. Which is more likely: that it will snow on exactly one day or that it will snow on exactly three days?

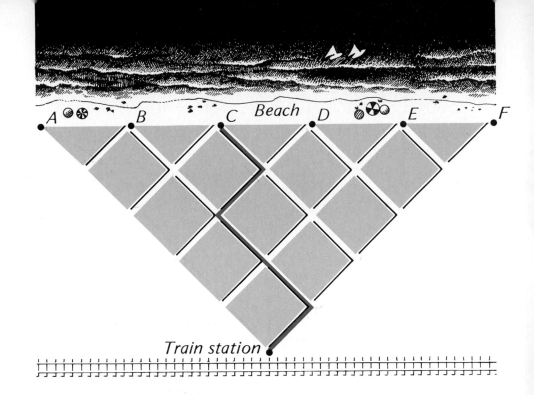

This map represents the streets of a beach town that lead from a train station to the beach.* An excursion train arrives at the station and the passengers head for the beach. No one knows whether to take the street at the left or right at each intersection, so we will assume that the probability of each person going in either direction is the same: $\frac{1}{2}$. One possible path is shown on the map.

The tree diagram on the facing page shows the possibilities of someone going left or right at each of the five intersections encountered between the station and the beach.

10. Refer to the diagram to copy and complete the following table. Round each probability to the nearest percent.

Number of intersections at which path goes to the left	5	4	3	2	1	0
Number of possibilities	1	▨	▨	▨	▨	▨
Probability	$\frac{1}{32}$	▨	▨	▨	▨	▨
Percent probability	3%	▨	▨	▨	▨	▨

---

*Adapted from a story in *Mathematics in Your World* by K. W. Menninger (Viking, 1962).

Chapter 8: THE MATHEMATICS OF CHANCE

First inter- section	Second inter- section	Third inter- section	Fourth inter- section	Fifth inter- section	Path

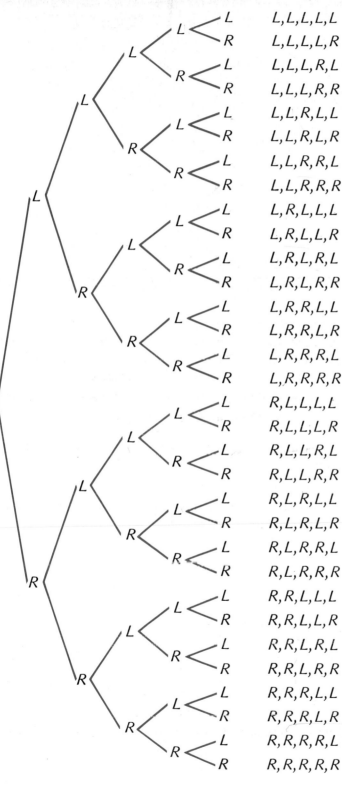

					L,L,L,L,L
					L,L,L,L,R
					L,L,L,R,L
					L,L,L,R,R
					L,L,R,L,L
					L,L,R,L,R
					L,L,R,R,L
					L,L,R,R,R
					L,R,L,L,L
					L,R,L,L,R
					L,R,L,R,L
					L,R,L,R,R
					L,R,R,L,L
					L,R,R,L,R
					L,R,R,R,L
					L,R,R,R,R
					R,L,L,L,L
					R,L,L,L,R
					R,L,L,R,L
					R,L,L,R,R
					R,L,R,L,L
					R,L,R,L,R
					R,L,R,R,L
					R,L,R,R,R
					R,R,L,L,L
					R,R,L,L,R
					R,R,L,R,L
					R,R,L,R,R
					R,R,R,L,L
					R,R,R,L,R
					R,R,R,R,L
					R,R,R,R,R

Look at the map of the town. Where along the beach will someone end up

11. who takes the street at the left at five intersections?

12. who takes the street at the left at three intersections?

13. If each person from the excursion train remains at the point at which he or she arrives at the beach, what do you think the arrangement of the people along the beach will be like? Explain.

Lesson 2: Binomial Probability

437

Czechoslovakian Academy of Sciences

Set II

The Austrian monk Gregor Mendel founded the science of genetics in the nineteenth century, when he experimented with the crossbreeding of plants. His discoveries, described in a report titled "Experiments with Plant Hybrids," were perhaps the first in which the theory of probability was applied to science.

In an experiment with garden peas, Mendel crossbred plants having round peas with plants having wrinkled peas. All the offspring had round peas. When these plants were bred with each other, 75% of the offspring had round peas and the remaining 25% had wrinkled peas.

To explain the appearance of both types of peas among these offspring, Mendel assumed that a pair of genes in each plant determined the type of pea that it produced. The pairs and types are shown below.

Genes	RR	RW	WR	WW
Type of pea	⊙	⊙	⊙	✿

The genes are labeled R for round and W for wrinkled in the diagram for simplicity. Note that a plant having a gene of each type produces round peas.

Each parent plant contributes one of its genes to each of its offspring. The crossbreeding of the first generation to produce the second is shown in this tree diagram.

First generation		*Second generation*
Gene from first plant	Gene from second plant	Offspring

```
        W       RW
    R <
        W       RW

        W       RW
    R <
        W       RW
```

1. According to this diagram, what is the probability of getting a plant with round peas in the second generation?

2. What is the probability of getting a plant with wrinkled peas?

The answers to these questions explain why all the offspring in the first stage of Mendel's experiment had round peas.

The crossbreeding of the second generation to produce the third is shown in this tree diagram.

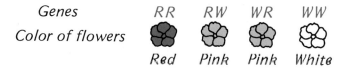

| *Second generation* | | *Third generation* |
| *Gene from first plant* | *Gene from second plant* | *Offspring* |

R   R   RR
R   W   RW
W   R   WR
W   W   WW

3. According to this diagram, what is the probability of getting a plant with round peas in the third generation?

4. What is the probability of getting a plant with wrinkled peas in the third generation?

Several decades after Mendel's work, the German botanist Karl Correns did an experiment with Japanese four-o'clocks. He cross-bred plants having red flowers with plants having white flowers. All the offspring plants had pink flowers. When the plants with pink flowers were bred with each other, 25% of the plants in the third generation had red flowers, 50% had pink flowers, and 25% had white flowers.

Suppose that a pair of genes in each plant determines the color of its flowers and that R and W represent red and white color genes respectively. The pairs and types are shown below.

| *Genes* | *RR* | *RW* | *WR* | *WW* |
| *Color of flowers* | Red | Pink | Pink | White |

Look again at the tree diagram on page 438.

5. According to the diagram, what is the probability of getting a plant with pink flowers in the second generation?

6. What is the probability of getting a plant with red or white flowers in the second generation?

Look again at the tree diagram on page 439.

7. According to this diagram, what is the probability of getting a plant with pink flowers in the third generation?

8. What is the probability of getting a plant with red flowers in the third generation? A plant with white flowers?

Set III

Several years ago a remarkable discovery was made about the patterns in which a coin can turn up heads and tails when it is tossed several times in a row.*

If a coin is tossed three times, there are eight equally probable outcomes, shown in the tree diagram below.

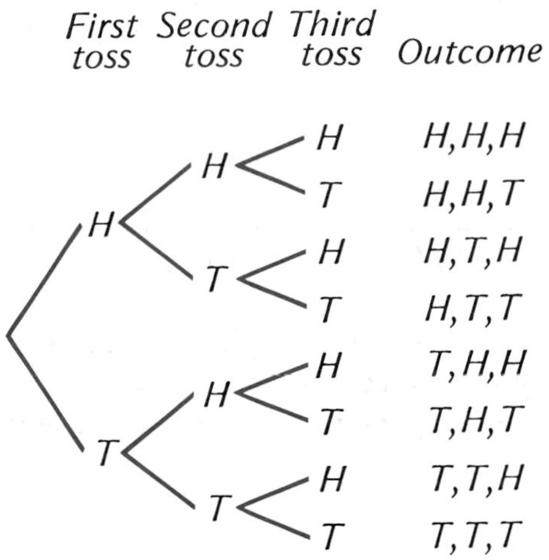

The discovery concerns a game in which each of two players bets on one of these outcomes. They then toss a coin repeatedly until one of the two outcomes occurs. The player who chose that outcome wins.

Suppose, for example, that the first player chooses H, T, H and the second player chooses H, H, T. If the coin is flipped and turns up T, T, H, H, T, the second player wins.

The remarkable thing about the game is that, *regardless of what outcome the first player chooses, the second player can choose one that is more likely to occur first.*

---

* The discovery was made by Walter Penney and is discussed by Martin Gardner in his "Mathematical Games" column, *Scientific American,* October 1974.

The table below shows what the second player should choose for each choice that the first player might make.

If the first player chooses	the second player should choose
H, H, H	T, H, H
H, H, T	T, H, H
H, T, H	H, H, T
H, T, T	H, H, T
T, H, H	T, T, H
T, H, T	T, T, H
T, T, H	H, T, T
T, T, T	H, T, T

Try playing the game with someone, following the strategy in this table. Play the game at least ten times and keep a record as shown below of what happens.

Other player	Me	What happened	Winner
H,T,H	H,H,T	T,T,H,H,T	Me

(If you do not have someone to play the game with, play it by yourself, making up choices for an imaginary opponent.) Regardless of what the other person does, the odds are in your favor so that you should win most of the time.

Lesson 3

Pascal's Triangle

All eight children in this remarkable family are girls. Because the probabilities of a child being a boy or girl are about the same, it would seem that in a family of this size, four girls and four boys would be much more likely. Exactly how do the probabilities of a family having eight girls and a family having four girls and four boys compare?

Although this could be figured out from a tree diagram, there is an easier way. It is based on a pattern of numbers so old that no one knows who first discovered it. The pattern, called Pascal's triangle, is named after the great seventeenth-century mathematician Blaise Pascal. One of the originators of probability theory, Pascal wrote a book about the triangle and its properties. Pascal's triangle looks like the figure at the top of the next page.

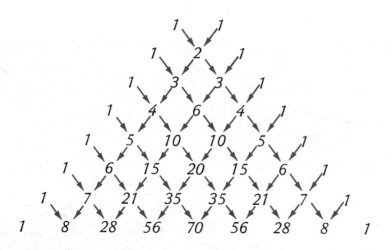

Each number within the triangle is found by adding the pair of numbers in the row above it at the left and right. The triangle can be continued indefinitely by writing a 1 at both ends of each new row and then adding each pair of numbers in the preceding row.

Each row of the triangle contains the numbers of ways of getting each possible outcome for a situation in which the probabilities are binomial. The fourth row, for example, contains the numbers of ways in which a coin can turn up when it is tossed four times.*

Number of heads	4	3	2	1	0
Number of ways	1	4	6	4	1

Because the total number of ways in this row is

$$1 + 4 + 6 + 4 + 1 = 16,$$

the probabilities for tossing a coin four times are:

Number of heads	4	3	2	1	0
Probability	$\frac{1}{16}$	$\frac{4}{16}$	$\frac{6}{16}$	$\frac{4}{16}$	$\frac{1}{16}$

The question about the probabilities for a family with eight children can be answered by looking at the eighth row of the triangle.

---

* The ways are illustrated in the tree diagram on page 432.

Number of girls	8	7	6	5	4	3	2	1	0
Number of ways	1	8	28	56	70	56	28	8	1

The total number of ways in this row is

$$1 + 8 + 28 + 56 + 70 + 56 + 28 + 8 + 1 = 256,$$

and so the probabilities for different numbers of girls and boys in a family of eight children are:

Number of girls	8	7	6	5	4	3	2	1	0
Probability	$\frac{1}{256}$	$\frac{8}{256}$	$\frac{28}{256}$	$\frac{56}{256}$	$\frac{70}{256}$	$\frac{56}{256}$	$\frac{28}{256}$	$\frac{8}{256}$	$\frac{1}{256}$

The probability of eight girls is $\frac{1}{256}$, or less than 1%. The probability of four girls and four boys is $\frac{70}{256}$, or about 27%.

Exercises

Set I

1. Copy the eight rows of Pascal's triangle shown on page 443 and then add two more rows.

2. At the left side of your triangle, write the numbers of the rows in a column. At the right side, write the sums of the numbers in each row in another column.

The top of your triangle should look like this.

Number of row						Sum of numbers in row
1		1	1			2
2	1	2	1			4
3	1	3	3	1		8

3. What kind of sequence do the numbers in the column of sums form?

In the figure below, the numbers in the first four rows of Pascal's triangle have been replaced by the possible ways in which families can have from one to four children.

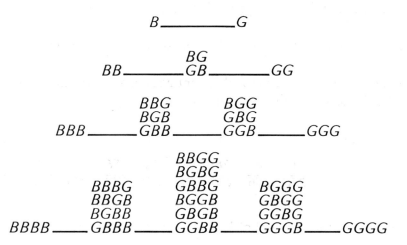

The third row can be used to find the probabilities of boys and girls in a family with three children.

Number of boys and girls	3–0	2–1	1–2	0–3
Number of ways	1	3	3	1
Probability	$\frac{1}{8}$	$\frac{3}{8}$	$\frac{3}{8}$	$\frac{1}{8}$

4. Use the fourth row to complete the following table for a family with four children.

Number of boys and girls	4–0	3–1	2–2	1–3	0–4
Number of ways					
Probability					

Use this table and the table for three children above to answer the following questions.
Which is more likely:

5. that a family with three children has three boys, or that a family with four children has three boys and a girl?

6. that a family with three children has two girls and a boy, or that a family with four children has two girls and two boys?

Suppose that a true-or-false test has ten questions and that you know the answers to five of them. The possible outcomes of guessing the answers to the other five questions are shown in the figure below.

```
                RRRWW   WWWRR
                RRWRW   WWRWR
                RRWWR   WWRRW
                RWRRW   WRWWR
                RWRWR   WRWRW
        RRRRW   RWWRR   WRRWW   WWWWR
        RRRWR   WRRRW   RWWWR   WWWRW
        RRWRR   WRRWR   RWWRW   WWRWW
        RWRRR   WRWRR   RWRWW   WRWWW
RRRRR——WRRRR——WWRRR——RRWWW——RWWWW——WWWWW
```

7. To which row of Pascal's triangle does this figure correspond?

8. Use it to complete the following table for the answers being guessed. Express each probability to the nearest percent.

Number of correct answers	5	4	3	2	1	0
Number of ways	▦	▦	▦	▦	▦	▦
Probability	▦	▦	▦	▦	▦	▦
Percent probability	▦	▦	▦	▦	▦	▦

Use your table to answer the following questions.

9. What numbers of correct answers are least likely to be guessed?

10. What numbers of correct answers are the most likely to be guessed?

11. To pass the test, at least seven of the questions must be answered correctly. What is the probability that you will pass the test?

Two tennis players are perfectly matched in ability so that the probability of either player winning a game that they play is the same: $\frac{1}{2}$.

If they play a set of six games, there are seven outcomes possible for the number of games each player has won: 6-0, 5-1, 4-2, 3-3, 2-4, 1-5, 0-6. These outcomes, however, are not equally likely.

12. Refer to Pascal's triangle to complete the following table for the six games. Express each probability to the nearest percent.

Final score	6–0	5–1	4–2	3–3	2–4	1–5	0–6
Number of ways							
Probability							
Percent probability							

Use your table to answer the following questions.

13. What scores are the least likely at the end of the six games?

14. What score is the most likely?

15. What is the probability that, at the end of the six games, there is a tie so that additional games have to be played to determine the winner?

16. Which is more likely: that at the end of the six games there is a tie or that one player has won by winning four of the games?

In the United States, the probability that a newborn baby will live to be at least 74 years old is $\frac{1}{2}$. Refer to Pascal's triangle to determine the following probabilities for a group of eight people born on the same day.
    What is the percent probability that

17. exactly four of the people will be living at the age of 74?

18. at least six of the people will be living at the age of 74?

Set II

This painting shows an unlikely situation: a set of pennies is being tossed onto a table in which every coin has turned up heads.

1. Use Pascal's triangle to complete the following table of probabilities of getting various numbers of heads when ten coins are tossed. Round each probability to the nearest percent.

Number of heads	10	9	8	7	6	5	4	3	2	1	0
Number of ways	1	10									
Percent probability	0%	1%									

2. Graph the information in your table, representing the numbers of heads on the x-axis and the percent probabilities on the y-axis. Convenient scales for the axes are shown at the left. After you have plotted the points, use a ruler to connect them in order with straight line segments.

3. If ten coins are tossed, what number of heads is most likely to turn up?

4. Would it be better to bet *in favor of* or *against* that number turning up? Explain.

Percent probability

30
20
10

O   5   10

Number of heads

Here is a graph of the probabilities of getting various numbers of heads when twenty coins are tossed.

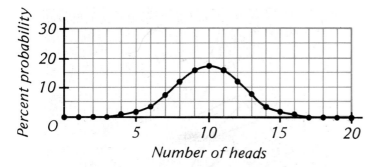

5. In what way is this graph similar to the one that you drew?

6. In what ways is it different?

Refer to the graph above to answer the following questions.

7. If twenty coins are tossed, what number of heads is most likely to turn up?

8. What is the probability of that number turning up? (Make an estimate from the graph.)

9. Which would be more surprising: equal numbers of heads and tails turning up when *ten* coins are tossed or when *twenty* coins are tossed?

10. What happens to the probability of equal numbers of heads and tails turning up as the number of coins being tossed increases?

Set III

Pascal's triangle has other interesting properties in addition to those related to probability. In fact, it has so many that Pascal said that more properties had been omitted from his book on the triangle than had been included.

1. By multiplying, find the values of 11^2, 11^3, and 11^4.

2. What do these numbers have to do with Pascal's triangle?

3. Now find the value of 11^5. Does it fit the pattern?

One sloping row of numbers in Pascal's triangle is shown in color in the figure below.

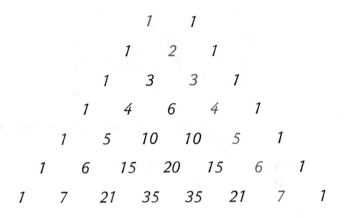

4. Starting at the top of this row, what is the sum of:

        its first two numbers?

        its first three numbers?

        its first four numbers?

        its first five numbers?

        its first six numbers?

5. What do you notice about these answers?

    Another sloping row of numbers is shown in color in the figure below.

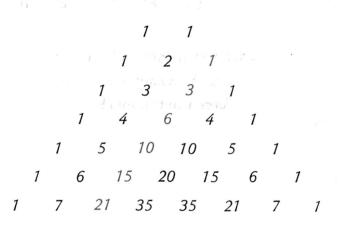

6. Starting at the top of this row, what is the sum of:

its first and second numbers?

its second and third numbers?

its third and fourth numbers?

its fourth and fifth numbers?

its fifth and sixth numbers?

7. These sums form a certain number sequence. What sequence is it?

The numbers in the figure below have been separated into a set of sloping rows.

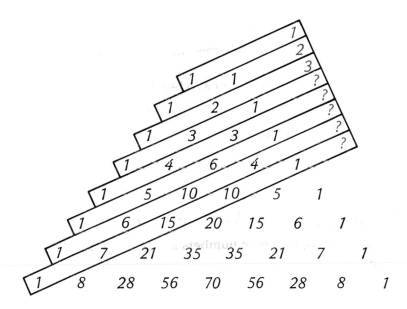

The sums of the numbers in the first three rows are 1, 2, and 3.

8. What are the sums of the numbers in the next five rows?

9. These eight sums form a certain number sequence. What sequence is it?

Dice Games and Probability

Red die	White die	Sum
1	1	2
	2	3
	3	4
	4	5
	5	6
	6	7
2	1	3
	2	4
	3	5
	4	6
	5	7
	6	8
3	1	4
	2	5
	3	6
	4	7
	5	8
	6	9
4	1	5
	2	6
	3	7
	4	8
	5	9
	6	10
5	1	6
	2	7
	3	8
	4	9
	5	10
	6	11
6	1	7
	2	8
	3	9
	4	10
	5	11
	6	12

By permission of Johnny Hart and Field Enterprises, Inc.

Although games with dice have been played for thousands of years, the game of craps did not originate until about 1890. It has since become the most popular dice game in the world.

According to the rules of the game, the person throwing the dice wins if they come up either 7 or 11. If the dice come up 2 ("snake eyes"), 3, or 12, then the person immediately loses. How do the chances of getting each of these sums compare?

To determine the probabilities in this game, we will consider a pair of dice, one of which is white and the other red. Because there are six ways in which each die can turn up, there are

$$6 \times 6 = 36 \text{ ways}$$

in which the dice can turn up together. They are shown in the tree diagram at the left.

A more convenient way of listing the sum of each of the 36 ways is shown in the table below.

		Red die					
		1	2	3	4	5	6
White die	1	2	3	4	5	6	7
	2	3	4	5	6	7	8
	3	4	5	6	7	8	9
	4	5	6	7	8	9	10
	5	6	7	8	9	10	11
	6	7	8	9	10	11	12

Each way is equally likely, and so the probability of getting any sum can be found by dividing the number of ways of getting that sum by 36. There are six ways in which a 7 can be thrown, and so the probability of rolling a 7 is

$$\frac{6}{36} = \frac{1}{6}.$$

There are only two ways in which an 11 can be thrown, and so the probability of rolling an 11 is

$$\frac{2}{36} = \frac{1}{18}.$$

Getting "snake eyes" is rather unlikely because there is only one way in which a 2 can be thrown; the probability is

$$\frac{1}{36}.$$

The percent probabilities of getting these sums are shown in the table below. (Each is rounded to the nearest percent.)

Sum	2	7	11
Probability	$\frac{1}{36}$	$\frac{1}{6}$	$\frac{1}{18}$
Percent probability	3%	17%	6%

These percentages tell us that, if a pair of dice is rolled 100 times, 2 will come up about three times, 7 about seventeen times, and 11 about six times.

Exercises

Set I

When two dice are thrown, they can turn up in 36 different ways resulting in sums ranging from 2 to 12.

1. Use the information in the table on page 453 to copy and complete the following one.

Sum of 2 dice	2	3	4	5	6	7	8	9	10	11	12
Number of ways	1	▓	▓	▓	▓	6	▓	▓	▓	2	▓

In craps, the dice thrower wins on the first throw if a sum of 7 or 11 comes up.

2. How many ways are there altogether of getting these sums?

3. What is the probability, both as a fraction and as a percentage, that the dice thrower will win on the first throw?

The dice thrower loses if a sum of 2, 3, or 12 comes up.

4. How many ways are there altogether of getting any of these sums?

5. What is the probability, both as a fraction and as a percentage, that the dice thrower will lose on the first throw?

By rolling a 4, 5, 6, 8, 9, or 10, a dice thrower neither wins nor loses and must roll again.

6. How many ways are there altogether of getting any of these sums?

7. What is the probability, both as a fraction and as a percentage, that the dice thrower will get one of these sums?

Having rolled one of these sums, the dice thrower must roll the same number again before rolling a 7 in order to win.

8. Which is more likely: that the same number will come up or that a 7 will?

9. Which of the sums, 4, 5, 6, 8, 9, 10, are least likely to turn up?

10. Which are most likely to turn up?

It is unlikely that anyone would be fooled by the doctored dice shown in this illustration from *Mad* magazine. If all the faces of one die are numbered 4 and all the faces of the other die are numbered 3, the probability of rolling a 7 with them is 100%.

*Doctored dice
from* Mad *magazine*

First die

	4	4	4	4	4	4
3	7	7	7	7	7	7
3	7	7	7	7	7	7
3	7	7	7	7	7	7
3	7	7	7	7	7	7
3	7	7	7	7	7	7
3	7	7	7	7	7	7

Second die (label on left side)

Professional cheats use misspotted dice that are much harder to detect. One type, called "tops and bottoms," consists of two dice numbered with only three different numbers on each die, each number being repeated on the die's opposite face.

11. Copy and complete the following table of sums for two of these dice, one numbered 3, 4, 5 and the other numbered 1, 5, 6.

		First die					
		3	4	5	3	4	5
Second die	1	4	?	?	?	?	?
	5	?	?	?	?	?	?
	6	?	?	?	?	?	?
	1	?	?	?	?	?	?
	5	?	?	?	?	?	?
	6	?	?	?	?	?	?

What is the probability that someone throwing these dice will

12. win on the first throw by rolling a 7 or 11?

13. lose on the first throw by rolling a 2, 3, or 12?

14. neither win nor lose on the first throw by rolling a 4, 5, 6, 8, 9, or 10?

15. Having rolled one of these sums on the first throw, what is the probability that the dice thrower will win by getting that sum again before 7 turns up?

Set II

Galileo became interested in probability when he was asked by some gamblers about the chances in a game played with three dice.

Because each die can turn up in 6 ways, three dice can turn up in

$$6 \times 6 \times 6 = 216$$

different ways resulting in sums ranging from 3 to 18. The 216

ways and their sums are shown in the following tables.

First die: 1

Second die

	1	2	3	4	5	6
1	3	4	5	6	7	8
2	4	5	6	7	8	9
3	5	6	7	8	9	10
4	6	7	8	9	10	11
5	7	8	9	10	11	12
6	8	9	10	11	12	13

(Third die, rows 1–6)

First die: 2

Second die

	1	2	3	4	5	6
1	4	5	6	7	8	9
2	5	6	7	8	9	10
3	6	7	8	9	10	11
4	7	8	9	10	11	12
5	8	9	10	11	12	13
6	9	10	11	12	13	14

(Third die, rows 1–6)

First die: 3

Second die

	1	2	3	4	5	6
1	5	6	7	8	9	10
2	6	7	8	9	10	11
3	7	8	9	10	11	12
4	8	9	10	11	12	13
5	9	10	11	12	13	14
6	10	11	12	13	14	15

(Third die, rows 1–6)

First die: 4

Second die

	1	2	3	4	5	6
1	6	7	8	9	10	11
2	7	8	9	10	11	12
3	8	9	10	11	12	13
4	9	10	11	12	13	14
5	10	11	12	13	14	15
6	11	12	13	14	15	16

(Third die, rows 1–6)

First die: 5

Second die

	1	2	3	4	5	6
1	7	8	9	10	11	12
2	8	9	10	11	12	13
3	9	10	11	12	13	14
4	10	11	12	13	14	15
5	11	12	13	14	15	16
6	12	13	14	15	16	17

(Third die, rows 1–6)

First die: 6

Second die

	1	2	3	4	5	6
1	8	9	10	11	12	13
2	9	10	11	12	13	14
3	10	11	12	13	14	15
4	11	12	13	14	15	16
5	12	13	14	15	16	17
6	13	14	15	16	17	18

(Third die, rows 1–6)

   Notice that there is only one way in which the sum of 3 can be thrown: each die must turn up a 1. It appears at the upper left in the first table. There is also only one way in which a sum of 18 can be thrown: each die must turn up a 6. It appears at the lower right of the last table.

1. Refer to the tables on the preceding page to copy and complete the following one.

Sum of three dice	3	4	5	6	7	8	9	10	11	12	13	14	15	16	17	18
Number of ways	1	3	6													

Check to see that the numbers of ways add up to 216.

2. One question the gamblers asked Galileo was why, when three dice are thrown, a sum of 10 turns up more often than a sum of 9. Use your table to explain why.

3. What sums are least likely to turn up when three dice are thrown?

4. What sums are most likely?

5. Graph the information in your table, representing the sums of the three dice on the x-axis and the numbers of ways of getting them on the y-axis. Convenient scales for the axes are shown here.

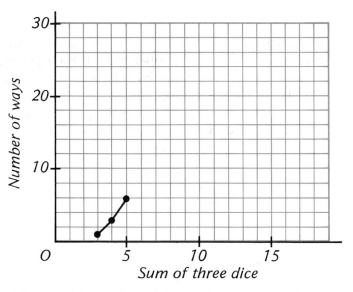

After you have plotted the points, use a ruler to connect them in order with straight line segments. The first part of the graph is shown above.

Some dice games are played with four or five dice.

6. In how many different ways can four dice turn up?

7. What sums are least likely to turn up when four dice are thrown?

Chapter 8: THE MATHEMATICS OF CHANCE

8. What sum is most likely to turn up?

9. In how many different ways can five dice turn up?

10. What sums are least likely to turn up when five dice are thrown?

11. What sums are most likely?

Set III

A dice game invented by Bradley Efron, a mathematician at Stanford University, has such a strange set of outcomes that even an experienced gambler would find them hard to believe.*

The faces of four dice are numbered as shown in the table below.

Die A   1, 2, 3, 9, 10, 11

Die B   0, 1, 7, 8, 8, 9

Die C   5, 5, 6, 6, 7, 7

Die D   3, 4, 4, 5, 11, 12

Each of two players chooses a die and the two dice are thrown. The player who gets the larger number wins. If both players get the same number, the tie is broken by rolling the dice again.

Suppose that one player chooses die A and the other player chooses die B. The player who has the greater probability of winning can be determined from the table of possible outcomes shown below. The A's indicate the outcomes in which die A wins,

Die A

	1	2	3	9	10	11
0	A	A	A	A	A	A
1	T	A	A	A	A	A
7	B	B	B	A	A	A
8	B	B	B	A	A	A
8	B	B	B	A	A	A
9	B	B	B	T	A	A

Die B (left vertical label)

*Described in Martin Gardner's "Mathematical Games" column, *Scientific American*, December 1970.

the B's the outcomes in which die B wins, and the T's the outcomes in which there is a tie. Because there are more outcomes in which A wins than B, the player with die A has the greater probability of winning.

1. What is the probability, both as a fraction and as a percentage, that A will beat B on the first throw?

2. Make a table showing the possible outcomes if one player chooses die B and the other player chooses die C.

3. Which player is more likely to win?

4. Make a table showing the possible outcomes if one player chooses die C and the other player chooses die D.

5. Which player is more likely to win?

6. Make a table showing the possible outcomes if one player chooses die D and the other player chooses die A.

7. Which player is more likely to win?

8. What is strange about the outcomes of this game?

Georges de la Tour, *Le Tricheur à l'As de Carreau* [The cheater with the ace of diamonds]; courtesy Musée du Louvre

Lesson 5

Independent and Dependent Events

Playing cards are thought to have originated in China, perhaps as early as the seventh century. The designs on the cards now used throughout the English-speaking world were introduced in France in the sixteenth century, about the time that the picture of the card players above was painted.

The probabilities of getting various types of hands in card games were determined soon after the mathematics of chance began. To illustrate the methods used, we will consider a simple example. Suppose that four cards are taken from a deck: two aces, a king, and a queen. One ace and the king are put in one pile and the other ace and the queen in a second pile. If a card is drawn at random from each pile, what is the probability that both cards will be the aces?

The tree diagram below illustrates the possibilities. Notice that

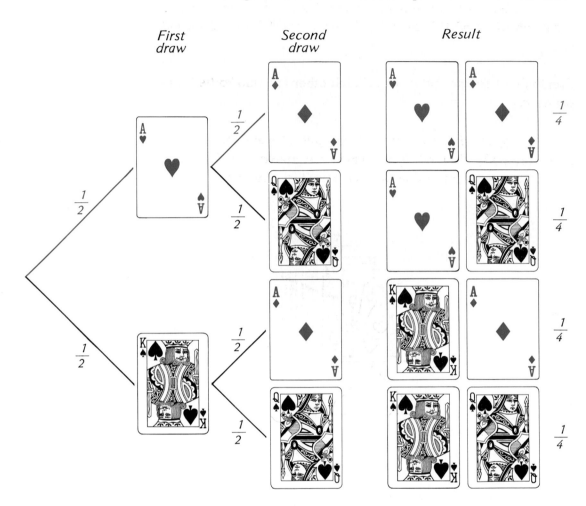

| First draw | Second draw | Result |

the probability of getting an ace in the first draw is $\frac{1}{2}$, the probability of getting an ace in the second draw is $\frac{1}{2}$, the probability of getting both aces is $\frac{1}{4}$, and

$$\frac{1}{2} \times \frac{1}{2} = \frac{1}{4}.$$

▶ To find the probability of several things happening in succession, multiply the probabilities of the individual happenings.

This rule works if the happenings do not influence each other. For example, what happens in drawing a card from the first pile has no effect on what happens in drawing a card from the second pile.

▶ Events that have no influence on each other are said to be **independent.**

Now suppose that all four cards are put into one pile and the pile thoroughly shuffled. If two cards are drawn from this pile, what is the probability that both cards will be the aces?

This is a different situation. What happens in the first draw—that is, whether or not the card is an ace—affects what happens in the second draw.

▶ Events that are influenced by other events are said to be **dependent.**

The tree diagram on the next page shows that the probability of getting the two aces can again be found by multiplying. The prob-

First draw Second draw Result

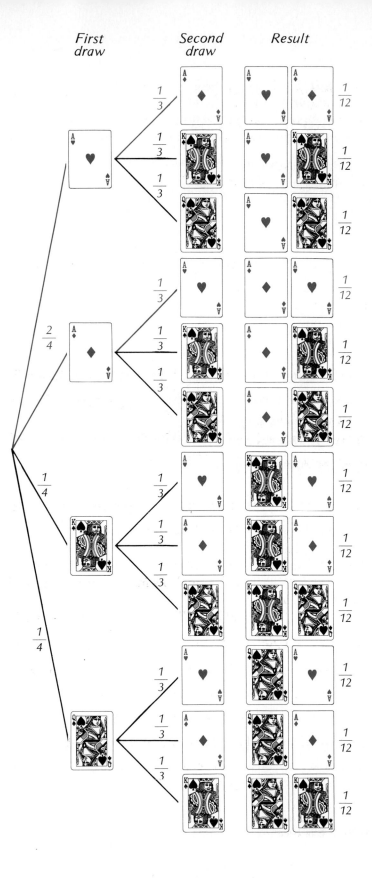

464

ability that the first card drawn will be an ace is $\frac{2}{4}$ because there are two aces in the pile of four cards. After the first ace has been drawn, there is one ace left in the three cards remaining in the pile, and so the probability that the second card is an ace is $\frac{1}{3}$. Multiplying these probabilities,

$$\frac{2}{4} \times \frac{1}{3} = \frac{2}{12},$$

we find that the probability that both cards are the aces is $\frac{2}{12}$ or $\frac{1}{6}$.

Exercises

Set I

A girl is told by her boyfriend that she is "one in a billion." She has a dimple in her chin, probability $\frac{1}{100}$, eyes of different colors, probability $\frac{1}{1,000}$, and is absolutely crazy about mathematics, probability $\frac{1}{10,000}$.

1. Do these events seem to be independent or dependent?

2. Show why the girl is "one in a billion."

In a carnival game called Spot the Spot, the player has to drop five disks onto a red circle.* The disks must cover it completely for the player to win.

The probability that a skilled player can drop one of the disks onto the exact place on the red circle that it must occupy is about $\frac{1}{3}$.

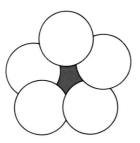

What is the probability that such a player will be able to drop

3. two of the disks exactly where they belong?

4. three of the disks exactly where they belong?

5. four of the disks exactly where they belong?

6. all five of the disks exactly where they belong so that the player wins the game?

*John Scarne, *Scarne's New Complete Guide to Gambling* (Simon and Schuster, 1974), pages 604–607.

Lesson 5: Independent and Dependent Events

465

The second Earl of Yarborough is said to have bet 1,000 to 1 that a bridge hand containing no tens, jacks, queens, or kings would never occur. It did, and he lost the bet. Such a hand is now called a "Yarborough."

The probability of being dealt a "Yarborough" is

$$\frac{36}{52} \times \frac{35}{51} \times \frac{34}{50} \times \frac{33}{49} \times \frac{32}{48} \times \frac{31}{47} \times \frac{30}{46} \times \frac{29}{45} \times \frac{28}{44} \times \frac{27}{43} \times \frac{26}{42} \times \frac{25}{41} \times \frac{24}{40}.$$

7. Why are there thirteen fractions in this product?

8. What do the numbers 36 and 52 in the first fraction represent?

9. What do the numbers 35 and 51 in the second fraction represent?

The product of the thirteen fractions is approximately equal to $\frac{1}{275}$.

10. How frequently is a "Yarborough" likely to occur?

The actor Sean Connery once bet on the number 17 three times in succession in a roulette game in the St. Vincent casino. All three times the number 17 came up and Mr. Connery won $20,000.

11. Are the events that the number 17 comes up each of the three times independent or dependent?

The wheel had 37 compartments, numbered 0 and 1 through 36.* What is the probability that the number 17 would come up

12. on a single spin?

13. on two successive spins?

14. on three successive spins?

In the United States, the probability that a man will live to the age of 75 is about $\frac{3}{10}$ and the probability that a woman will live to the age of 75 is about $\frac{1}{2}$.

On the assumption that these probabilities are true for a person's grandparents, find each of the following both as a fraction and as a percentage.

---

* European roulette wheels have 37 compartments, unlike those in the United States, which have 38.

Chapter 8: THE MATHEMATICS OF CHANCE

15. What is the probability that both grandparents on a person's father's side will live to the age of 75?

16. What is the probability that both grandparents on a person's mother's side will live to the age of 75?

17. What is the probability that all four of a person's grandparents will live to the age of 75?

18. Do you think it is correct to assume that all four events are independent?

Set II

The following bet is from a book titled *Never Give a Sucker an Even Break.* *

Take a small opaque bottle and seven olives, two of which are green, five black. The green ones are considered the "unlucky" ones. Place all seven olives in the bottle, the neck of which should be of such a size that it will allow only one olive to pass through at a time. Ask the sucker to shake them and then wager that he will not be able to roll out three olives without getting an unlucky green one amongst them. If a green olive shows, he loses.

1. Are the events that each successive olive to come out of the bottle is black independent or dependent?

* Written by John Fisher (Pantheon Books, 1976).

2. What is the probability that the first olive that comes out of the bottle is black?

3. Would it be better to bet in favor of it being black or against it?

4. What is the probability that the second olive also is black?

5. Would it be better to bet in favor of it being black or against it?

6. What is the probability that the third olive also is black?

7. Would it be better to bet in favor of it being black or against it?

8. What is the probability that all three olives are black?

9. Would it be better to bet in favor of all three being black or against it?

The probabilities that someone living in the United States has blood of a given type is shown in the table below.

Blood type	Probability
O	45%
A	41%
B	10%
AB	4%

10. What is the sum of these probabilities?

11. Does this sum seem reasonable?

What is the probability, to the nearest percent, that both a husband and his wife have blood of type

12. O?

13. A?

14. B?

15. AB?

16. What is the sum of the four probabilities you have just determined?

17. Why is this sum not equal to 100%?

In an experiment with animal intelligence, a group of rats and a group of fish were given two choices: X or Y.* The probability of

---

*M. E. Bitterman, "The Evolution of Intelligence," *Scientific American*, January 1965.

Chapter 8: THE MATHEMATICS OF CHANCE

being rewarded for selecting choice X was $\frac{7}{10}$ and the probability of being rewarded for selecting choice Y was $\frac{3}{10}$.

After some practice, the rats began selecting choice X all the time.

18. What was their probability, both as a fraction and as a percentage, of being rewarded for a single choice?

After some practice, the fish began selecting choice X $\frac{7}{10}$ of the time and choice Y $\frac{3}{10}$ of the time.

19. What was their probability, both as a fraction and as a percentage, of being rewarded for a single choice?

20. Which group of animals behaved more intelligently?

Set III

In a court case on a charge of overtime parking several years ago, a policeman observed the positions of the valves of the tires on one side of a parked car as shown in this figure.* He recorded them as

being at "one o'clock" and "six o'clock." Later, when the allowed time had expired, he observed that the car was still in the same parking space with the valves of the two tires in the same positions as before. The owner of the car claimed that he had left the parking space before the time had expired and returned to it later. He said that the valves being in the same position as before must have been a coincidence.

---

* H. Zeisel, "Statistics as Legal Evidence" in *International Encyclopedia of the Social Sciences*, edited by D. L. Sills (Macmillan, 1968).

1. If the position of a tire valve is recorded to the nearest "hour," how many different positions are possible?

2. What is the probability that one of the valves would be in its original position if the owner of the car was telling the truth?

3. Do you think that the event of each valve returning to its original position is independent or dependent of the others?

4. If the events are independent, what would the probability be of both valves returning to their original positions?

5. If the events are dependent, what effect might that have on the probability of both valves returning to their original positions?

The judge acquitted the defendant but said that he would have been convicted if all four wheels had been checked and found to have been in their original positions.

6. If the events are independent, what would the probability be of all four valves returning to their original positions?

7. Do you think this result, which the judge accepted, applies to the situation? Explain.

The Birthday Problem: Complementary Events

By permission of Johnny Hart and Field Enterprises, Inc.

An amazing problem in probability is that of the "Coinciding Birthdays." The problem concerns the number of people needed in a group in order for the odds to favor two of them sharing the same birthday. The answer, just 23 people, is quite surprising.

The solution of the "Coinciding Birthdays" problem is based on the probabilities of *complementary events*. An example of such probabilities is shown in the tree diagram below. If two coins are tossed,

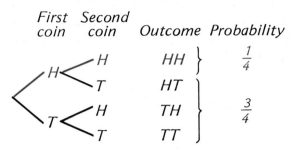

First Second
coin coin   Outcome Probability

H —— H   HH } $\frac{1}{4}$
H —— T   HT
T —— H   TH } $\frac{3}{4}$
T —— T   TT

the probability of both coins turning up heads is $\frac{1}{4}$. The probabil-

ity that both coins do *not* turn up heads is $\frac{3}{4}$. These two probabili-

ties add up to 1:

$$\frac{1}{4} + \frac{3}{4} = \frac{4}{4} = 1$$

▶The event that something happens and the event that it does not happen are called **complementary.**

▶Probabilities of complementary events always add up to 1.

This means that the probability of one of the complementary events can be found by subtracting the probability of the other one from 1. In the "Coinciding Birthdays" problem, either there are two people in a group who share the same birthday or no two people do. Because these events are complementary, the probability that two people share the same birthday can be found by subtracting the probability that no two people share the same birthday from 1.

What is the probability that no two people in the group share the same birthday? The first person's birthday can be any day.*

The probability that the second person's birthday is different is $\frac{364}{365}$,

the probability that the third person's birthday is different is $\frac{363}{365}$,

the probability that the fourth person's birthday is different is $\frac{362}{365}$, and so on.

---

*For simplicity, we will ignore leap year and assume that the probability of someone being born on any one of the 365 days of an ordinary year is the same.

Chapter 8: THE MATHEMATICS OF CHANCE

Because the probability of several things happening in succession is found by multiplying the probabilities of all of them, the probability that the second person's birthday is different *and* the third person's birthday is different *and* the fourth person's birthday is different, and so on, is

$$\frac{364}{365} \times \frac{363}{365} \times \frac{362}{365} \times \cdots$$

This product, then, is the probability that *no* two people in the group share the same birthday. The probability that two people in the group *do* share the same birthday is

$$1 - \frac{364}{365} \times \frac{363}{365} \times \frac{362}{365} \times \cdots$$

How this probability varies with the number of people in the group will be considered in some of the exercises in this lesson.

Exercises

Set I

A basketball player has a free-throw shooting average of 0.700. This means that the probability that he will score on a free throw is $\frac{7}{10}$, or 70%.

The player is entitled to two free throws. The tree diagram below shows the possible outcomes.

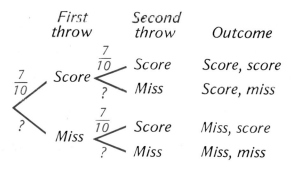

What is the probability, both as a fraction and as a percentage,

1. that he will *not* score on a free throw?

2. that he will score on both free throws?

3. that he will not score on either free throw?

4. that he will score on exactly one of the free throws?

The American artist Rube Goldberg (1883–1970) was so well known for his cleverly ridiculous inventions that, according to one dictionary,* his name has come to mean "having a fantastically complicated, improvised appearance" and "deviously complex and impractical."

The invention for sharpening pencils shown here makes use of an opossum and a woodpecker and includes an emergency knife in case either animal "gets sick and can't work."

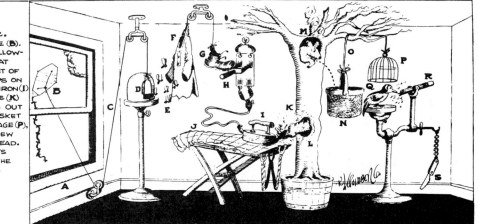

From *Rube Goldberg vs. the Machine Age* by Reuben L. Goldberg. © 1968 King Features Syndicate, Inc. Reprinted by permission of Hastings House, Publishers, Inc.

Suppose that the probability that the opossum will get sick is $\frac{3}{10}$ and the probability that the woodpecker will get sick is $\frac{1}{10}$. What is the probability that

5. the opossum will not get sick?

6. the woodpecker will not get sick?

7. *both* the opossum and woodpecker will get sick?

8. neither the opossum nor the woodpecker will get sick?

9. Are the events that both will get sick and neither one will get sick complementary? Explain.

* *The Random House Dictionary of the English Language*, unabridged edition, 1966.

Chapter 8: THE MATHEMATICS OF CHANCE

The makers of Cracker Jack began putting a prize in each box in 1912. Suppose that the company is using one hundred different prizes at present and that the probability of any one of the prizes being included in a box is the same.

Someone opens a box of Cracker Jack and finds a particular prize. What is the probability that, when that person opens another box,

10. he will find the same prize?

11. he will find a different prize?

The probabilities of getting the same prize from different numbers of boxes of Cracker Jack are shown in the graph below.

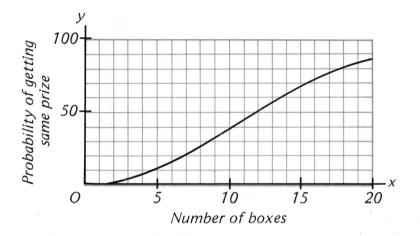

Would you expect to find a different prize in each box if you opened

12. five boxes?

13. ten boxes?

14. fifteen boxes?

Set II

The Arthur Murray Studios once offered $25 worth of dancing lessons to anyone who had a "lucky dollar." A "lucky dollar" was a bill whose serial number included a 2, 5, or 7. If you have any dollar bills with you, check to see whether or not you would have been a winner.

Courtesy of the U.S. Department of the Treasury

1. The probability that any given digit is a lucky one is $\frac{3}{10}$. Explain.

2. What is the probability that any given digit is *not* a lucky one?

There are eight digits in the serial number of a dollar bill.

3. Write an expression to represent the probability that all eight digits will not be lucky ones.

Expressed as a percentage, the value of this expression is 6%.

4. What is the probability that at least one of the digits in the serial number is a lucky one?

5. Out of one hundred people with dollar bills, how many would be winners?

The probabilities that there are two people in a group who share the same birthday and that no two people do are shown below and at the top of the next page. Each is rounded to the nearest percent.

Number of people in group	Probability that all birthdays are different	Probability that there are two people who share the same birthday
2	100%	0%
4	98%	2%
6	96%	4%
8	93%	7%
10	88%	12%
12	83%	17%
14	78%	22%
16	72%	28%
18	65%	35%
20	59%	41%
22	52%	48%
24	46%	54%
26	40%	60%
28	35%	65%
30	29%	71%
32	25%	75%
34	20%	80%
36	17%	83%
38	14%	86%
40	11%	89%

Chapter 8: THE MATHEMATICS OF CHANCE

Number of people in group	Probability that all birthdays are different	Probability that there are two people who share the same birthday
42	9%	91%
44	7%	93%
46	5%	95%
48	4%	96%
50	3%	97%

6. Graph the information in the table, representing the number of people in the group on the x-axis and the probabilities on the y-axis. Convenient scales for the axes are shown below.

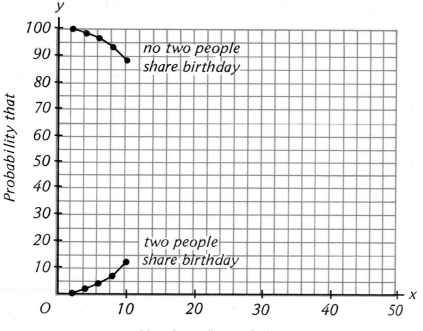

Connect each set of points in order. The first part of the graph is shown in the figure.

7. What is the significance of the point in which the two curves intersect?

Suppose that the graph were continued to the right. What would eventually happen to the curve representing the probability

8. that there are two people who share the same birthday?

9. that no two people share the same birthday?

The birthdays of the first thirty presidents of the United States are included in the list below.

President	Birthday
1. Washington	Feb. 22
2. J. Adams	Oct. 30
3. Jefferson	Apr. 13
4. Madison	Mar. 16
5. Monroe	Apr. 28
6. J. Q. Adams	July 11
7. Jackson	Mar. 15
8. Van Buren	Dec. 5
9. W. H. Harrison	Feb. 9
10. Tyler	Mar. 29
11. Polk	Nov. 2
12. Taylor	Nov. 24
13. Fillmore	Jan. 7
14. Pierce	Nov. 23
15. Buchanan	Apr. 23
16. Lincoln	Feb. 12
17. A. Johnson	Dec. 29
18. Grant	Apr. 27
19. Hayes	Oct. 4
20. Garfield	Nov. 19
21. Arthur	Oct. 5
22. Cleveland	Mar. 18
23. B. Harrison	Aug. 20
24. McKinley	Jan. 29
25. T. Roosevelt	Oct. 27
26. Taft	Sept. 15
27. Wilson	Dec. 28
28. Harding	Nov. 2
29. Coolidge	July 4
30. Hoover	Aug. 10

10. Before looking through this list, would you bet *in favor of* or *against* two of them having the same birthday? Explain.

11. Now go through the list carefully. Do two share the same birthday? If so, which two?

Chapter 8: THE MATHEMATICS OF CHANCE

Suppose that a telephone book is opened to a page at random and any set of consecutive telephone numbers (such as that from a Turkish telephone book shown here) is chosen from that page. What is the probability that the last two digits of two of the telephone numbers in the set are the same? This problem is identical with the one about getting the same prize twice from a series of Cracker Jack boxes, described in Set I after exercise 9.

12. What aspect of this problem makes it identical with the Cracker Jack problem?

Because the probabilities in the two problems are the same, the graph for the Cracker Jack problem can be used for the telephone digits problem.

13. How should the *x*-axis be relabeled?

14. How should the *y*-axis be relabeled?

15. If you looked at twenty consecutive telephone numbers, would you expect to find the last two digits of two of them to be the same?

Choose a set of consecutive telephone numbers from the telephone book for your city.

16. How many did you look through before finding two numbers that end in the same two digits? If possible, show the set of numbers.

352 Varaşoğlu - Varol

Varaşoğlu Osman Suat
  Merter Sitesi 23 73 61
Varavir Şevki Kyalı Bağdat C 181   52 12 64
Varaylı Zeki Ev Slmy Eczane S 50   33 24 32
Varbarbut Avram
  Ev Şiş Sıracevizler C 51 .... 48 78 22
Varbarut Yako
  Ev Şiş Sıracevizler C 35 ...... 40 83 52
Varboz Muammer
  Ev Fth Sarıgüzel C 111/1 ..... 23 71 21
Varcın Kutsiye
  Ev Fth Sofular C 163 .... 23 97 74
  Ev Aks Atatürk Bulv 146 ...... 28 26 98
Vardal Dursun Ali
  Ayvans Demir C 49/2 .... 21 70 77
Vardal Enver Ev Atıkalı Lodos S 2   25 30 35
Vardal İhsan
  Ev Gayrett Yıldızposta C 66 01 32
Vardal Kenan
  Mcköy Kervangeçmez S 47 61 23
Vardal Muammer
  Ev Kuşt Bestekârşakirağa S 22 .. 46 15 07
Vardal Nebile Ev Yköy Naima S 73 91 16
Vardal Nebile
  Ev Gayrett Yıldızposta C 66 14 64
Vardal Necati Ev Ayvans Çember S   23 84 31
Vardal Refik Ev Üsk Kuşatçı S 1/3 33 02 52
Vardal Sabri
  Ev Bahçelie Basın Sit Bl B 2 ... 71 42 94
Vardal Veysel Bkapı Liman H 101   27 95 67
Vardar Abdülriza
  Krköy Nohut Çk 3/2 .... 45 84 25
Vardar Abdürrezak Tarlab Çukur S 6 43 05 73

Set III

In the seventeenth century, a wealthy Frenchman known as the Chevalier de Mérè was fond of betting that, if a die is rolled four times, the number six will turn up at least once.

1. If a die is rolled once, what is the probability that it will *not* turn up six?

2. What is the probability that, if a die is rolled four times, it will not turn up six on any of the rolls? Express it both as a fraction and as a percentage, rounded to the nearest percent.

3. What is the probability to the nearest percent that, if a die is rolled four times, the number six will turn up at least once?

4. Would it be better to bet *in favor of* or *against* this happening?

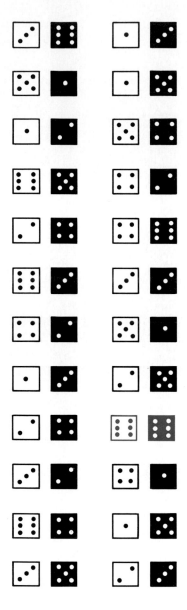

After winning a lot of money with this bet, de Méré switched to a different bet. He bet that if two dice are rolled 24 times, a sum of twelve will turn up at least once.

5. If a pair of dice is rolled once, what is the probability that it will *not* turn up twelve?

6. Write an expression to represent the probability that if two dice are rolled 24 times, they will not turn up twelve on any of the rolls.

Expressed as a percentage, the value of this expression is 51%.

7. What is the probability that if two dice are rolled 24 times, a sum of twelve will turn up at least once?

8. Would it be better to bet *in favor of* or *against* this happening?

Chapter 8 / Summary and Review

In this chapter we have become acquainted with:

Probability: the measure of chance (*Lessons 1 and 4*) The probability of an event in which a set of equally likely things can occur can be calculated as a ratio.

$$\text{Probability of an event} = \frac{\text{number of ways in which the event can occur}}{\text{total number of ways}}$$

Probabilities can be expressed as numbers on a scale ranging from 0 to 1 or as percentages on a scale ranging from 0% to 100%. They can also be expressed as odds.

If the number of ways in which an event can occur is x and the number of ways in which it does not occur is y, then the *odds against the event* are said to be in the ratio of "y to x."

Binomial probability and Pascal's triangle (*Lessons 2 and 3*) The probabilities in any situation in which there are two possible outcomes for each part are called binomial.

Pascal's triangle is a pattern of numbers from which binomial probabilities can be easily determined. Each number within the triangle is found by adding the pair of numbers directly above it at the left and right. The triangle begins like this:

$$
\begin{array}{ccccccc}
 & & & 1 & & 1 & & & \\
 & & 1 & & 2 & & 1 & & \\
 & 1 & & 3 & & 3 & & 1 & \\
1 & & 4 & & 6 & & 4 & & 1
\end{array}
$$

Independent and dependent events (*Lesson 5*) To find the probability of several things happening in succession, multiply the probabilities of all of them happening.

Events that have no influence on each other are said to be independent; events that are influenced by other events are dependent.

Complementary probabilities (*Lesson 6*) The event that something happens and the event that it does not happen are called complementary.

Probabilities of complementary events always add up to 1.

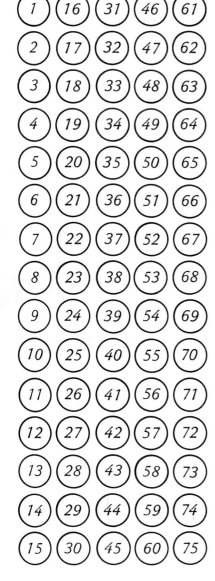

Exercises

Set I

Probabilities about smoking are listed below.

A smoker wants to quit	66%
A smoker has tried to quit	84%
A smoker will succeed in quitting	21%

1. Do most people who smoke want to quit?

2. What is the probability that someone who smokes does not want to quit?

3. What is the probability that someone who smokes has never tried to quit?

4. Do most people who try to quit smoking succeed?

5. What is the probability that someone who tries to quit smoking will not succeed?

A box contains 75 disks for a bingo game. The disks are numbered from 1 through 75. Find the probability, both as a fraction and as a percentage, that the first disk drawn from the box contains

6. a one-digit number.

7. an even number.

8. a number greater than 60.

9. a number smaller than 80.

Chapter 8: THE MATHEMATICS OF CHANCE

This figure appeared at the front of *Ssu Yuan Yü Chien,* a book written in China by Chu Shih-Chieh in 1303.

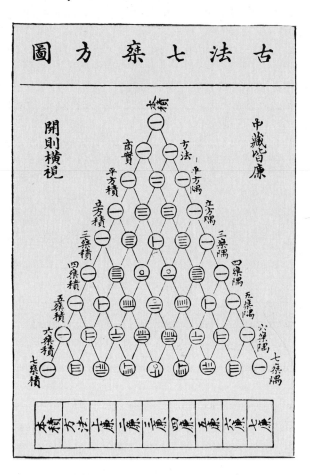

10. What is it?

11. Translate the symbols in the circles into more familiar symbols.

12. What type of probabilities can be determined from it?

A popular song of the 1920s began like this:

Keep your sunny side up, up!
Hide the side that gets blue.
If you have nine sons in a row
*Baseball teams make money, you know!**

13. If we assume that the probability of a child being a boy is $\frac{1}{2}$, then the probability of a family having two sons in a row is $\frac{1}{4}$. Explain.

14. What is the probability of a family having nine sons in a row?

Set II

To test for extrasensory perception (ESP), a deck of 25 cards is usually used in which there are five cards marked with each of the symbols shown below.

1. If the deck is shuffled and a card drawn at random, what is the probability that someone could guess the symbol on it simply by luck?

2. What is the probability that someone could guess three cards in succession if each card is put back and the deck reshuffled before the next card is drawn?

3. Suppose that this was done once with 500 people and that several of the people named all three cards correctly. Does this prove that they have ESP? Explain.

Automobile insurance rates vary according to the probability that the driver will be in an accident in the course of a year. The probabilities for drivers in the United States are shown in the table at the left.

Age group	Probability of an accident
Under 20	0.37
20–24	0.33
25–34	0.21
35–44	0.14
45–74	0.13
Over 74	0.17

According to this table, what is the probability, as a percentage, that a driver under the age of 20

4. will have accidents in two successive years?

5. will not have an accident in two successive years?

6. Should an insurance company bet *in favor of* or *against* a driver

under the age of 20 not having an accident in two successive years?

What is the probability, as a percentage, that a driver between the ages of 45 and 74

7. will have accidents in two successive years?

8. will not have an accident in two successive years?

9. Would you bet *in favor of* or *against* a driver between the ages of 45 and 74 not having an accident in two successive years?

It would be very surprising if all thirteen cards of a bridge hand were of the same suit. The probability of this occurring is 0.0000000006%.

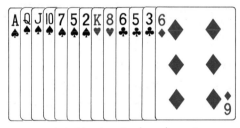

A bridge hand whose longest suit contains seven cards

The probabilities of being dealt longest suits containing other numbers of cards are listed in the table below.*

Longest suit	Probability
12	0.0000003%
11	0.00004%
10	0.002%
9	0.04%
8	0.5%
7	4%
6	17%
5	44%
4	35%
3	0%

10. What is the most probable longest suit in a bridge hand?

---

*From *How to Figure the Odds* by Oswald Jacoby (Doubleday, 1947).

11. Why is the probability of the longest suit in a bridge hand containing only three cards 0%?

Would it be surprising for a bridge player who played thirty hands during an evening to be dealt a hand whose longest suit contained

12. seven cards?

13. eight cards?

Set III

The Counterfeiter's Club is having a raffle and each of the 25 members buys a ticket. The 25 stubs are put into a box from which one will be drawn to determine which member will win the prize.

1. What is the probability that Fraudulent Fred, one of the members, will win the prize?

2. Fred decides to improve his chances by forging seven copies of his stub and sneaking them into the box. Now what is the probability that he will win?

3. Now what is the probability that any one of the other members will win?

Just before the drawing, Fred finds out that, after the first stub has been drawn to determine the winner, a second stub will be drawn to decide who will get the booby prize.

4. What is the probability that Fred will win both prizes, so that his forgery is discovered?

5. The first stub has been drawn and it is not one of Fred's. What is the probability that he will still get the booby prize?

Chapter 8 / Problems for Further Exploration

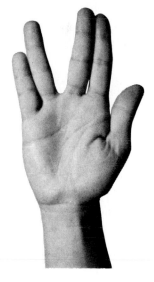

Lesson 1

1. Several years ago, the young son of a doctor in Springfield, Massachusetts, discovered a simple test for left-handedness. He got the idea from observing the V-shaped sign that Mr. Spock gave on the "Star Trek" television series.

   The test requires spreading apart the fingers of each hand. The fingers are stretched to form a large V with two fingers on each branch of the V. The possible outcomes of this test and the corresponding predictions are shown below.

Hand whose fingers stretch farther	Handedness of person
Right hand	Left-handed
Left hand	Right-handed
No difference	Right-handed

   When an article about the test appeared in a national magazine,* 3,225 readers tried it and reported the results summarized in the following table.

Handedness	Number of people	Number for whom test predicted correctly	incorrectly
Right-handed	2,880	2,097	783
Left-handed	345	289	56

---

* *Current Science*, February 20, 1974.

From this information, find the probability to the nearest percent that a person in this group is

a) right-handed.

b) left-handed.

What is the probability that a person the test predicts to be

c) right-handed actually is right-handed?

d) left-handed actually is left-handed?

e) What is the probability that the test predicts the handedness of a person correctly?

2. Four people—A, B, C, and D—have appointments for interviews for a job. Only one of them will be hired. It is the policy of the company to hire the first person who is better than the preceding candidate and cancel any interviews that remain. One person has to be hired and so, if each candidate is worse than the one before, the last person to be interviewed gets the job.

Suppose that A is the best candidate, B the second best, C the third best, and D the worst. If their interviews are scheduled in the order A, C, B, D, according to the policy just described candidate B gets the job and candidate D's interview is canceled.

a) Make a list of the 24 different possible orders in which the interviews can be scheduled. Circle the letter of the applicant who gets the job in each case.

b) Use your list to find the probability, as a percentage, of each candidate getting the job.

What is the probability that

c) two interviews are required to fill the position?

d) three interviews are required?

e) four interviews are required?

Lesson 2

1. Two handball players, A and B, decide to play each other until one player has won three games. Each player is equally good so that the probability of either one winning a game is $\frac{1}{2}$. How many games are they likely to play?

Although the winner may be decided in three, four, or five games, it is easiest to consider all the possible orders of winners for five games. In the tree diagram on the next page, the possible winners of games that are not played are written in parentheses.

First game	Second game	Third game	Fourth game	Fifth game	Order of winners

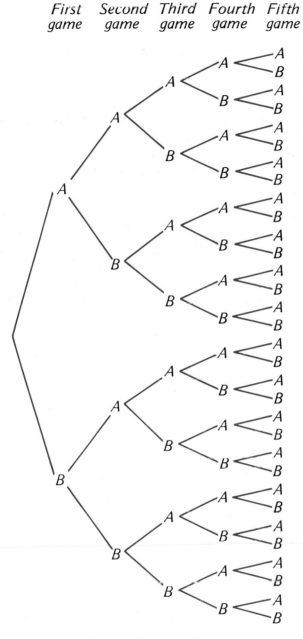

Order of winners

AAA(AA)
AAA(AB)

a) Copy and complete the list of possible outcomes.

What is the probability, as a fraction and as a percentage, that player A will win in

b) three games?

c) four games?

d) five games?

e) What are the probabilities for player B?

f) Do these results seem reasonable? Explain.

2. Here is another heredity experiment that can be explained with probability. Several blue parakeets are mated with some yellow ones and every one of the offspring is green. When the all-green generation is mated, the colors of its offspring are green, blue, yellow, and white.

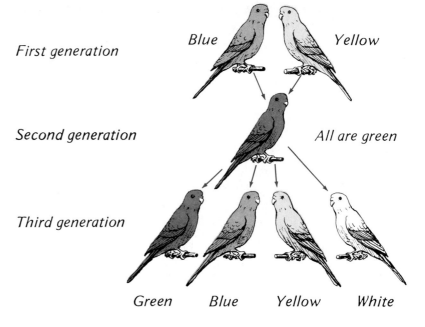

First generation   Blue   Yellow

Second generation   All are green

Third generation

Green   Blue   Yellow   White

To explain the appearance of parakeets with four different colors, it is necessary to suppose that pairs of two genes determine the color. Again we will represent each gene with a letter. Suppose that the genes of the first generation are those given in the table at the left. Then the breeding of this first generation will result in a second generation of all green parakeets, as shown in the tree diagram below. (We will assume that, if the genes of a parakeet include both B and Y, it is green, and that W

Color of parakeet	Genes
Blue	BB-WW
Yellow	YY-WW

First generation		Second generation	
Gene from first parent	Gene from second parent	Genes of offspring	Color of offspring
BW	YW	BW-YW	Green
	YW	BW-YW	Green
BW	YW	BW-YW	Green
	YW	BW-YW	Green

does not affect the color unless it appears without the other letters.)

The breeding of the all-green generation of parakeets to produce the green, blue, yellow, and white parakeets in the third generation is shown in this tree diagram.

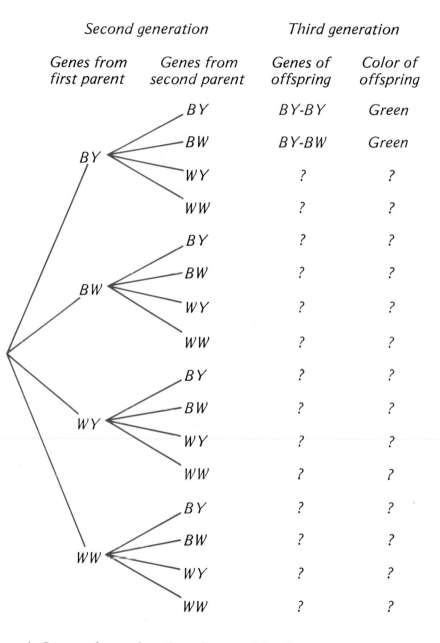

Second generation		Third generation	
Genes from first parent	Genes from second parent	Genes of offspring	Color of offspring
BY	BY	BY-BY	Green
	BW	BY-BW	Green
	WY	?	?
	WW	?	?
BW	BY	?	?
	BW	?	?
	WY	?	?
	WW	?	?
WY	BY	?	?
	BW	?	?
	WY	?	?
	WW	?	?
WW	BY	?	?
	BW	?	?
	WY	?	?
	WW	?	?

a) Copy and complete the columns of the diagram representing the genes and color of the parakeets in the third generation.

b) What color of parakeet is most common in the third generation? What color is least common?

c) Copy and complete the following table of probabilities for the third generation.

Color of parakeet:	green	blue	yellow	white
Probability:	▨	▨	▨	▨

Lesson 3

1. There is a remarkable connection between Pascal's triangle and a problem that you encountered earlier.* The problem concerns choosing two or more points on a circle and connecting every pair of points with a straight line segment. For a given number of points, what is the greatest number of regions that can be formed in this way?

The figures below illustrate solutions to the problem for circles on which from two to eight points have been chosen.

---

* See page 29.

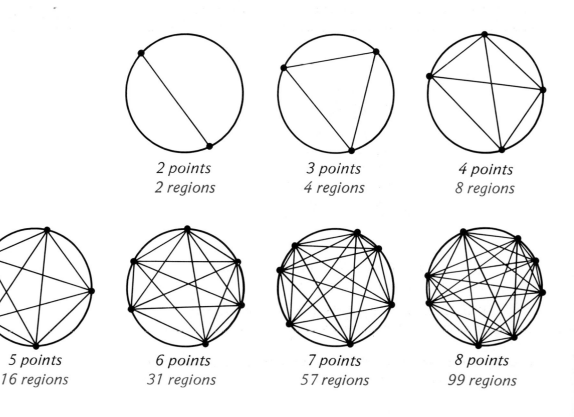

2 points	3 points	4 points
2 regions	4 regions	8 regions

5 points	6 points	7 points	8 points
16 regions	31 regions	57 regions	99 regions

Chapter 8: THE MATHEMATICS OF CHANCE

Compare the numbers of regions with the numbers in Pascal's triangle below.

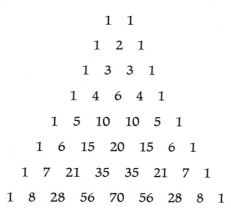

1   1

1   2   1

1   3   3   1

1   4   6   4   1

1   5   10   10   5   1

1   6   15   20   15   6   1

1   7   21   35   35   21   7   1

1   8   28   56   70   56   28   8   1

a) What is the connection between them?
b) How many regions do you think can be formed by choosing nine points on a circle? Explain.

2. The pattern of circles at the right shows the arrangement of the numbers in the first five rows of Pascal's triangle. If a circle is added to the top of the pattern and some of the circles shaded according to the following rules, the pattern looks like this:

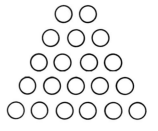

Rule 1. Shade the circle at the top and the circle at each end of every row.

Rule 2. Shade every circle that is below a pair of circles, of which exactly one is shaded.

a) Copy the figure, add ten more rows of circles, and shade them according to the rules above.

Compare the figure that you have drawn with Pascal's triangle shown on the preceding page. According to the first rule, the circle at each end of every row is shaded. In Pascal's triangle, all these circles correspond to 1's.

b) What type of numbers do the shaded circles correspond to in general?

c) Use rule 2 to explain why this is true.

Lesson 4

1. "Double-deuces" and "double-fives" are misspotted dice in which one number is repeated on the opposite face. A double-deuce has 2 twos and no fives. A double-five has 2 fives and no twos.

a) Make a table of sums for a pair of double-deuces and for a pair of double-fives.

b) Copy and complete the following table comparing the numbers of ways in which a pair of normal dice, a pair of double-deuces, and a pair of double-fives can turn up.

Sum of two dice	2	3	4	5	6	7	8	9	10	11	12
Number of ways with two normal dice	1	2	3	4	5	6	5	4	3	2	1
two double-deuces	▥	▥	▥	▥	▥	▥	▥	▥	▥	▥	▥
two double-fives	▥	▥	▥	▥	▥	▥	▥	▥	▥	▥	▥

c) How do the probabilities of winning on the first throw by rolling a 7 or 11 compare with each type of dice?

d) How do the probabilities of losing on the first throw by rolling a 2, 3, or 12 compare?

Suppose that you are throwing the dice and you neither win nor lose on the first throw.

e) What sums would you hope to get if you are playing with normal dice?

f) If you got one of those sums on the first throw, which would be more likely: that you win by getting it again or that you lose by getting a 7 first?

g) What sums would you hope to get if you are playing with double-deuces?

h) If you got one of those sums on the first throw, which would be more likely: that you win by getting it again or that you lose by getting a 7 first?

i) What sums would you hope to get if you are playing with double-fives?

j) If you got one of those sums on the first throw, which would be more likely: that you win by getting it again or that you lose by getting a 7 first?

2. The game of Chuck-a-Luck is played with three dice. Before the dice are thrown, each player bets on one or more of the numbers 1, 2, 3, 4, 5, or 6. If a number on which a bet is placed turns up on one die, the bettor wins the amount of the bet. If it turns up on two of the dice, the bettor wins twice the amount of the bet. If it turns up on all three dice, the bettor wins three times the amount of the bet.

Suppose that, before the dice are thrown, a player bets $1 on each number. If the dice turn up three different numbers—say, 1, 5, and 6—the player wins $3 ($1 each for betting on 1, 5, and 6) and loses $3 ($1 each for betting on 2, 3, and 4).

Because the rules of the game allow betting on all six numbers, it seems as if a player who does this will neither win nor lose money in the long run.

a) Suppose that two of the dice turn up the same number—say, 2, 4, and 4. How much does the player win and lose?

b) Suppose that all three dice turn up the same number—say, 5, 5, and 5. How much does the player win and lose?

The table at the right shows 36 of the 216 different ways in which the three dice can turn up. The ways in which the three dice turn up different numbers are shown as white circles, the ways in which two of the dice turn up the same number are shown as black circles, and the way in which all three dice turn up the same number is shown as a brown circle.

c) Make five more tables showing the rest of the ways in which the dice can turn up.

In how many different ways can

d) the three dice turn up different numbers?

e) exactly two of the dice turn up the same number?

f) all three dice turn up the same number?

Suppose that a player plays Chuck-a-Luck 216 times, each time betting $1 on each number.

g) If the dice turn up each of the 216 possible ways, will the player win, lose, or break even?

h) If the player does not break even, how much money does he win or lose?

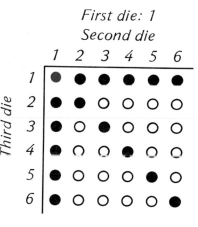

First die: 1

Second die

	1	2	3	4	5	6
1	●	●	●	●	●	●
2	●	●	○	○	○	○
3	●	○	●	○	○	○
4	●	○	○	●	○	○
5	●	○	○	○	●	○
6	●	○	○	○	○	●

Third die

MONOPOLY ® game equipment © 1935, 1946, 1961 Parker Brothers, Beverly, Massachusetts 01915.
MONOPOLY is Parker Brothers' registered trademark for its real estate trading game equipment.
Used by Permission.

Lesson 5

1. According to the rules of MONOPOLY®, a player who lands in jail has three ways to get out. He can use a "get out of jail free" card, pay a fine of $50, or throw doubles. A player who decides to stay in jail cannot remain there more than three turns.

   Suppose that someone has just landed in jail, does not have a "get out of jail free" card, and does not want to pay $50.

   a) What is the probability that he will get out of jail on his next turn by rolling doubles?
   b) What is the probability that he will *not* get out of jail on his next turn by rolling doubles?
   c) What is the probability that he will not get out of jail on three turns in a row so that he has to pay the $50?
   d) Would you bet *in favor of* or *against* a player being able to get out of jail by rolling doubles?
   e) What if the rules allowed a player to remain in jail for *four* turns? What effect would this have on the likelihood of the player being able to get out by rolling doubles?

Chapter 8: THE MATHEMATICS OF CHANCE

2. A poker player has a hand consisting of three clubs, a heart, and a diamond. He is considering trying for a flush in clubs (five clubs) by discarding the heart and diamond and drawing two more cards.

   Find the probability, both as a fraction and as a percentage, that
   a) the first card drawn is a club.
   b) the second card drawn is a club.
   c) the player will get a flush.
   d) Do you think it would be a good idea for the player to try for the flush? Explain.

Lesson 6

1. In a game of Hangman, one player chooses a seven-letter word in which each letter is different.
   a) If the other player guesses letters at random, what is the probability that his first guess will not be any one of the seven letters?
   b) If he is wrong on the first guess, what is the probability that he is also wrong on the second guess?
   c) What is the probability that no letter has been guessed correctly after two guesses?
   d) How many letters must be guessed in order for it to be more likely than not that at least one of the letters has been guessed correctly?

2. In the novel *The Bridge on the River Kwai* by Pierre Boule, an air force officer talking about the chances of three commandos successfully parachuting into the jungle says:

   > If they do only one jump, you know, there's a fifty percent chance of injury. Two jumps, it's eighty percent. The third time, it's dead certain they won't get off scot free.*

   If the chance of injury on one jump is 50%, the other probabilities stated by the officer are incorrect. Find the actual probability that at least one person will be injured if
   a) two commandos make the jump.
   b) three commandos make the jump.
   c) Would an injury be "dead certain" if many commandos made the jump? Explain.

---

*This exercise is based on a discussion of the problem by Darrell Huff in his book *How to Take a Chance* (Norton, 1959).

Chapter 9

AN INTRODUCTION TO STATISTICS

© 1960 United Feature Syndicate, Inc.

Lesson 1

Organizing Data: Frequency Distributions

Many people think of the word "statistics" in the sense that sports fans use it—as meaning numbers or tables of numbers. Mathematicians, however, use the word in a more general sense. The subject of statistics is a branch of mathematics that deals with the collection, organization, and interpretation of numerical data. We will begin our survey of statistics by considering some of the ways in which numerical data can be organized.

The winners and the number of games played in each World Series from 1903 through 1980 are listed on the facing page. The number of games played in a series varied from 4 to 8. (The rule limiting the series to 7 games was established in 1922.)

Has there been a trend in these numbers through the years? If there has been, it might be more easily discovered by arranging them another way.

Year	Winner, Games played	Year	Winner, Games played
1903	Boston Red Sox, 8	1943	New York Yankees, 5
1905	New York Giants, 5	1944	St. Louis Cardinals, 6
1906	Chicago White Sox, 6	1945	Detroit Tigers, 7
1907	Chicago Cubs, 5	1946	St. Louis Cardinals, 7
1908	Chicago Cubs, 5	1947	New York Yankees, 7
1909	Pittsburgh Pirates, 7	1948	Cleveland Indians, 6
1910	Philadelphia Athletics, 5	1949	New York Yankees, 5
1911	Philadelphia Athletics, 6	1950	New York Yankees, 4
1912	Boston Red Sox, 8	1951	New York Yankees, 6
1913	Philadelphia Athletics, 5	1952	New York Yankees, 7
1914	Boston Braves, 4	1953	New York Yankees, 6
1915	Boston Red Sox, 5	1954	New York Giants, 4
1916	Boston Red Sox, 5	1955	Brooklyn Dodgers, 7
1917	Chicago White Sox, 6	1956	New York Yankees, 7
1918	Boston Red Sox, 6	1957	Milwaukee Braves, 7
1919	Cincinnati Reds, 8	1958	New York Yankees, 7
1920	Cleveland Indians, 7	1959	Los Angeles Dodgers, 6
1921	New York Giants, 8	1960	Pittsburgh Pirates, 7
1922	New York Giants, 5	1961	New York Yankees, 5
1923	New York Yankees, 6	1962	New York Yankees, 7
1924	Washington Senators, 7	1963	Los Angeles Dodgers, 4
1925	Pittsburgh Pirates, 7	1964	St. Louis Cardinals, 7
1926	St. Louis Cardinals, 7	1965	Los Angeles Dodgers, 7
1927	New York Yankees, 4	1966	Baltimore Orioles, 4
1928	New York Yankees, 4	1967	St. Louis Cardinals, 7
1929	Philadelphia Athletics, 5	1968	Detroit Tigers, 7
1930	Philadelphia Athletics, 6	1969	New York Mets, 5
1931	St. Louis Cardinals, 7	1970	Baltimore Orioles, 5
1932	New York Yankees, 4	1971	Pittsburgh Pirates, 7
1933	New York Giants, 5	1972	Oakland Athletics, 7
1934	St. Louis Cardinals, 7	1973	Oakland Athletics, 7
1935	Detroit Tigers, 6	1974	Oakland Athletics, 5
1936	New York Yankees, 6	1975	Cincinnati Reds, 7
1937	New York Yankees, 5	1976	Cincinnati Reds, 4
1938	New York Yankees, 4	1977	New York Yankees, 6
1939	New York Yankees, 4	1978	New York Yankees, 6
1940	Cincinnati Reds, 7	1979	Pittsburgh Pirates, 7
1941	New York Yankees, 5	1980	Philadelphia Phillies, 6
1942	St. Louis Cardinals, 5		

Years 1903–1928

```
8 7 4 8 7
5 5 5 7 7
6 6 5 8 7
5 8 6 5 4
5 5 6 6 4
```

Years 1929–1953

```
5 7 4 6 5
6 6 7 7 4
7 6 5 7 6
4 5 5 7 7
5 4 5 6 6
```

Years 1954–1978

```
4 6 7 5 5
7 7 7 5 7
7 5 4 7 4
7 7 7 7 6
7 4 7 7 6
```

The first 75 numbers separated into three sets of 25 each are listed at the left. To see what patterns these numbers may have, we organize each set in a **frequency distribution.** Each frequency distribution is made by listing the numbers 4 through 8 (that is, the different numbers of games played) in a column. We then go through each set of numbers in the order in which it is written and tally the number of times each number of games played appears in the set. The tally marks are entered in the second column of each table. The numbers of tally marks, called the *frequencies,* are then listed in the third column of each table.

Years 1903–1928			Years 1929–1953			Years 1954–1978		
No. of games	Tally marks	Frequency	No. of games	Tally marks	Frequency	No. of games	Tally marks	Frequency
4					3	4						4	4						4
5	⍫|||	8	5	⍫||	7	5	||||	4
6	⍫	5	6	⍫||	7	6	|||	3
7	⍫	5	7	⍫||	7	7	⍫ ⍫ ||||	14
8						4	8		0	8		0
	Total =	25		Total =	25		Total =	25

The information in these tables can be more easily understood if it is presented in the form of graphs called **histograms.** The *frequencies* in a frequency distribution are represented by *bars* in a histogram.

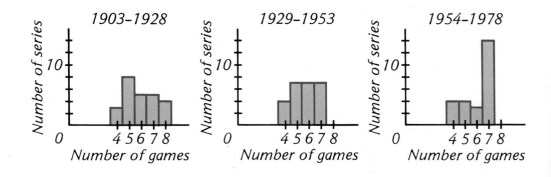

These histograms suggest that the likelihood of a World Series being played in the minimum number of games has not changed through the years. They also suggest that the likelihood of a World

Chapter 9: AN INTRODUCTION TO STATISTICS

Series lasting the maximum number of games has increased. The degree of reasonableness of these conclusions can be determined by more advanced statistical methods.

The subject of statistics has become an important tool with which to learn more about our physical, biological, social, political, and economic worlds. H. G. Wells once said:

> *Statistical thinking will one day be as necessary for efficient citizenship as the ability to read and write.*

Exercises

Set I

This photograph shows Bob Beamon making his record-breaking long jump in the 1968 Olympics, a record that one sportswriter predicted would "last into the twenty-first century."

The Olympic records in the long jump are listed in the table below.

Photograph by E. D. Lacey

Year	Winner	Meters
1896	Ellery Clark	6.35
1900	Alvin Kraenzlein	7.18
1904	Myer Prinstein	7.34
1908	Francis Irons	7.48
1912	Albert Gutterson	7.60
1920	William Pettersson	7.15
1924	De Hart Hubbard	7.44
1928	Edward Hamm	7.73
1932	Edward Gordon	7.63
1936	Jesse Owens	8.06
1948	William Steele	7.82
1952	Jerome Biffle	7.57
1956	Gregory Bell	7.83
1960	Ralph Boston	8.12
1964	Lynn Davies	8.07
1968	Robert Beamon	8.90
1972	Randy Williams	8.24
1976	Arnie Robinson	8.35
1980	Lutz Dombrowski	8.54

1. Make a frequency distribution of these distances by grouping them in intervals of 0.25 meters. The first column in your table

should look like this.

Distance in meters
6.01–6.25
6.26–6.50
6.51–6.75
6.76–7.00
7.01–7.25
7.26–7.50
7.51–7.75
7.76–8.00
8.01–8.25
8.26–8.50
8.51–8.75
8.76–9.00

2. Make a histogram of the distances, labeling the axes as shown here.

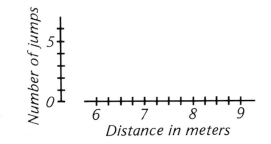

How many of the jumps were

3. less than 7 meters?

4. between 7 and 8 meters?

5. more than 8 meters?

This histogram shows the number of hours that adults sleep each night.

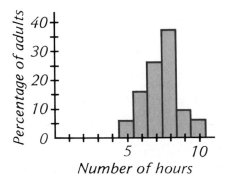

6. Approximately what percentage of adults sleep 8 hours each night?

7. Which is more common: for an adult to sleep *more* than 8 hours each night or *less* than 8 hours each night?

8. How does the number of adults who sleep 5 hours a night seem to compare with the number who sleep 10 hours a night?

These histograms show the distributions in ages of the populations of Mexico and the United States. The last bar in each graph represents people 80 years of age and older.

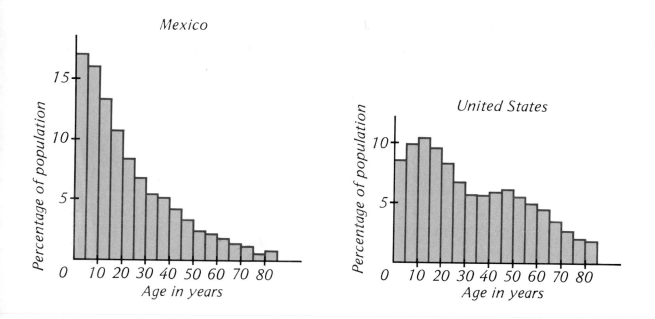

9. In general, how do the ages of the people in Mexico compare with the ages of the people in the United States?

10. How does the percentage of children in Mexico younger than 5 years of age compare with the percentage of children in the United States in the same age group?

Approximately what percentage of the population of

11. Mexico is younger than 20 years of age?

12. the United States is younger than 20 years of age?

13. Mexico is 60 years of age or older?

14. the United States is 60 years of age or older?

The percentage of licensed drivers in each age group in the United States and the percentage in each age group of the drivers in fatal accidents are given in the following table.

Age of driver	Percentage of licensed drivers	Percentage of drivers in fatal accidents
15–19	9	18
20–24	13	21
25–29	13	14
30–34	11	11
35–39	9	7
40–44	8	6
45–49	8	5
50–54	7	5
55–59	6	4
60–64	6	3
65–69	4	2
70–74	3	2
75–	3	2

15. Make a histogram of the ages of licensed drivers in the United States. Label the axes as shown here.

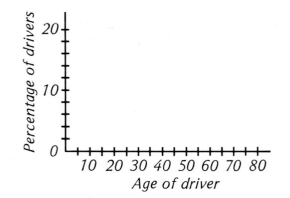

16. Make a histogram of the ages of drivers in fatal accidents.

17. What age group accounts for the highest percentage of the fatal accidents?

18. In what age group are the drivers most likely to be in fatal accidents?

19. What three age groups account for the lowest percentage of the fatal accidents?

20. In what two age groups are the drivers least likely to be in fatal accidents?

Set II

The amounts of money paid on $2 bets on the winners in two weeks of horse races at Hollywood Park are listed below.*

First week						Second week				
14.40	10.60	5.00	5.80	13.60		10.00	3.80	14.20	5.20	14.80
7.40	10.40	4.40	10.20	5.00		4.60	6.00	20.20	9.40	3.80
8.80	9.80	5.20	8.20	5.40		6.20	9.00	12.80	10.40	7.60
9.00	8.80	9.00	6.00	8.80		4.20	36.80	2.60	6.80	22.80
7.20	4.20	4.20	12.60	8.20		13.00	4.80	14.00	5.80	9.20
9.80	13.00	13.00	4.60	56.60		4.00	7.00	14.80	4.80	20.80
5.80	3.60	4.40	3.00	5.60		23.60	3.00	11.00	4.00	8.20
6.80	24.80	9.20	21.80	16.40		40.40	16.00	18.20	4.60	9.00
10.80	4.20	14.40	12.40	10.40		5.40	7.80	9.40	7.40	11.60

1. Make a frequency distribution of the payoffs on the winners of the first week by grouping them together in intervals of $3.00. The first column in your table should look like the one at the right.

   In how many of the races in the first week were the payoffs

2. less than $9?

3. $21 or more?

4. Make a comparable frequency distribution of the payoffs on the winners of the second week.

   In how many of the races in the second week were the payoffs

5. less than $9?

6. $21 or more?

7. Does the distribution of payoffs seem to change considerably from week to week or does it seem to remain approximately the same?

Payoff
0–2.99
3.00–5.99
6.00–8.99
9.00–11.99
12.00–14.99
15.00–17.99
18.00–20.99
21.00–23.99
24.00 and up

---

* June 17–21 and 24–28, 1981.

This histogram represents the distribution of the earth's surface area by elevation above (+) and below (−) sea level.

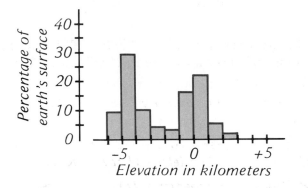

8. Within what interval of elevation is most of the earth's dry land?

9. Within what two intervals of elevation is most of the ocean floor?

Approximately what percentage of the earth's surface is

10. above sea level?

11. below sea level?

Hawaiian, a Polynesian language, is considered one of the most musical in the world. Although this language has been spoken for more than a thousand years, its written form was devised only as recently as 1822. Here is a sample of the language, a story about a menehune and a shark:

AKA AME KA MANO*

O Aka he menehune uuku momona.
I kekahi la ua luu o Aka,
Ua nahu ka mano i kona manamana wawae nui.
Alaila ua huhu loa ka mano,
No ka mea ua pololi loa oia.

Ua huhu pu no ka menehune,
No ka mea he hoaloha o Aka no lakou.
Nolaila hana lakou i hinai koali,
A ua hoopiha ia me ka maunu.

---

* From *Where the Red Lehua Grows* by Jane Comstock Clarke (Honolulu Star-Bulletin, Ltd., 1943).

Chapter 9: AN INTRODUCTION TO STATISTICS

Me keia ua paa ia lakou ka mano.
Ua huki ia iluna o ke one e make.
Ua hoike mai na iwi keokeo
"Mai hoopa i ka menehune."

12. The first paragraph contains 117 letters. Write the alphabet in a column and make a frequency distribution of the letters in this paragraph.

13. State some conclusions about the Hawaiian language that seem reasonable on the basis of your frequency distribution.

14. The second paragraph contains 28 words. Make a frequency distribution of the last letters of these words.

15. What does your frequency distribution suggest?

16. Is this conclusion supported by the other two paragraphs? Explain.

17. The third paragraph also contains 28 words. Make a frequency distribution of the first letters of these words.

18. What does your frequency distribution suggest?

19. Is this conclusion supported by the other two paragraphs? Explain.

Set III

On March 12, 1966, a civil-service examination was given in Chicago for the position of city engineer.* There were 223 applicants for 15 jobs. Their test scores are shown in the margin.

1. Make a frequency distribution of the scores, numbering the first column from 26 (the lowest score) to 95 (the highest.)

2. On the basis of this evidence, the examiners were charged with rigging the scores. Why?

---

* This problem is from *Statistics* by David Freedman, Robert Pisani, and Roger Purves (Norton, 1978).

42	49	74	47	78
51	37	92	68	39
90	30	81	95	74
27	48	76	68	39
52	84	64	42	62
34	43	44	56	80
54	34	48	60	91
39	59	37	45	46
58	31	42	50	31
84	56	47	50	46
33	66	91	91	37
49	43	54	43	55
32	69	45	48	71
58	51	83	46	29
37	34	67	55	40
84	93	75	39	65
84	48	59	33	49
52	93	90	67	59
92	30	43	56	69
45	84	61	58	39
93	47	45	54	92
95	37	69	83	61
80	56	60	45	57
84	57	37	46	58
83	72	69	33	53
30	58	42	43	44
37	43	63	27	36
51	61	48	47	44
66	71	69	53	46
27	35	69	33	44
58	60	33	62	52
73	43	55	69	45
48	44	32	61	57
26	84	58	48	58
41	62	80	39	52
80	46	61	53	63
82	30	60	27	75
54	47	39	60	51
67	69	44	45	81
35	48	83	47	31
81	82	61	43	54
36	53	93	51	74
67	59	76	66	57
56	60	49	36	52
53	90	42		

Lesson 2

The Breaking of Ciphers and Codes: An Application of Statistics

In one of the Sherlock Holmes stories, "The Adventure of the Dancing Men," the methods of statistics are applied to deciphering a series of mysterious messages. The messages were written as rows of little stick figures in which each figure stands for a letter of the alphabet. The first message looked like this:

As you can see, some of the figures are holding flags, and Sherlock Holmes concluded that the flags were used to indicate the ends of words. This message, then, consists of four words. Of the fifteen figures that it contains, one appears four times:

Holmes, knowing that the letter E is used more frequently than any other letter in English, decided that this figure represented E. This seems especially likely because two of these figures are holding flags and E is the last letter of many words.

Later in the story more messages appeared and Holmes, by putting a number of clues together, was able to decipher what they said. For example, the first word of the longest message

is five letters long and begins and ends with the letter E. The name of the woman to whom it was sent was Elsie, and so it seemed probable that this word was ELSIE. Therefore, the figures

stand for L, S, and I respectively.

Reasoning in a similar fashion with some other messages, Holmes decided that the first one said

AM HERE ABE SLANEY

and that the other message shown here said

ELSIE, PREPARE TO MEET THY GOD.

Holmes trapped the villain, Abe Slaney, by writing a message to him with the stick figures. Slaney, assuming that only Elsie could read the cipher, thought that the message was from her and so he went to her home, where he was arrested.*

The use of coded messages today has advanced far beyond detective stories and puzzles. In our age of electronic communications, coding has become very important not only in national security, but also in protecting private information stored in computers. The U.S. Department of State receives and sends several million coded words every week. The National Security Agency, which develops and breaks codes, has more than ten thousand employees and, because modern codes are so complex, is thought to have more computers than any other organization in the world. Today, cryptology, the science of code writing and breaking, makes use of many mathematical ideas, especially in algebra and statistics.

Exercises

Set I

The messages in "The Adventure of the Dancing Men" are written in a *simple substitution cipher*, a cipher in which each letter is always represented by the same symbol. Here is a list of the letters and symbols used in the two messages included in this lesson.

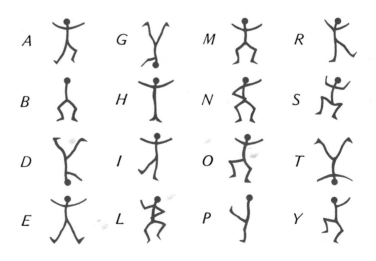

---

*If you are interested in reading the entire story, "The Adventure of the Dancing Men" is included in the series of adventures titled *The Return of Sherlock Holmes* by Sir Arthur Conan Doyle.

Chapter 9: AN INTRODUCTION TO STATISTICS

Most substitution ciphers do not have any pattern in the order of symbols used for the various letters of the alphabet.

1. Does there seem to be any pattern in the symbols in this "dancing men" cipher or do they appear to be chosen at random?

2. What does the following message say?

Here is the message that Sherlock Holmes sent to Abe Slaney.

3. Although it begins with a symbol that does not appear in the list, you should be able to figure out what it says.

The frequencies of the letters in any written language are always about the same in every large sample of it. Here is a graph of the frequencies (in percent) of the letters in ordinary English.*

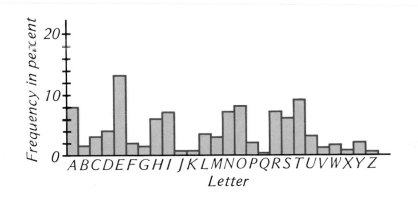

* After "Modern Cryptology" by David Kahn. Copyright © 1966 by Scientific American, Inc. All rights reserved.

4. What two letters are used the most frequently in English?

5. What two letters are used the least frequently?

From the graph, we see that the letter E appears about 13 times in a typical sample of 100 letters. In the first 100 letters of *A Christmas Carol* by Charles Dickens, the letter E appears 13 times.

6. From the graph, how many times would you expect the letter A to appear in the first 300 letters of the twenty-third psalm?

It actually appears 22 times (in the King James Version).

7. How many times would you expect the letter C to appear in the first 1,000 letters of Lincoln's Gettysburg Address?

It actually appears 31 times.

AAAAABBCCCCC
DDDDEEEEEEEFFFGG
GHHHHIIIIIIIJJKKLLLL
LMMMMMNNNNNNO
OOOOOPPPQQRRR
RRSSSSSSTTTTTUU
UVVWWXXYYZZ

A company that produces commercial lettering sheets says that the "letters are provided in the frequency they are most commonly used." The adjoining figure shows the alphabet as it appears on one of the company's sheets. The sheet contains 100 letters in all.

Look at this figure and the graph of the frequencies of the letters in ordinary English on page 513.

8. Do the numbers of the most frequently used letters in English,

E, T, A, O, I, N, and R,

that appear on this sheet seem correct?

9. Explain.

10. Do the numbers of the most rarely used letters in English,

V, J, K, X, Q, and Z,

that appear on this sheet seem correct?

11. Explain.

12. How do you think the sheet might be improved?

A sentence sometimes used as a typing exercise is:

Pack my box with five dozen liquor jugs.

13. Write the alphabet in a column and make a frequency distribution of the letters in this sentence.

Chapter 9: AN INTRODUCTION TO STATISTICS

14. Why is the sentence used as a typing exercise?

15. How does the frequency of E in the "liquor jugs" sentence compare with the frequency of E in ordinary English?

16. How does the frequency of Q in the "liquor jugs" sentence compare with the frequency of Q in ordinary English?

17. If the "liquor jugs" sentence were written in code, would it be easy or difficult to decipher?

Set II

Ciphers and codes are particularly important when countries are at war. This telegram was sent to Tokyo from the Japanese consulate in Honolulu on the evening before the attack on Pearl Harbor.

Written in a cipher in which the words were run together and the letters grouped in blocks of five, the telegram reported on the ships of the United States fleet that were in port.

1. Here is another cipher in which the words are not separated so that there are no clues about their lengths. For convenience, the message has been written in groups of five letters each.

```
B F C O N    A N Y K F    I X K U S    I X H U C
O N G F B    C N I A C    H C N A A    K S T N I
C O N P H    B F Y K I    N K M C O    N R H W H
F N A N F    H T L C O    B A W S H    L N I H F
B P W K U    C H F C U    K S N B F    C O N H P
N U B Y H    F T B Y C    K U L H C    C O N I N
Y B A B T    N D H C C    S N K M P    B I X H L
```

a) Copy the cipher on your paper, leaving two blank lines between each line of letters. Also write the letters of the alphabet in a column so that you can record what each letter stands for as you figure it out.

A graph of the frequencies of the letters in the cipher is shown below.

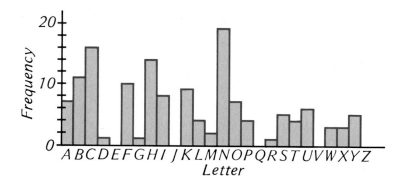

b) The most frequent letter is probably E, so write an E underneath each place this letter appears in the cipher.

c) The second most frequent letter is probably T. Write a T underneath each place this letter appears.

d) A very common word in English is THE. If THE appears several times in the cipher, you should be able to guess what letter represents H.

Chapter 9: AN INTRODUCTION TO STATISTICS

e) A strong clue in solving a cipher is knowledge of certain words that are likely to appear in it. For example, this cipher mentions the Second World War. If you can find a place where the words THE SECOND WORLD WAR can appear, you will know what several more letters represent.

f) Solving for the remaining letters in the cipher is left to you. What does the cipher say?

The machine in this photograph can be used both to translate information into coded form and to translate it back again. Manufactured in Switzerland, it has been used by many governments and commercial organizations.

The printed tapes shown below are similar to those produced by the machine.

Hagelin cryptograph; courtesy of Crypto AG

A

```
THISW ISWAW SECRE TWMES SAGEW WRITT ENWON WAWHA GELIN WCR

HUTIH QLIYZ OQYRR SVOED TRUDK XRWAF UVHQG AAOZT QRUYO YRR
```

B

```
HUTIHQLIYZOQYRRSVOEDTRUDKXRWAFUVHQGAAOZTQRUYOYRRLJZSUTDHM

THIS IS A SECRET MESSAGE  RITTEN ON A HAGELIN CRYPTOGRAPH
```

The first line of tape A shows what is typed on the machine and the second line shows it in coded form. (The machine automatically breaks the message into five-letter groups.)

2. What is the purpose of the W's in the first line?

3. Is each letter on the first line always replaced by the same letter in the coded form?

The first line of tape B shows what is typed on an identical machine when the code is received and the second line shows the deciphered message produced by the machine.

4. Why do you suppose the word "written" on the second line turns out misspelled?

5. Do you think a cipher produced by this machine could be solved by making a frequency distribution of the letters contained in it?

Set III

The following mystery story is from *Games* magazine.* The idea is not to guess "who did it," because it is the author "who's done it." The problem is to figure out exactly what he has done.

A SIN OF OMISSION

Around midnight, a sly-looking man slips into a luxury city building. A woman occupant, watching his actions from a fourth-floor window, grows suspicious and dials 911 for a patrol car. This lady complains, "A man in a brown suit, with shaggy hair, a slight build, and a criminal air is prowling through my lobby."

Fairly soon two young cops, Smith and Jarvis, pull up. Looking for an unknown vagrant, Smith spots Jim Oats walking out a front door. Oats, a minor burglar, is bold as brass, arrogant, and calm. Smith grabs him by his collar.

"O.K., Oats," snarls Smith, "what brings you to this location?"

Fixing his captor with a chilly look and frosty indignation, Oats quips, "I can go on a short stroll. Lift your filthy hands off my shirt. I'm not guilty of anything."

Smith drops his hands limply. This haughty air is too much for him to swallow. Angrily Smith says, "What a story. I'm nobody's fool, you punk. I just wish I could put you back in jail, but I can't obtain any proof against you. You know all about why I'm at this building—a station log full of burglary, arson, and muggings."

"Now, now," Oats laughs, "think of my rights. How can you talk this way?" Smith's probing hands start to frisk Oats for guns, narcotics, anything unlawful or contraband. Nothing shows up—only a small bound book. "What's this?" Smith asks.

Oats, tidying up his clothing, pluckishly says, "That's my political study of voting habits in this district. Why don't you look at my lists? I work for important politicians now—guys with lots of clout." An ominous implication lurks in this last thrust.

"Don't talk down to us," Smith snaps. But studying Oats's book, Jarvis finds nothing unusual. Smith finally hands him back his lists. Our cops can't hold him. Jarvis admits Oats can go. Just as a formality, Jarvis asks him, "Did you commit any criminal act in this building? Anything at all of which a courtroom jury could find you guilty?"

"No," Oats says flatly. "No way," and jauntily skips off. Halting six blocks away, Oats digs a tiny picklock from his sock and a diamond ring from his shaggy hair.

---

* *Games* magazine, November/December 1977.

Lesson 3

Measures of Location

The best-selling book *Working*, by Studs Terkel, contains interviews with many different people about their jobs. In one of the interviews, a steelworker remarks:

> *I don't know where they got the idea that we make so much. The lowest class payin' job there, he's makin' two dollars an hour if he's makin' that much. It starts with jobs class-1 and then they go up to class-35. But no one knows who that one is. Probably the superintendent. So they put all these class jobs together, divide it by the number of people workin' there, and you come up with a fabulous amount. But it's the big bosses who are makin' all the big money and the little guys are makin' the little money.**

To illustrate the situation with the salaries described, consider a company with just fifteen people. Suppose that their annual salaries are those shown in the figure at the right.

The sum of the fifteen salaries is $360,000. If this number is divided by 15, the number of people, the result is

$125,000

$60,000

$24,000 EACH

$15,000 EACH

$12,000

$10,000 EACH

$$\frac{\$360,000}{15} = \$24,000.$$

This number, called the *mean*, or *arithmetic average*, is the amount of money that each person in the company would make if everyone had the same salary and the total remained the same.

▶ The **mean** of a set of numbers is found by adding them and dividing the result by the number of numbers added.

---

* Steve Dubi, quoted in *Working* by Studs Terkel (Pantheon Books, 1974).

Although it would be correct to say that the "average" salary in the company is $24,000, in this case this number gives a distorted picture of the situation. If $24,000 is average, then most of the salaries in the company are below average.

Perhaps a more realistic way of illustrating the situation in this company would be to use the salary in the middle. This number, $12,000, is the *median.* Just as many people have salaries above this number as have salaries below it.

▶ The **median** of a set of numbers is the number in the middle when the numbers are arranged in order of size.

Another look at the salaries reveals that more people in the company make $10,000 than any other number. This number is called the *mode.*

▶ The **mode** of a set of numbers is the number that occurs most frequently, if there is such a number.

In general, a measure of location for a set of numbers should give us an idea of a typical number in the set. Which measure of location—*mean, median,* or *mode*—is the most appropriate for a set of numbers depends on the numbers in the set and the purpose that the measure is to serve.

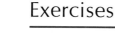

By permission of Johnny Hart and Field Enterprises, Inc.

Exercises

Set I

B. C. and Thor are playing golf. The scores on the first nine holes are shown in the table below.

| | \multicolumn{9}{c}{*Hole*} | |
	1	*2*	*3*	*4*	*5*	*6*	*7*	*8*	*9*	*Total*
B. C.	4	7	5	2	4	7	3	6	7	45
Thor	3	5	4	2	3	7	3	5	16	48

On the ninth hole, Thor got stuck in a sand trap and lost the game.

1. What was B. C.'s mean, or average, score on the nine holes?

2. What was Thor's mean, or average, score?

3. Which player did better on most of the holes?

4. Is this obvious from their mean scores?

B. C.'s scores on the nine holes arranged in order from smallest to largest are:

$$2 \quad 3 \quad 4 \quad 4 \quad 5 \quad 6 \quad 7 \quad 7 \quad 7$$

5. What was his median score?

6. Arrange Thor's scores on the nine holes in order from smallest to largest.

7. What was his median score?

8. What was the mode of B. C.'s scores?

9. What was the mode of Thor's scores?

10. Which measure of location—the mean, median, or mode—do you think gives the best comparison of the abilities of the two players?

The greatest recorded snowfall in a 24-hour period occurred at Silver Lake, Colorado, in April, 1921. It was 192 centimeters.

11. What was the mean hourly snowfall during that period?

12. Does this mean that that number of centimeters of snow fell during each hour of that period?

13. What was the least amount of snow that could have fallen there in a single hour?

14. What was the greatest amount of snow that could have fallen there in a single hour?

A useful fact that can sometimes be used to simplify the calculation of a mean, or average, can be discovered from the following exercises.

Find the mean of each of these sets of numbers.

15. 4, 6, 7, 15

16. 104, 106, 107, 115

17. Exactly how do the numbers in the second set compare in size with those in the first set?

18. How does the mean of the numbers in the second set compare with the mean of the numbers in the first set?

19. What shortcut does this suggest for finding the mean of the second set of numbers?

In 1861, the British scientist Sir William Crookes discovered the element thallium. He made ten measurements of its atomic weight, getting the numbers shown here.

203.63   203.63   203.64   203.64   203.64

203.64   203.64   203.65   203.65   203.67

20. Use the shortcut suggested by exercises 15 through 19 to find the mean of these measurements. Round it to the nearest hundredth.

Set II

In his *History of the Peloponnesian War*, the Greek historian Thucydides described the efforts of the Plateans, who wanted to climb over their enemy's walls. He wrote:

> They made ladders equal in height to the enemy's wall, getting the measure by counting the layers of bricks at a point where the . . . wall . . . happened not to have been plastered over. Many counted the layers at the same time, and while some were sure to make a mistake, the majority were likely to hit the true count. . . . The measurement of the

*ladders, then, they got at in this way, reckoning the measure from the thickness of the bricks.**

1. Which measure of location—mean, median, or mode—did the Plateans use in deciding how long to build their ladders?

A motor-vehicle bureau survey has revealed that, thirty years ago, each car on the road contained an average of 3.2 persons.[†] Twenty years ago, occupancy had declined to 2.1 persons per car. Ten years ago, the average was down to 1.4 persons. If this trend continues, ten years from now every third car going by will have nobody in it.

2. What sort of measure of location are the numbers in this survey?

According to probability theory, the expected numbers of girls in sixteen families having four children each are

0 1 1 1 1 2 2 2 2 2 2 3 3 3 3 4

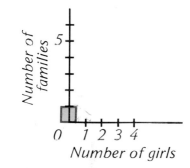

3. Make a histogram of this information, labeling the axes as shown at the right. The bar shown represents the only family having no girls.

4. What kind of symmetry do the five bars in the histogram have?

5. Find the mean, median, and modal numbers of girls per family.

The *actual* numbers of girls in a sample[‡] of sixteen families having four children each are

2 2 2 2 0 2 3 2 3 2 3 1 1 4 3 2

6. Make a histogram of this information, labeling the axes as in exercise 3.

7. Find the mean, median, and modal numbers of girls per family.

8. How does the shape of the histogram suggest that not all of these measures of location are the same number?

---

* *Thucydides*, translated by Charles Forster Smith (Harvard University Press, 1965.)

† Adapted from an anecdote in *Mathematical Circles Revisited* by Howard W. Eves (Prindle, Weber and Schmidt, 1971.)

‡ The first sixteen families having four children listed in *Who's Who 1980* (London: Black).

Ten of his friends decided to throw a surprise birthday party for Uncle Fletcher. The mean age of the first eight people to arrive is 21 years. The ninth person to arrive, Cora Bucksaddle, is 30 years old.

9. What does the mean age become after she joins the party?

The tenth person to arrive, Ole Chinbunny, is also 30 years old.

10. What does the mean age become after he joins the party?

11. Given that Ole and Cora are the same age, why doesn't Ole's joining the party raise the average age by the same amount that Cora's did?

When the guest of honor, Uncle Fletcher, joins the party, the mean average age becomes 30 years.

12. What birthday is Uncle Fletcher celebrating?

Set III

Redistribution of the Wealth in Oilaria*

The Sheik of Oilaria has proposed the following share-the-wealth program for his sheikdom. The population is divided into five economic classes. Class 1 is the poorest, class 2 is the next-poorest, and so on, to class 5, which is the richest. The plan is to average the wealth by pairs, starting with classes 1 and 2, then 2 and 3, then 3 and 4, and finally 4 and 5. Averaging means that the total wealth of the two classes is redistributed evenly to everyone in them.

The Sheik's Grand Vizier approves the plan but suggests that averaging begin with the two richest classes, then proceed down the scale instead of up.

Suppose that there are the same number of people in each economic class and that the following table shows the amount of money possessed by each class before the wealth is redistributed.

Class	Amount of money in billions
1	1
2	3
3	4
4	7
5	13

*This problem by Walter Penney originally appeared in Martin Gardner's "Mathematical Games" column in *Scientific American*, December 1979.

Figure out how much money each of the five classes would have if the wealth were redistributed according to

1. the Sheik's plan.

2. the Grand Vizier's plan.

3. Which plan would the poorest class prefer?

4. Explain why this choice would be preferred by the poorest class even if the amount of money possessed by each class was unknown.

5. Which plan would the richest class prefer?

6. Explain why this choice would be preferred by the richest class even if the amount of money possessed by each class was unknown.

"Oh! Oh!"

Drawing by John Gallagher

Lesson 4

Measures of Variability

The basketball team from Dribble High has arrived for its game with Wembly and its prospects of winning look pretty good. Although Wembly's coach had been told that the average height of the eleven players on the Dribble team was 180 centimeters, he did not know about Joe Dunkshot, its most promising player. This is the first game of the season and Dribble's coach had been keeping Joe a secret.

A measure of location, by itself, does not tell everything that someone might want to know about a set of numbers. In addition to a measure of location, it is often helpful to have information about how the numbers vary. One *measure of variability* is the *range*.

Here is a list of the heights, in centimeters, of the eleven members of the Dribble team, in order from shortest to tallest:

168   170   171   176   178   178   180   180   181   183   215

This list shows that the players vary in height from 168 centimeters to 215 centimeters. The *range* in heights is

$$215 \text{ cm} - 168 \text{ cm} = 47 \text{ cm}.$$

▶ The **range** in a set of number is the difference between the largest and the smallest numbers in the set.

Although the range in a set of numbers is easy to figure out, it is determined by only two of the numbers and does not give us any information about how the rest of the numbers in the set vary. Another *measure of variability* is the *standard deviation*.

▶ The **standard deviation** of a set of numbers is determined by finding:

1. the mean, or average, of the numbers,

2. the difference between each number in the set and the mean,

3. the squares of these differences,

4. the mean of the squares, and

5. the square root of this mean.

The calculation of the standard deviation in heights of the members of the Dribble team is shown below.

Height	Difference from mean	Square of difference
168	12	144
170	10	100
171	9	81
176	4	16
178	2	4
178	2	4
180	0	0
180	0	0
181	1	1
183	3	9
+ 215	35	+1,225
1,980		1,584

$$\text{Mean} = \frac{1,980}{11} = 180 \qquad \begin{array}{l}\text{Mean of} \\ \text{squares}\end{array} = \frac{1,584}{11} = 144 \qquad \begin{array}{l}\text{Square root} \\ \text{of mean}\end{array} = \sqrt{144} = 12$$

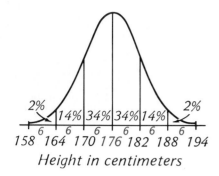

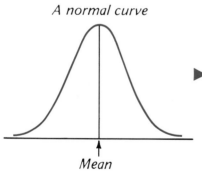

A normal curve

Mean

The mean height of the eleven members of the Dribble team is 180 centimeters and the standard deviation in heights is 12 centimeters.

For comparison, consider the variation in the heights of the approximately 2,000,000 seventeen-year-old boys in the United States. The mean height is 176 centimeters and the standard deviation is 6 centimeters. If a histogram were made of the heights and a smooth curve drawn through the midpoints of the tops of the bars, it would look like the adjoining figure.

The standard deviation has been used as the unit for numbering the horizontal axis. The curve is bell-shaped with its peak above the mean, 176 centimeters. From the graph, we see that the heights of 68% of the seventeen-year-old boys in the United States are within one standard deviation of the mean; in other words, 68% of the boys are between 170 centimeters and 182 centimeters tall. The heights of 96% are within two standard deviations of the mean (between 164 centimeters and 188 centimeters), and nearly 100% are within three standard deviations of the mean (between 158 centimeters and 194 centimeters). The graph is a *normal curve* and the percentages are significant because many large sets of numbers are distributed in the same way.

▶ A **normal curve** is a bell-shaped curve that closely matches the distribution of many large sets of numbers. For such sets of numbers, 68% of the numbers are within one standard deviation of the mean, 96% are within two standard deviations of the mean, and nearly 100% are within three standard deviations of it.

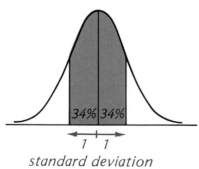

standard deviation
from mean

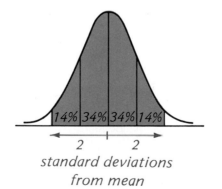

standard deviations
from mean

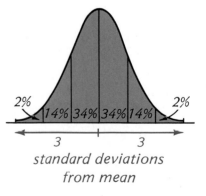

standard deviations
from mean

Chapter 9: AN INTRODUCTION TO STATISTICS

Exercises

Set I

The times, in seconds, that it takes ten runners to run an 800-meter race are

114   116   116   119   120   120   121   121   125   128.

1. What is the range in times?

2. Find the mean time.

3. Copy and complete the table at the right.

4. Find the mean of the numbers in the last column of the table.

5. Find the standard deviation of the times by taking the square root of this average.

6. What percentage of the ten times are one standard deviation or less from the mean?

7. What percentage are two standard deviations or less from the mean?

Time	Difference from mean	Square of difference
114	6	36
116		
116		
119		
120		
120		
121		
121		
125		
128		

Suppose that the time of the fastest runner was decreased by 10 seconds and that the time of the slowest runner was increased by 10 seconds so that the ten times are

104   116   116   119   120   120   121   121   125   138.

8. Now what is the range in times?

9. What is the mean time?

10. Copy and complete the table at the right.

11. Find the mean of the numbers in the last column of the table.

12. Find the standard deviation in the times by taking the square root of this average.

Time	Difference from mean	Square of difference
104	16	256
116		
116		
119		
120		
120		
121		
121		
125		
138		

On the basis of your answers to exercises 1 through 12, if the smallest and largest numbers in a set of numbers are increased and decreased by equal amounts, what do you think happens to

13. the range of the numbers?

14. the mean of the numbers?

15. the standard deviation of the numbers?

7	7	7	7	9
10	10	13	13	14
14	14	14	14	14
17	17	17	17	17
17	18	20	20	20
21	21	21	21	23
24	25	27	27	27
28	29	29	34	34
35	35	38	40	42
42	42	44	45	49

The scores of the winning teams in the fifty Rose Bowl games played from 1931 to 1980 are arranged from smallest to largest in the list at the left.

16. What is the range of the scores?

   The sum of the numbers in the list is 1,150.

17. Find the mean score of the winning team in these games.

   The standard deviation of these numbers is 11.

18. How many of the scores are one standard deviation or less from the mean?

19. What percentage of the scores are one standard deviation or less from the mean?

20. How many of the scores are two standard deviations or less from the mean?

21. What percentage of the scores are two standard deviations or less from the mean?

Set II

The normal curve first appeared in a paper by the French mathematician Abraham de Moivre in 1733. Since then, it has been found to have many useful applications.

   It is related, for example, to binomial probabilities. Compare the histograms below showing the relative frequencies of various numbers of heads when a set of coins is repeatedly tossed with the normal curve shown at the top of the next page.

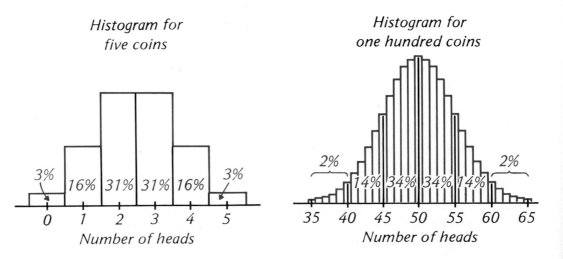

Histogram for five coins

Histogram for one hundred coins

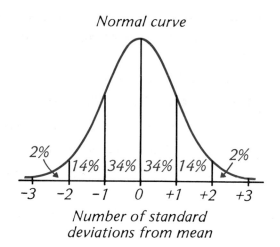

Normal curve

2% 14% 34% 34% 14% 2%

-3 -2 -1 0 +1 +2 +3

Number of standard
deviations from mean

The first histogram at the bottom of the preceding page shows that, when five coins are repeatedly tossed, 2 or 3 heads turn up

$$31\% + 31\% = 62\%$$

of the time.

1. When one hundred coins are repeatedly tossed, what percentage of the time do between 45 and 55 heads turn up?

2. When a large set of numbers fits a normal curve, what percentage of the numbers are within one standard deviation of the mean?

3. When five coins are repeatedly tossed, what percentage of the time do between 1 and 4 heads turn up?

4. When one hundred coins are repeatedly tossed, what percentage of the time do between 40 and 60 heads turn up?

5. When a large set of numbers fits a normal curve, what percentage of the numbers are within two standard deviations of the mean?

6. When five coins are repeatedly tossed, what percentage of the time do between 0 and 5 heads turn up?

7. When one hundred coins are repeatedly tossed, approximately what percentage of the time do between 35 and 65 heads turn up?

8. When a large set of numbers fits a normal curve, approximately what percentage of the numbers are within three standard deviations of the mean?

9. What happens to the shape of a "coin-tossing histogram" as the number of coins increases?

The Scholastic Aptitude Test (SAT) is the test most commonly required for college entrance. It was originally designed so that the distribution of scores looked like this.

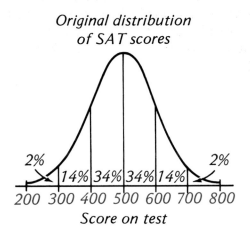

10. What does the curve in this graph look like?

11. What is the mean score?

The scale of scores extends three standard deviations below the mean and three standard deviations above the mean.

12. What is the standard deviation of the scores?

What percentage of the scores were between

13. 500 and 600?

14. 500 and 700?

15. 500 and 800?

Some colleges do not admit applicants whose scores are less than 600.

16. According to this distribution, what percentage of students would be expected to have scores of 600 or more?

In recent years, the mean score on the SAT has declined, although the standard deviation has remained about the same.

Boxes of breakfast cereal are often labeled as follows:

This package is sold by weight, not volume. Some settling of contents may have occurred during shipment and handling.

Although such boxes are labeled with a specific weight, their weights actually vary slightly because of chance variations in manufacturing.

The frequency distribution at the right shows the net weights of a sample of 800 boxes of cereal.

17. Make a histogram of the weights, labeling the axes as shown below. After you have made the histogram, mark the midpoint of the top of each bar with a dot and connect the dots with a smooth curve. (The first part of the graph is shown here as an example.)

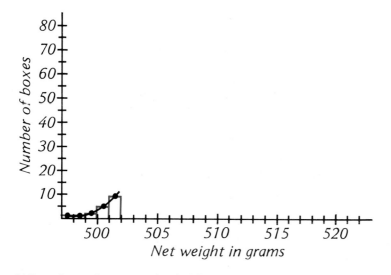

Net weight in grams	Number of boxes
497.0–497.9	1
498.0–498.9	1
499.0–499.9	2
500.0–500.9	5
501.0–501.9	9
502.0–502.9	14
503.0–503.9	21
504.0–504.9	31
505.0–505.9	44
506.0–506.9	53
507.0–507.9	67
508.0–508.9	73
509.0–509.9	80
510.0–510.9	79
511.0–511.9	74
512.0–512.9	66
513.0–513.9	54
514.0–514.9	42
515.0–515.9	32
516.0–516.9	20
517.0–517.9	14
518.0–518.9	10
519.0–519.9	5
520.0–520.9	2
521.0–521.9	0
522.0–522.9	1

18. What does the curve look like?

The peak of the curve corresponds to the mean net weight of the boxes in the sample.

19. What does the mean net weight of the boxes appear to be?

Calculation from the 800 individual weights of the standard deviation in weight reveals it to be 4.0 grams.

20. Use the numbers in the frequency distribution to determine the number of boxes that are within one standard deviation of the mean. (Use the eight intervals from 506.0 g to 513.9 g.)

21. What percentage of the 800 boxes are within one standard deviation of the mean?

22. How many boxes are within two standard deviations of the mean? (Use the sixteen intervals from 502.0 g to 517.9 g.)

23. What percentage of the 800 boxes are within two standard deviations of the mean?

24. How many boxes are within three standard deviations of the mean? (Use the twenty-four intervals from 498.0 g to 521.9 g.)

25. What percentage of the 800 boxes are within three standard deviations of the mean?

The numbers of children of the first 25 presidents of the United States are shown in this histogram.

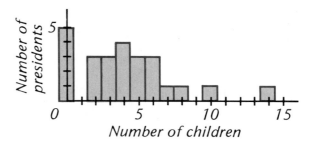

The mean of the numbers is 4.1 and the standard deviation is 3.3.*

26. How many of the first 25 presidents had a number of children within one standard deviation of the mean?

27. What percentage of these presidents had a number of children within one standard deviation of the mean?

28. How many of the first 25 presidents had a number of children within two standard deviations of the mean?

29. What percentage of these presidents had a number of children within two standard deviations of the mean?

30. How many of the first 25 presidents had a number of children within three standard deviations of the mean?

31. What percentage of these presidents had a number of children within three standard deviations of the mean?

Look again at the histogram of the numbers of children and at your answers to exercises 27, 29, and 31.

32. What is interesting about the answers?

---

* Each of these numbers has been rounded to the nearest tenth.

Chapter 9: AN INTRODUCTION TO STATISTICS

Set III

The curves in this figure show the variations in the weights of quarters that were minted and put into circulation at the same time.*

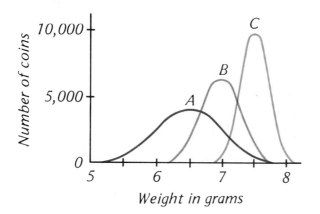

One curve shows the weight distribution when the coins were new and the other two show the distributions when they had been in circulation for five years and for ten years.

1. Which curve do you think shows the weights of the newly minted quarters, which curve the coins after five years, and which curve the coins after ten years?

2. What happens to the average weight of the coins as time passes?

3. What happens to the standard deviation of the weight of the coins as time passes?

---

* Adapted from a graph in the article "Scientific Numismatics" by D. D. Kosambi, *Scientific American*, February 1966.

Reprinted by permission; © 1974 by NEA, Inc.

" . . . As you can see, the profit picture for oil
companies isn't all THAT bright . . . "

Lesson 5

Displaying Data: Statistical Graphs

Year	Profits in millions
1970	403
1971	429
1972	458
1973	491
1974	515
1975	536
1976	574
1977	620
1978	677
1979	734
1980	811

Statistical information is often presented in graphical form. Graphs reveal patterns in tables of numbers in a very simple way—so simple, in fact, that the patterns can often be seen in a single glance.

Unfortunately, graphs can also be used to deceive. Although it seems unlikely that anyone would be gullible enough to be fooled by the trick used in the cartoon above, there are other ways in which a graph may be used to create a false impression of the data that it represents.

Here is an example. The annual profits of a certain oil company are shown in the table at the left. A graph of this information is shown below. We can tell from just a glance at it that the compa-

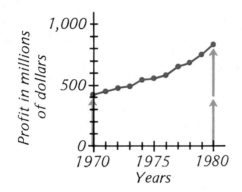

ny's profits steadily increased during the 1970s. Looking at it more carefully, we see that the line rises to roughly twice its original height, which corresponds to the fact that the company's profits at the end of the decade were about twice those at the beginning.

Now suppose that we remove the lower part of the graph, getting the result shown here. This simple change alters the picture

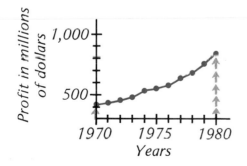

considerably. Someone looking at this graph and neglecting to notice the numbering on the vertical axis might conclude that the company's profits at the end of the decade were about *five* times those at the beginning.

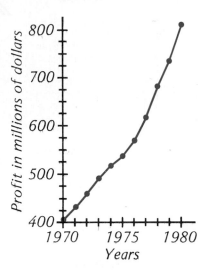

This effect can be exaggerated even further by renumbering the vertical axis to get the graph at the left. To someone who ignores the numbering, the increase in profits seems shocking. A graph of this type serves a useful purpose, however: it makes the changes easier to see.

One thing, then, that determines the impression given by a graph is the numbering of its axes. The scales are important and, if they are removed, such a graph becomes almost meaningless.

Although graphs are used to present statistical information because of the simple picture that they can give of the patterns it contains, they can also be deliberately drawn to give a false impression. And even a graph that is meant to give an honest representation can be misinterpreted.

Exercises

Set I

Temperatures during the year vary considerably from city to city. The following tables give the normal low and high Fahrenheit temperatures in Baghdad, Acapulco, and Melbourne for each month of the year.

Baghdad			Acapulco			Melbourne		
Month	Low	High	Month	Low	High	Month	Low	High
Jan.	39	60	Jan.	70	85	Jan.	57	78
Feb.	42	64	Feb.	70	87	Feb.	57	78
Mar.	48	71	Mar.	70	87	Mar.	55	75
Apr.	57	85	Apr.	71	87	Apr.	51	68
May	67	97	May	74	89	May	47	62
June	73	105	June	76	89	June	44	57
July	76	110	July	75	89	July	42	56
Aug.	76	110	Aug.	75	89	Aug.	43	59
Sept.	70	104	Sept.	75	88	Sept.	46	63
Oct.	61	92	Oct.	74	88	Oct.	48	67
Nov.	51	77	Nov.	72	88	Nov.	51	71
Dec.	42	64	Dec.	70	87	Dec.	54	75

1. Make a line graph for Baghdad to show how the low and high temperatures vary during the year, labeling the axes as shown

at the right. The temperatures for the months of January, February, and March are shown as an example.

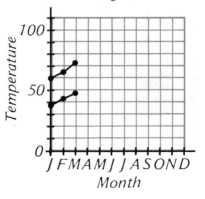

Baghdad

2. Make a line graph showing the variations of low and high temperatures in Acapulco during the year, labeling the axes in the same way.

3. Make a line graph showing the variations of low and high temperatures in Melbourne during the year, labeling the axes in the same way.

 Refer to your line graphs to determine which city has

4. the least variation in temperature during the year.

5. the greatest variation in temperature during the year.

6. What advantage do your line graphs have over the tables in presenting the information about the temperature variations in the three cities?

7. Graph the Melbourne information again, using a temperature scale ranging from 40 to 90 as shown at the right.

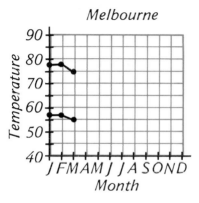

Melbourne

 Compare this graph with the other Melbourne graph.

8. In what way does changing the temperature scale from 0°–110° to 40°–90° change the impression given of the variation in temperatures?

9. Why does the change in the temperature scale do this?

This bar graph appeared several years ago in an ad of an insurance company.

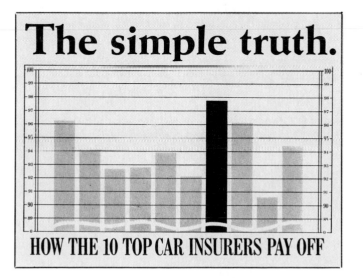

10. Does it really show the simple truth?

The bars represent the "payoffs" of ten insurance companies, the tallest bar representing that of the company running the ad.

11. Roughly how many times as tall as the shortest bar in the graph is the tallest one?

12. Make a list of the "payoff" numbers shown for the ten companies, rounding each one to the nearest whole number. For example, the rounded "payoff" numbers for the first three bars are 96, 94, and 93.

13. Graph this information to give a more honest picture of the situation. Number the vertical scale in 10's from 0 to 100.

14. Why do you suppose the more honest graph was not used in the ad?

The shaded slices of these circle (or "pie") graphs show the parts of the population of the United States in 1900 and 1980 that were less than nineteen years old.

Part of the U.S. population
less than nineteen years old

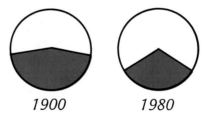

1900   *1980*

15. In which year was a greater percentage of the population less than nineteen years old?

16. Does it follow that there were more people in the United States less than nineteen years old in that year than in the other? Explain.

Set II

By permission of Johnny Hart and Field Enterprises, Inc.

Picture graphs, frequently used to make comparisons, can be easily misinterpreted. Suppose, for example, that Peter has twice as many clams as B. C. and that we decide to show this with two bags of clams, one drawn so that it is twice as tall as the other.

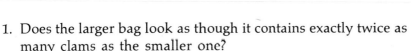

1. Does the larger bag look as though it contains exactly twice as many clams as the smaller one?

Rather than comparing the heights of the two bags, someone might compare the *areas* covered by them. Notice that the larger bag is twice as wide as the smaller one, in addition to being twice as tall.

2. How do you think their areas compare?

Hint:

Lesson 5: Statistical Graphs   541

The number of clams that each bag contains is determined by its *volume*.

3. How do you think their volumes compare?

Hint:

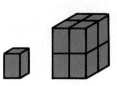

A different picture graph could be drawn, also using bags of clams, which makes it perfectly clear that Peter has just twice as many clams as B. C.

4. What would it look like?

This graph appeared several years ago in a bulletin for the employees of a large company. It compared the number of absences on four consecutive Mondays, the last one preceding a legal holiday.

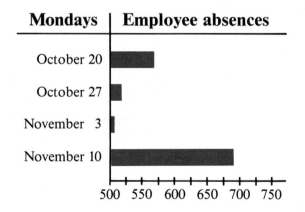

5. What impression does the graph seem intended to convey?

6. The actual numbers of absences on these four days were

566, 517, 501, and 689.

Do the lengths of the four bars make this obvious?

7. Why does the graph give the impression that it does?

8. Graph the number of absences again, renumbering the scale along the bottom as follows: 0, 150, 300, 450, 600, 750.

Chapter 9: AN INTRODUCTION TO STATISTICS

This graph shows the change in the U.S. population per square mile of land as recorded by each census since 1790. Although

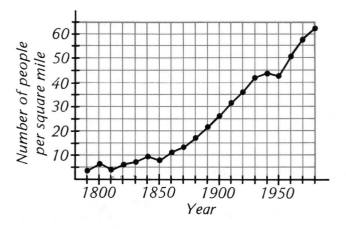

almost every census has shown an increase in the population per square mile, three censuses have shown a decrease.

9. Which years are they?

10. Does the graph show that the population of the United States decreased at these censuses? Explain.

Blood sugar is a source of energy. A study of nutrition has revealed that, if you eat a high-carbohydrate breakfast, your blood sugar quickly reaches a peak and then drops rapidly. If you eat a high-protein breakfast, your blood sugar rises more slowly to a lower peak, but it also drops more slowly, never getting as low as with a high-carbohydrate breakfast.

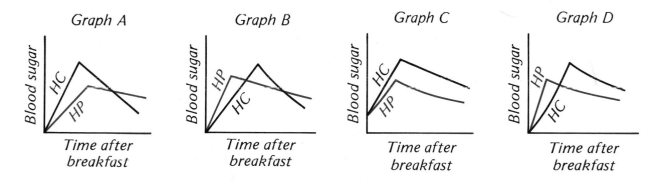

11. Which graph is consistent with this information?

Is it possible to tell from any of the graphs

12. that the blood-sugar peak that results from a high-carbohydrate breakfast is higher than that produced by a high-protein breakfast?

13. exactly how the blood-sugar peaks that result from the two types of breakfasts compare in size?

The first graph below shows what happened to the average price of a quart of milk between 1939 and 1979 and the second graph shows what happened to the average time that a person had to work to earn a quart of milk during the same period.

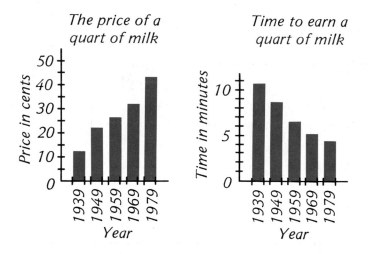

14. Which graph presents a more encouraging picture?

In 1949, a quart of milk cost almost twice as much as it did in 1939.

15. Roughly how did the price of a quart of milk in 1979 compare with the price in 1939?

16. Roughly how did the time required to earn a quart of milk in 1979 compare with the time in 1939?

17. The graphs appear to contradict each other, yet both are correct. How is this possible?

18. What do the answers to exercises 15 and 16 imply about the average wages in 1979 compared with those in 1939?

Set III

Which face looks happier? Your answer to this question may depend on your age.

The graph below shows the results of an experiment in which people of different ages were shown drawings of two faces, identical except for the size of the pupils, and asked to choose the happier one.* The dark-brown bars represent the percentages of peo-

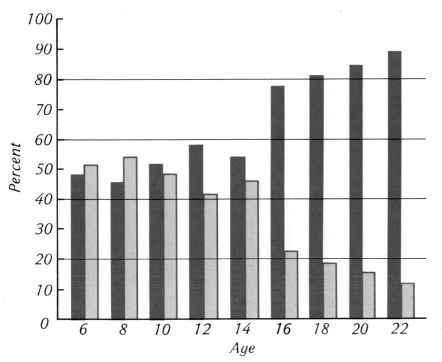

ple who chose the face with the large pupils and the light-brown bars represent the percentages of people who chose the face with the small pupils.

1. Do you notice any pattern in the graph? If so, what is it?

2. Is there a connection between the height of each light-brown bar and the height of the corresponding dark-brown bar? If so, what is it?

3. Why do you suppose both bars are shown for each age?

4. What conclusion do you suppose was drawn from this experiment?

---

* "The Role of Pupil Size in Communication" by Eckhard H. Hess. Copyright © 1975 by Scientific American, Inc. All rights reserved.

'I can't say that I go for this kind of TV poll . . .'

Lesson 6

Collecting Data: Sampling

The popularity of television programs in the United States is determined by sampling. The Nielsen ratings are based on a sample of 1,200 homes out of the approximately 80,000,000 homes with television sets. According to the Nielsen Company,*

> *To do a proper job, a sample needs to reflect, with a proper degree of accuracy, the universe it measures. It is, in effect, a scale model of this universe. The Nielsen TV samples, for example, include all types of households and neighborhoods.*

▶ A **sample** is a group of items chosen to represent a larger group called the *population* or *universe*.

---

* "Audience Research," *T.V. Guide Almanac,* compiled and edited by Craig T. and Peter G. Norback and the editors of *TV Guide* Magazine (Ballantine Books, 1980).

The word *population* in statistics does not refer to people inhabiting a place but rather to a group of things (people or animals or objects) being studied. Samples are used because they can supply accurate information at less cost and in less time than would be required by a complete survey of the population. In fact, for very large populations, a complete survey may be impossible. For example, the average annual rainfall in New York City is determined from samples recorded at weather stations because it would be impossible to collect and measure all the rain that fell in that city in a year.

Samples are sometimes the only way that information can be obtained about even a limited population. If a fireworks company tested every one of the firecrackers that it produced to find out whether it would explode, it would not have any firecrackers left to sell.

For a sample to accurately represent a population, it must be *sufficiently large*. If the sample is too small, there is a good chance that the items included in it are not typical of the population. Statisticians have methods of determining just how large a sample should be to be reasonably dependable.

It is also important that the sample be a *random* one. In other words, the items in it should be chosen so that *every item in the population has an equal chance of being included in the sample*. A sample that is not chosen in this way may be biased and give a distorted picture of the population that it is supposed to represent.

There is another reason for wanting samples to be random. It is impossible to be absolutely certain of the nature of a population on the basis of samples of it because the samples will vary. If the samples are random, however, then the methods of probability can be applied to them to get an idea of how accurately they represent the population. Probability theory, then, is fundamental to the sampling process and, hence, a very important part of statistics.

Exercises

Set I

Several years ago, the Chesapeake and Ohio Railroad Company made a study of sampling.* From a *sample* of freight bills for a six-month period, the company estimated that another railroad company owed it $64,568. By checking *all* the freight bills of that period, the company found that the other company actually owed it $64,651.

1. How much money would the Chesapeake and Ohio Railroad Company have lost by trusting the number from the sample?

Checking the sample of freight bills cost the company $1,000; checking all the freight bills cost the company $5,000.

2. How much money would the Chesapeake and Ohio Railroad have saved by relying on the sample instead of checking all of the bills?

3. If these results are typical, which method would benefit the company more: taking a sample of the freight bills for each six-month period or checking all of them? Explain.

---

* "How Accountants Save Money by Sampling," by John Neter, in *Statistics: A Guide to the Unknown*, edited by Judith M. Tanur (Holden-Day, 1972).

Chapter 9: AN INTRODUCTION TO STATISTICS

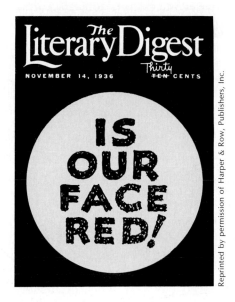

Reprinted by permission of Harper & Row, Publishers, Inc.

In 1936, a magazine called the *Literary Digest* sent a questionnaire to a sample of telephone subscribers and automobile owners asking them who they would vote for in the presidential election of that year. On the basis of the 2,376,523 replies, the magazine predicted that Landon would defeat Roosevelt, a prediction so far off-target that it may have been partly responsible for the magazine going out of business the following year. The table below shows how the *Literary Digest's* prediction compared with the outcome of the election.

Candidate	Percentage of votes predicted	Percentage of votes received
Landon	57	38
Roosevelt	43	62

4. Do you think that the sample of voters chosen was large enough?

5. Was it random? Explain.

6. Why do you suppose the poll could have been so wrong?

A method sometimes used to estimate the number of fish in a lake is to catch a sample of the fish, tag them, and throw them back. Later, another sample is caught and the number of tagged fish counted.

Suppose that the second sample consisted of 50 fish, of which 10 were tagged.

7. What fraction of the fish in this sample were tagged?

8. If we assume that this sample is typical of the population of fish in the lake, what fraction of the fish in the lake are tagged?

Suppose that the first sample consisted of 100 fish.

9. How many fish do the samples suggest are in the lake?

In using this method, it is assumed that each fish has an equal chance of being caught and that this is true for both of the samples.

10. Give a reason why this may not be true.

An article in the magazine *California Highways** reported that a survey of the clothing worn by pedestrians killed in traffic at night revealed that about 80% of the victims were wearing dark clothes and 20% light-colored clothing. The conclusion was drawn that pedestrians are safer at night if they wear something white so that drivers can see them more easily.

11. What conclusion could be drawn instead?

12. What additional survey would be helpful in deciding which conclusion is correct?

Set II

The effectiveness of the Salk vaccine against polio was tested in 1954 with approximately 2,000,000 children. In one of the towns in which it was tested, there were approximately 1,000 children, half of which were given the vaccine and half of which were not.

A polio epidemic passed through the state in which the town was located and not one of the children innoculated with the vaccine caught the disease.

1. Did this prove that the vaccine was effective?

None of the children who had not been innoculated caught the disease either.

2. What do you suppose this indicated about the size of the sample?

---

*Cited in *How to Take a Chance* by Darrell Huff (Norton, 1959).

To find out how many owners of cars are making payments on their cars, a group of poll-takers stationed themselves at the exits of the parking lot of an amusement park. After questioning 800 drivers leaving the lot, they reported that 92% of the owners of cars had paid for them in full.

3. Is this a random sample of car owners? Explain why or why not.

4. If the sample were not random, would it matter? Explain why or why not.

In any poll, there is always the possibility that some of the answers given are not true.

5. Why might this be likely in this survey?

6. What effect would these answers have on the figure reported?

Several years ago Pepsi-Cola based an advertising campaign on a "blind taste test" comparing Pepsi and Coca-Cola.*

A sample of Coca-Cola drinkers was served Pepsi in a glass marked M and Coke in a glass marked Q. More than half of them preferred the cola in the glass marked M.

7. What do you think Pepsi-Cola claimed that this test proved?

8. What do you think Coca-Cola claimed that this test proved about the two letters used?

Coca-Cola then did a test in which Coca-Cola was served in glasses marked M and in glasses marked Q.

9. What do you think they discovered?

After Coca-Cola advertised the results of this test, Pepsi changed the letters on its glasses to L and S, with Pepsi in the

* *Time* magazine, July 26, 1976.

glasses marked L and Coke in the glasses marked S. More than half of the people tested preferred the cola in the glass marked L.

10. What would have been a more convincing way of changing the test?

In an investigation to determine whether there is a connection between children's ability in arithmetic and the size of their feet, a random sample of 200 children was chosen from a large elementary school. Each child was given an arithmetic test and the length of his or her feet was measured. The results are shown in the table at the left.

Length of feet in centimeters	Average score on test
15.0–16.9	40%
17.0–18.9	54%
19.0–20.9	67%
21.0–22.9	78%
23.0–24.9	84%

Do these results show that

11. the bigger a child's foot is, the better he or she tends to be at arithmetic?

12. big feet cause a child to be good at arithmetic?

13. What is the cause of the pattern in this table?

The managers of a summer camp were curious about the numbers of children in the families of the children that attend it. They asked each of the 100 children attending the camp one week how many children were in his or her family. The results are summarized in the frequency distribution below.

Number of children in family	Number of children reporting
1	17
2	23
3	22
4	15
5	9
6	8
7	0
8	6

From this information, they computed the mean number of children per family.

14. What did they get?

The actual mean number of children in the families of the 100 children attending that week was 2.4.

15. What was wrong with the procedure used by the management to find the average?

Several years ago, two psychologists in Los Angeles did an experiment to test people's honesty. They used 375 envelopes that were addressed to one of the psychologists at his home. Of these envelopes:

75 were empty and had typed on them: "This is a research study. Drop this envelope in the nearest postbox. Thank you for your cooperation."

150 contained blank sheets of folded paper but did not have a message typed on the outside.

150 seemed to contain money (two coins and a bill that were actually fake).

The psychologists dropped each envelope on a sidewalk near a mailbox. A third of the envelopes were left in wealthy sections of the city, a third in middle-income areas, and a third in poor neighborhoods. Each one was marked on the inside to show the area in which it had been left.

The numbers of each type of envelope that were returned by the people who found them are shown in the table below.

Type of envelope	Number dropped on sidewalks	Number returned
Empty envelopes labeled "research study"	75	68
Envelopes containing blank sheets of paper	150	120
Envelopes seeming to contain money	150	102

Which type of envelope had

1. the highest rate of return?

2. the lowest rate of return?

A purpose of the experiment was to find out how many people would do something dishonest, such as taking a letter or money that did not belong to them, if there was very little chance of being caught.

3. Why, then, do you think empty envelopes labeled "This is a research study" were used in the experiment?

Some of the envelopes that contained either blank sheets of paper or the fake money were opened and resealed by the people who found them before they were returned.

4. How do you think these envelopes were counted?

The table below shows the number of each type of envelope returned from each area.

Type of envelope	Area of city		
	Poor	Middle	Wealthy
Empty envelopes labeled "research study"	22	24	22
Envelopes containing blank sheets of paper	43	37	40
Envelopes seeming to contain money	28	33	41

There seems to be one significant pattern in this table.

5. What is it?

Although the sample used in this experiment was both sufficiently large and random, the design of the experiment prevents any conclusion about people's honesty from being drawn from the pattern.

6. Why? (*Hint:* The problem concerns temptation.)

"*Tonight, we're going to let the statistics speak for themselves.*"

Chapter 9 / Summary and Review

In this chapter we have become acquainted with the branch of mathematics that deals with the collection, organization, and interpretation of numerical facts. We have studied:

Organizing data: frequency distributions (*Lessons 1 and 2*) A frequency distribution is a convenient way of organizing data to reveal what patterns they may have. The data can be condensed by grouping them together in intervals.

Measures of location (*Lesson 3*) Three measures of location are commonly used:

The *mean*, or average, of a set of numbers is found by adding them and dividing the result by the number of numbers added.

The *median* is the number in the middle when the numbers are arranged in order of size.

The *mode* is the number that occurs most frequently, if there is such a number.

Measures of variability (*Lesson 4*) The *range* of a set of numbers is the difference between the largest and the smallest numbers in the set.

The *standard deviation* of a set of numbers is determined by finding:

1. the mean of the numbers,
2. the difference between each number in the set and the mean,
3. the squares of these differences,
4. the mean of the squares, and
5. the square root of this mean.

The standard deviation is useful because of its relation to the normal curve.

A *normal curve* is a bell-shaped curve that closely matches the distribution of many large sets of numbers. For such sets of num-

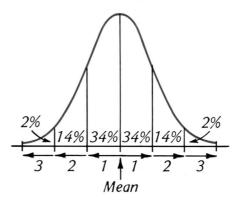

bers, 68% of the numbers are within one standard deviation of the mean, 96% are within two standard deviations of the mean, and nearly 100% are within three standard deviations of it.

Displaying data: statistical graphs (*Lesson 5*) Statistical graphs are sometimes drawn in a way that gives a false impression. In bar and line graphs, it is especially important to look at the numbering of the scales. Picture graphs can be misleading if it is not clear whether heights or areas or volumes should be compared.

Collecting data: sampling (*Lesson 6*) A *sample* is a group of items chosen to represent a larger group called the *population*. For a sample to be representative, it is important that it be sufficiently large and that it be random. For a sample to be random, every item in the population must have an equal chance of being included in it.

Chapter 9: AN INTRODUCTION TO STATISTICS

Exercises

Set I

The cruising ranges in miles of 36 cars are shown in the list below.*

370 510 460 325 420 315 270 300 400
515 350 425 420 380 460 395 420 485
480 300 505 355 690 375 340 300 505
285 360 325 305 340 295 485 495 510

1. Make a frequency distribution of these numbers by grouping them in intervals of 100 miles. Number the first column in your table like this.

Distance in miles
_____

200–299
300–399
400–499
500–599
600–699

2. Make a histogram of the distances, labeling the axes as shown at the right.

How many of the cars have cruising ranges of

3. less than 400 miles?

4. 400 miles or more?

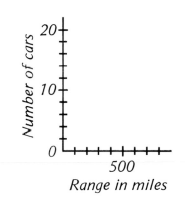

In the introduction to their book *American Averages,*[†] Mike Feinsilber and William B. Mead wrote:

> *Did you know that about half the people in America are below average?*
> *Not you. Not us.*
> *Them.*

5. Which measure of location is being referred to?

6. For which measure of location is it possible that more than half of the people in America could be "below average"?

_____

* "The 1981 Cars" in *Consumer Reports,* April 1981.
† Dolphin Books, 1980.

A drug company once included a graph like the one shown here in one of its ads claiming that its product acts "twice as fast as aspirin."

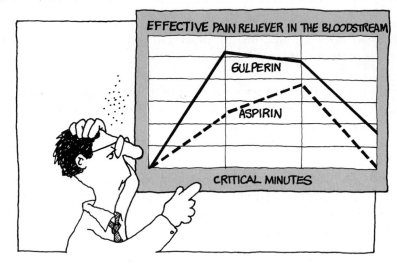

7. Do the lines in the graph seem to show that the drug acts twice as fast as aspirin?

8. What is missing from the graph?

9. Is it possible to tell from the graph how long the effects of the drug will last?

The lines do not seem to be accurately drawn over all of the time shown.

10. Explain why not.

These histograms show the age distribution of males and females living in the United States in 1980.

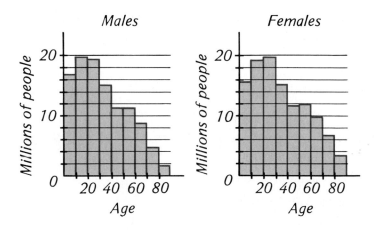

Chapter 9: AN INTRODUCTION TO STATISTICS

11. Which age interval has the most males?

12. Which age interval has the most females?

13. How does the number of males younger than 20 compare with the number of females younger than 20?

14. How does the number of males older than 50 compare with the number of females older than 50?

There is a legend that many years ago someone wanted to find out how tall the emperor of China was. To ask to measure the emperor was out of the question; so this person decided to poll a million of his countrymen to learn what they thought the height of the emperor was.

After the poll was completed, the investigator based his guess on the mean height named, even though not one of the people asked had ever seen the emperor.

Suppose that a graph of the guesses looked like this.

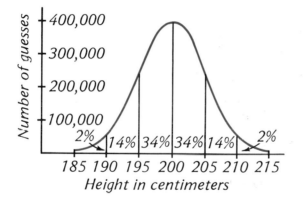

15. What kind of curve does this appear to be?

16. What was the mean, or average, guess of the emperor's height?

17. What was the standard deviation from the mean?

What percentage of the guesses were within

18. one standard deviation of the mean?

19. two standard deviations of the mean?

20. three standard deviations of the mean?

Approximately how many of the million people polled thought that the emperor was

21. at least 200 centimeters tall?

22. more than 215 centimeters tall?

Every day thousands of airline passengers take trips in which they transfer from one airline to another. The airlines decide how much money should be exchanged between them because of these transfers on the basis of just a small sample of them.

23. Why do you think the airlines use a sample of the transfers rather than all of them?

Federally owned land

California   *Oregon*

The shaded slices of the circle graphs at the left show the parts of California and Oregon owned by the federal government.

24. In which state does the government own a greater percentage of the land?

25. Does it follow that the government owns more land in that state than in the other? Explain.

Set II

If a runner had a race with a roller skater, the skater would probably win. The world record times for three different distances are shown in the table below.

Distance	Runner	Roller skater
400 meters	44 seconds	35 seconds
800 meters	102 seconds	73 seconds
1,500 meters	212 seconds	135 seconds

1. Which one of these three bar graphs comparing the times for the distances is misleading? Why?

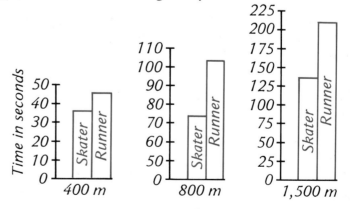

2. What should be noticed in comparing the graphs for 400 meters and 1,500 meters?

Seven people were asked to measure the length of a dollar bill in centimeters. The answers that they reported are listed here:

15   15.2   15.2   15.3   15.3   15.3   153

3. Which answer does not seem to be in centimeters?

4. Judging from the answers, what do you think the length of a dollar bill in centimeters is?

5. Which measure of location did you use? Explain.

6. Find the mean of the seven measurements.

7. Which measures of location are the least affected by the fact that one person's answer is so far off?

According to the 1980 census, more households in the United States consist of 2 people than any other number. A table of the relative sizes of households is shown at the right.

Number of persons in household	Percentage of households
1	23
2	30
3	18
4	16
5	8
6	3
7 or more	2

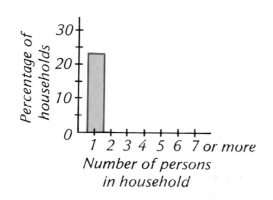

8. Make a histogram of this information, labeling the axes as shown here. After you have made the histogram, mark the midpoint of the top of each bar with a dot and connect the dots with a smooth curve.

9. Do the relative sizes of the households seem to lie along a normal curve?

The mean number of persons per household is about 2.7 and the standard deviation is about 1.5.

10. What percentage of households are within one standard deviation of the mean?

11. What percentage are within two standard deviations of the mean?

12. Are these close to the percentages that one would expect for a normal curve?

Voting records for the 1976 presidential election show that 54% of those old enough to vote did so. In a survey of eligible voters in 1977, 59% of the people in the sample said they voted.

13. Why do you suppose the figure obtained from the sample was higher than the figure obtained from the voting records?

14. What other properties of a sample might cause the information obtained from it to be inaccurate?

This picture graph accompanied a newspaper ad with the headline: "The Times has 2,244,500 readers every weekday—more than the next four area newspapers *combined!*"

One weekday issue	The Times	The Examiner	The Register	The Daily News	The Press-Telegram
Number of readers	2,244,500	624,000	485,500	350,500	310,500

15. How do the numbers of readers of the five papers seem to have been used in determining the sizes of the trucks?

16. Is the graph misleading? Explain.

Set III

A well-known signal of distress is ··· ⁻⁻⁻ ···, which is Morse code for the letters S O S. In the transmission of Morse code, each dash takes three times as long as each dot. The figure near the top of the next page shows the relative amounts of time for each part of the

S O S signal. To send an S takes 5 units of time and to send an O takes 11 units of time.

The table below shows the Morse code symbol for each letter of the alphabet.

Letter	Symbol		Letter	Symbol
A			N	
B			O	
C			P	
D			Q	
E			R	
F			S	
G			T	
H			U	
I			V	
J			W	
K			X	
L			Y	
M			Z	

1. Which letter takes just 1 unit of time to send?

2. Which letters take 3 units of time to send?

3. Which letters take 5 units of time to send?

Look at the graph of the frequencies in percent of the letters in ordinary English on page 513.

4. What do you notice about the frequencies of the letters that you named in exercises 1 through 3?

5. Which letters take the longest time (13 units) to send?

6. What do the frequencies of these letters have in common?

7. Which letters take 11 units of time to send?

The symbol for one of these letters seems like a very poor choice.

8. Which letter?

9. Why?

Chapter 9 / Problems for Further Exploration

Lesson 1

1. In 1787–1788, a series of essays was published to persuade the citizens of New York to ratify the proposed constitution of the United States.* These essays, called the Federalist papers, were written anonymously by Alexander Hamilton, John Jay, and James Madison. It is now known that 51 of them were written by Hamilton, 14 by Madison, 12 by either Hamilton or Madison, and 5 by Jay.

   To determine the authorship of the 12 disputed papers, historians have studied the frequencies in which certain words appear in them. One word used was "by." The numbers of times that "by" was used in 48 essays by Hamilton, 50 essays† by Madison, and the 12 disputed papers are summarized in the table below.

Number of times per 1,000 words	48 essays by Hamilton	50 essays by Madison	12 disputed papers
1.0–2.9	2	0	0
3.0–4.9	7	0	0
5.0–6.9	12	5	2
7.0–8.9	18	7	1
9.0–10.9	4	8	2
11.0–12.9	5	16	4
13.0–14.9	0	6	2
15.0–16.9	0	5	1
17.0–18.9	0	3	0

---

* The information in this exercise is from the article "Deciding Authorship," by Frederick Mosteller and David L. Wallace, in *Statistics: A Guide to the Unknown,* edited by Judith M. Tanur (Holden-Day, 1972).

† This number includes essays in addition to the Federalist papers.

From this table, we see that "by" was used between 1.0 and 2.9 times per 1,000 words in 2 of the 48 essays by Hamilton:

$$\frac{2}{48} \times 100\% = \frac{200\%}{48} \approx 4\%$$

a) Copy the table, replacing each number of essays with the corresponding percentage.
b) Refer to your table to make three histograms to represent the appearance of "by" in the essays by Hamilton, the essays by Madison, and the disputed papers. Label the axes of each histogram like this.

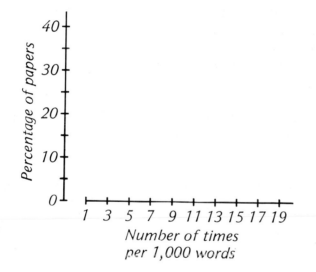

Refer to your histograms to answer the following questions.
c) Who used the word "by" more frequently: Hamilton or Madison?
d) What do the histograms seem to indicate about the authorship of the disputed papers? Explain your reasoning.

2. If three cards are dealt from a shuffled pack, which of the following is more likely: that all three are of different suits, that two cards are of the same suit, or that all three cards are of the same suit?

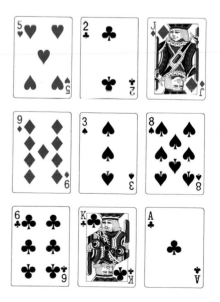

Greatest number of cards in any suit
3
2
1

a) Take a pack of playing cards, shuffle it, and deal the cards out in groups of three. Make a frequency distribution of the 17 groups that result, labeling the first column as shown at the left.

Reshuffle the cards and deal them out again, adding the results of the 17 newly formed groups to the distribution. Continue to do this until you have recorded the results for exactly 100 groups of three. (This means that you will shuffle the deck six times since $17 + 17 + 17 + 17 + 17 + 15 = 100$.)

Use what you have learned about probability to calculate the probability in percent of dealing three cards from a complete deck and getting

b) three cards of the same suit.

c) three cards of different suits.

d) two cards of the same suit. (*Hint:* Subtract the sum of the other two probabilities from 100%.)

e) How do the numbers in your frequency distribution compare with the probabilities that you calculated?

Lesson 2

1. The arrangement of the alphabet on the typewriter keyboard, designed by Christopher Sholes in 1872, is shown in the figure below.

Left hand ●
Right hand ●

The row of keys outlined in black is the "home row" on which the fingers rest.

The table below shows the typical number of times each key is used in typing 100 letters.

A	8	F	2	K	1	P	2	U	3	Z	0
B	1	G	1	L	3	Q	0	V	1		
C	3	H	6	M	3	R	7	W	1		
D	4	I	7	N	7	S	6	X	1		
E	13	J	1	O	8	T	9	Y	2		

Refer to this table and the picture of the keyboard above to determine, in typing 100 letters, the number of times of each of

the following occurring.
a) Using a key in the home row.
b) Using a key in the row above the home row.
c) Using a key in the row below the home row.
d) Using a finger of the left hand.
e) Using a finger of the right hand.

The American Simplified Keyboard, designed by August Dvorak in 1936, is available on some typewriters. It is illustrated below.

Left hand
Right hand

Refer to the table on page 566 and this picture to determine, in typing 100 letters on the American Simplified Keyboard, the number of times of each of the following occurring.
f) Using a key in the home row.
g) Using a key in the row above the home row.
h) Using a key in the row below the home row.
i) Using a finger of the left hand.
j) Using a finger of the right hand.

In what way is the American Simplified Keyboard an improvement over the keyboard designed by Sholes with respect to the frequency with which
k) the rows are used?
l) the left and right hands are used?

2. Cryptograms, ciphers that have the original punctuation and spacing between words, are published in many newspapers. Their popularity dates back to 1841, when Edgar Allan Poe wrote an article titled "Secret Writing" for a magazine of the time.

You may enjoy trying to decipher the following cryptograms, each of which is written in a different cipher. Because they are so brief, one word is given in each to help you get started.
a) NO BAGS ITSRUYE XNXU'Y CTLR TUB
   PCNQXSRU, YCRSR'E T MAAX PCTUPR
   YCTY BAG DAU'Y CTLR TUB.
                                    —PQTSRUPR XTB

(The fourth word is DIDN'T.)

b) A HARD HPTE SCHRF, DFIDXACHHO
   SVDT MVDO CUD MCRDT GO IDPIHD
   SVP CTTPO ZD.

   —JUDY CHHDT

   (One of the words is ANNOY.)

c) EVO CODE BNJ EY WOOL AVSRFTOZ
   VYUO SD EY UNWO EVO VYUO
   NEUYDLVOTO LRONDNZE—NZF ROE
   EVO NST YHE YM EVO ESTOD.

   —FYTYEVJ LNTWOT

   (One of the words is PLEASANT.)

Lesson 3

1. One way in which languages differ is in the numbers of syllables in their words. The histograms below show the frequencies
   of the numbers of syllables per word in English, German, Japanese, and Arabic.*

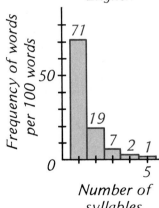
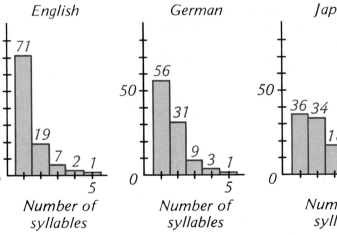
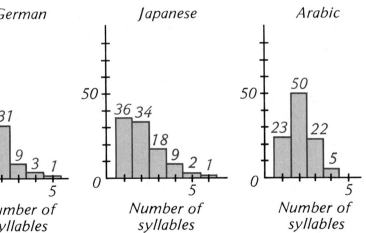

From the histogram for English, we see that the most frequent number of syllables per word is 1.

a) What is the most frequent number of syllables per word for
   each of the other languages?

From the histogram for English, we see that, if a typical sample of 100 words is arranged in order of increasing numbers of
syllables, the 50th (or 51st) word would have 1 syllable.

---

* From *The Way Things Work Book of the Computer* (Simon and Schuster, 1974.)

b) What is the median number of syllables per word for each of the other languages?

c) Find the mean number of syllables per word for English. Show your work. (*Hint:* Use the information contained in the histogram to determine the number of syllables in a typical sample of 100 words. For example, 71 words have 1 syllable each; 19 words have 2 syllables each, and so on.)

d) Find the mean number of syllables per word for each of the other languages. Show your work in each case.

2. Experiment: *Rolling a Die to Get All Six Numbers*

How many times, on the average, must a die be rolled so that each of its six numbers comes up at least once?

One way to get an idea of the number of times is to take a die and try it. For example, successive tosses of a die might turn up the following numbers:

3   5   2   2   1   2   3   6   4

In this case, it took 9 tosses to get all six numbers. A convenient way to keep track of the numbers is with a frequency distribution like the one at the right. The outcomes of the tosses are tallied in a column until there is at least one mark for each number; the total number of tosses is then written at the bottom of the column. The outcomes of additional trials can be recorded in additional columns.

Number on die	Example trial
1	|
2	|||
3	||
4	|
5	|
6	|
	9

a) Take a die and carry out 25 trials of this experiment, recording the results as shown above.

b) Make a frequency distribution of the results; label the first column "Number of tosses," and tally the numbers of tosses made in each of the 25 trials.

c) Use your frequency distribution to determine the median number of tosses necessary to get all six numbers.

d) Calculate the mean number of tosses necessary.

Lesson 4

1. A bakery sells small loaves of bread that supposedly weigh 250 grams each. A statistics student bought three of the loaves, took them home, and weighed them. Their weights were 247 grams, 252 grams, and 239 grams.

a) What was their mean weight?

Wondering about this result and about the variation in the weight, the student decided to buy one loaf of bread at the bakery each morning and weigh it. The weights of the loaves

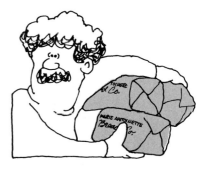

bought on 25 successive days are listed below.

238 242 249 236 253
258 234 245 249 244
247 248 244 242 239
240 246 251 249 254
244 239 243 247 244

Weight in grams
230–234
235–239
240–244
245–249
250–254
255–259

b) Make a frequency distribution of these weights by grouping them in intervals of 5 grams. Number the first column in your table as shown at the left.

c) Make a histogram of the weights, labeling the axes as shown at the left. After you have made the histogram, mark the midpoint of the top of each bar with a dot and connect the dots with a smooth curve.

d) What does the curve seem to be?

e) Use the curve to estimate the mean value of the weights of the loaves of bread.

On discovering this, the student complained to the baker who said he would correct the situation. The weights of the loaves bought on the 15 days that followed the complaint are listed below.

251 254 250 252 255
253 251 257 250 250
251 250 255 252 250

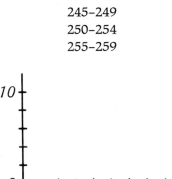

f) Make a frequency distribution of these weights, using the same intervals as in part b.

g) Make a histogram of the weights, labeling the axes as in part c.

h) On the basis of this evidence, how do you think the baker "corrected the situation"?

i) Explain.

2. Several years ago a letter appeared in Dear Abby's column from a woman who said she had been pregnant for 310 days before giving birth to her baby. This is considerably longer than the typical pregnancy.

Lengths of pregnancy of women having children are normally distributed, with a mean of 266 days and a standard deviation of 16 days.

Between what two lengths of time would you expect

a) about 68% of the pregnancies to last?

b) about 96% of the pregnancies to last?

c) almost 100% of the pregnancies to last?

Chapter 9: AN INTRODUCTION TO STATISTICS

The table at the right shows the percentages of the numbers in a normal distribution at various distances from the mean.

d) What percentage of children are born from pregnancies lasting 310 days or more?

e) About how many such children would be expected in 1,000 births?

f) Do you think it is reasonable to assume that the woman was pregnant as long as she claimed?

Number of standard deviations from mean	Percentage of numbers
0.25	19.7
0.50	38.3
0.75	54.7
1.00	68.3
1.25	78.9
1.50	86.6
1.75	92.0
2.00	95.5
2.25	97.6
2.50	98.8
2.75	99.4
3.00	99.7

Lesson 5

1. *Discrimination in the Wistful Vista School District**

The following table shows the numbers of men and women who applied for teaching positions one year in the Wistful Vista School District and the numbers who were hired.

Grade level	Men who applied	Men hired	Women who applied	Women hired
K–6	40	2	120	12
7–9	50	7	20	6
10–12	180	72	10	9
Total	270	81	150	27

From it, we see that 2 of the 40 men who applied for teaching positions in grades K–6 were hired:

$$\frac{2}{40} \times 100\% = \frac{200\%}{40} = 5\%$$

a) Make similar calculations and copy and complete the following table.

Grade level	Percentage of male applicants hired	Percentage of female applicants hired
K–6	5	▨
7–9	▨	▨
10–12	▨	▨
Total	▨	▨

b) Copy and complete the bar graph at the right comparing the percentages of men and women hired at each grade level.

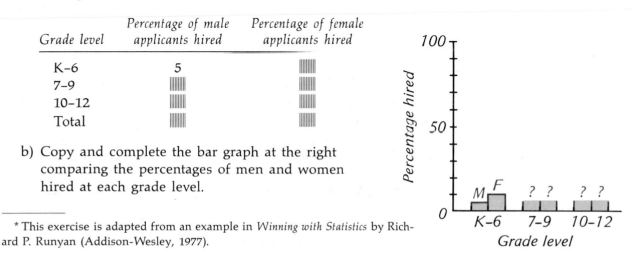

---

* This exercise is adapted from an example in *Winning with Statistics* by Richard P. Runyan (Addison-Wesley, 1977).

This seems to show that the Wistful Vista School District practices discrimination in hiring its teachers.

c) Explain.

d) Copy and complete the bar graph at the left comparing the percentages of men and women who were hired at all grade levels combined.

e) What does this graph seem to show?

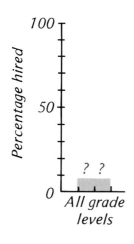

2. A common marketing practice is to offer a new product at an "introductory low price." In one experiment to find out if this increases sales in the long run, rolls of aluminum foil were introduced in seven stores of a large discount chain at a price of 59 cents and at seven other stores of the chain at a price of 64 cents. After three weeks, the price in all the stores became 64 cents.*

A graph of the sales for the first eight weeks is shown here.

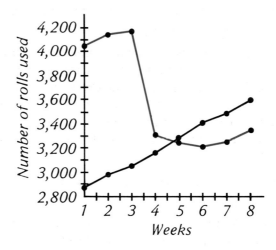

a) Which line do you think represents the sales of the foil introduced at 59 cents?

b) What happened to its sales in the course of the eight weeks of the experiment?

c) What happened to the sales of the foil introduced at 64 cents in the course of the eight weeks of the experiment?

Similar patterns in the sales of other items were observed in additional experiments. In each case, the prices were not adver-

*"Effect of Initial Selling Price on Subsequent Sales," by Anthony N. Doob, J. Merrill Carlsmith, Jonathan L. Freedman, Thomas K. Landauer, and Soleng Tom, Jr., *The Journal of Personality and Social Psychology,* 11(4):1969.

tized and the customers did not know that the product was being sold in other stores at a different price. Those buying it at the lower price were not told that it was an "introductory price" and did not know that the price would go up after three weeks.

d) Why do you suppose that the item that started out with the lower sales each time did better in the long run?

e) Judging from these experiments, does the practice of introducing a new product at less than its normal price seem like a good idea?

Lesson 6

1. How many words do you suppose are in this book? The only way to find the *exact* number would be to count them all. If we are content with knowing the *approximate* number, however, we might make an estimate of it on the basis of a sample.

a) Determine, as accurately as you can, the number of words in this book.

b) Explain the procedure that you used to make your estimate.

2. *The Case of the Missing Stockings**

   The owner of a company that manufactured nylon stockings had reason to believe that about $1,000,000 worth of the stockings were being stolen from the factory each year. He brought in detectives, put recording devices on the machines, and had the foremen questioned, but without success.

   Finally, a psychologist hired by the owner discovered the following facts:

   1) The amount of material used in making the stockings varied from one worker to another.

   2) The annual output of the company was estimated from a test run by one of the machine operators.

   3) This operator was the company's best worker.

   On the basis of these clues, what do you suppose was happening to the missing stockings? Explain your reasoning.

---

*From *Sampling* (originally titled *Sampling in a Nutshell*), by Morris James Slonim (Simon and Schuster, 1960).

573

Chapter **10**

TOPICS IN TOPOLOGY

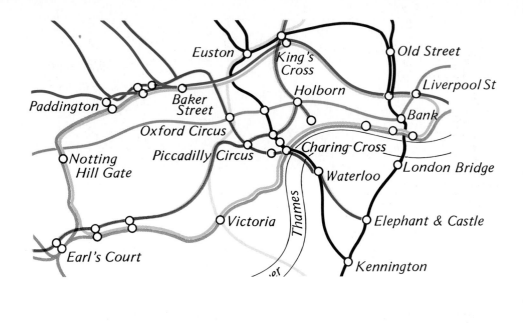

The Mathematics of Distortion

Beneath the city of London is an elaborate network of subways. Called the Underground, it consists of 279 stations connected by 410 kilometers of tracks.

Although the map above is an accurate representation of the subway lines and stations in central London, it is not the map used by passengers. For someone traveling on the Underground, all that matters is the way in which the stations are connected by the various lines. The map on the facing page, which shows these connections in a simple way, is the one posted in each Underground station.

It is easy to see how the second map was made. The artist treated the lines of the original map as if they were elastic bands and stretched and bent them into simpler shapes. The result is a figure having some properties different from the original one and some properties that have remained unchanged. The properties that have remained unchanged are *topological*.

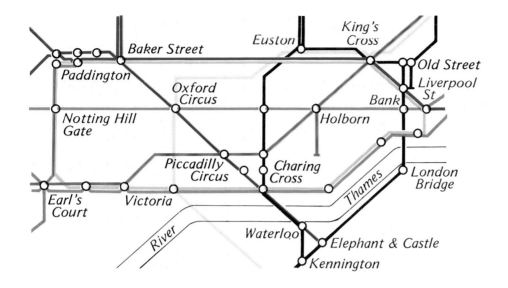

Topology, sometimes called "the mathematics of distortion," deals with very basic properties of geometric figures: properties that remain unchanged no matter how those figures are bent, stretched, or twisted. Look, for example, at the figures below. From

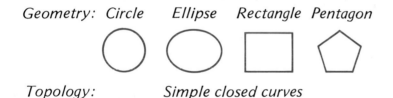

the standpoint of geometry, each figure is different and has its own name. From the standpoint of topology, they are all alike. The figures are examples of *simple closed curves*.

▶ A **simple closed curve** is a curve in which it is possible to start at any point and move continuously around the curve, passing through every other point exactly once before returning to the starting point.

The figures are also *topologically equivalent*.

▶ Figures are **topologically equivalent** if they can be twisted and stretched into the same shape without connecting or disconnecting any points.

Lesson 1: The Mathematics of Distortion

Because topology deals with very basic ideas, its influence on other areas of mathematics has been immense. Even though it is a comparatively young branch of mathematics, it has proved to have many practical applications and, because of its many surprises, is often a source of much enjoyment.

Exercises

Set I

The figures below are from a book on fingerprints.

Ridge types

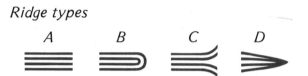

1. Which two figures are topologically equivalent?

2. Use the definition of topological equivalence in this lesson to explain why.

The figures below were devised by Marvin L. Minsky and Seymour A. Papert of the Massachusetts Institute of Technology as a test of visual perception.*

3. Place your paper over this page and trace each figure. Are the figures topologically equivalent?

4. What does each one consist of?

Perceptrons: An Introduction to Computational Geometry by Marvin L. Minsky and Seymour Papert (MIT Press, 1969).

Chapter 10: TOPICS IN TOPOLOGY

The first neon sign made its appearance at the Paris Motor Show in 1910. The symbols that are the easiest to make in neon are those that are topologically equivalent to a line. Examples are shown at the right.

Use the forms of the capital·letters shown below to answer the following questions.

ABCDEFGHIJK LMNOPQRSTUVWXYZ

5. Twelve of these letters are topologically equivalent to a line. Which letters are they?

Which letters are topologically equivalent to each of the following shapes?

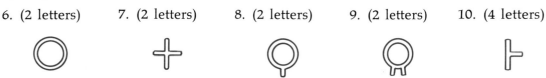

6. (2 letters) 7. (2 letters) 8. (2 letters) 9. (2 letters) 10. (4 letters)

The following topological problem was presented in a lecture by the German mathematician August Ferdinand Moebius in 1840.*

A king with five sons stated in his will that after his death his land was to be divided into five regions so that each region shared some of its border (more than just a point) with each of the others. Can the will be carried out?

Look at the adjoining map.

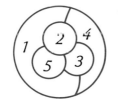

11. Does region 1 share some of its border with each of the other regions?

12. Does region 2 share some of its border with each of the other regions?

13. Does region 3 share some of its border with each of the other regions?

14. Does the map solve the problem? Explain why or why not.

15. Can you draw a different map that solves the problem?

* *Famous Problems of Mathematics* by Heinrich Tietze (Graylock Press, 1965.)

Lesson 1: The Mathematics of Distortion

Set II

A simple closed curve is a figure in which it is possible to start at any point and travel over every other point of the figure exactly once before returning to the starting point.

1. Which of the figures at the left are simple closed curves?

A basic idea of topology is that a simple closed curve in a plane divides the plane into two regions: an inside and an outside. This fact is called the *Jordan Curve Theorem,* after the nineteenth-century French mathematician Camille Jordan.

2. Which of the four figures at the left divide the plane into two regions?

Cardioid

Spiral

Octagon

Lemniscate

This simple closed curve appeared on a medieval wall tile.*

3. Is point A inside or outside the curve?

4. How many times does each straight line from point A to the outside cross the curve?

5. Is point B inside or outside the curve?

6. How many times does each straight line from point B to the outside cross the curve?

---

* *Handbook of Regular Patterns* by Peter S. Stevens (MIT Press, 1980).

Chapter 10: TOPICS IN TOPOLOGY

The number of times that a straight line joining a point to the outside of a simple closed curve crosses the curve depends on whether the point is inside or outside the curve.

7. Explain.

8. Determine whether each of the points C, D, and E is inside or outside the curve below.

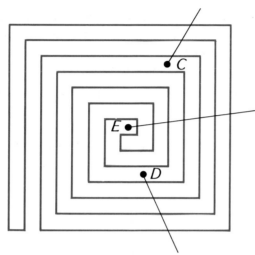

An old puzzle titled "Water, Gas, and Electricity" is to draw lines connecting the three utilities to each of three houses without any of the lines crossing each other.*

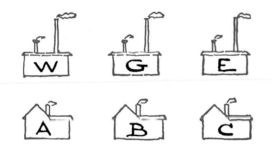

9. Copy the figure below showing the utilities and houses and see if you can connect them as indicated.

---

*Amusements in Mathematics by Henry Ernest Dudeney (Nelson and Sons, 1917; Dover, 1958).

The figure below illustrates a partial solution to the puzzle.

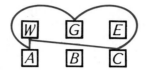

10. What do the four connections shown form?

It is now impossible to connect one of the utilities to one of the houses without crossing one of these connections.

11. Which ones?

12. Why is it impossible?

Regardless of how the connections are made, a situation of this sort always arises.

13. What does this indicate about the puzzle?

Topology is used in the design of the electric circuits used in microelectronics:

> The circuit elements are interconnected by a conducting film of evaporated metal that is photoengraved to leave the appropriate pattern of connections.*

These connections cannot cross each other or the circuits will short.

The following figures represent electric circuits. Copy each one and then figure out a way to connect, within each figure, each pair of squares labeled with the same letter so that none of the connections cross. If you think a figure does not have such a circuit, explain why not.

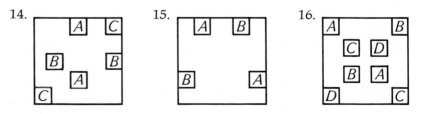

* "Microelectronics" by Robert N. Noyce, *Scientific American*, September 1977.

Chapter 10: TOPICS IN TOPOLOGY

Copyright Universiteitsmuseum, Utrecht; photograph by Jac. P. Stolp

This strange picture from the eighteenth century is titled *Two Men Boxing*. There is a mirror in the shape of a cone in the middle of it. When the picture is viewed in the mirror, it appears as the image in the center.

1. What topological relation do the pictures seem to have?

2. As the artist distorted the picture to produce the intended image, did he twist and stretch all parts of it by the same amount? Explain.

3. What does the mirror do to the picture in addition to removing the distortion?

More examples of topological distortions in art are included in the book, *Hidden Images*, by Fred Leeman (Abrams, 1976) and in Martin Gardner's "Mathematical Games" column in the January 1975 issue of *Scientific American*.

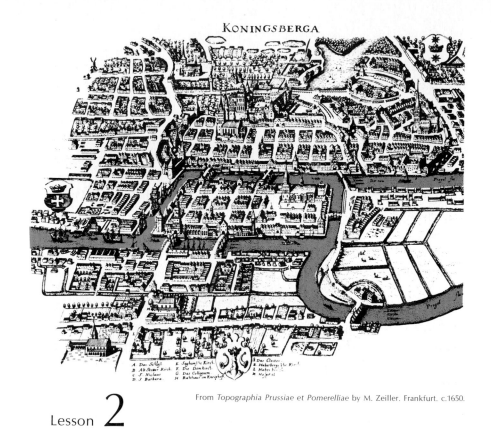

KONINGSBERGA

From *Topographia Prussiae et Pomerelliae* by M. Zeiller. Frankfurt. c.1650.

Lesson 2

The Seven Bridges of Königsberg: An Introduction to Networks

One of the problems that led to the development of topology is about seven bridges in the city of Königsberg in old Germany. The drawing above, from a book published in about 1650, shows that the center of Königsberg was on an island in the middle of a river. The river flowed around the island from the left and, on the right, separated into two branches. Seven bridges made it possible to travel from one part of the city to another.

The citizens of Königsberg wondered whether it was possible to travel around the city and cross each of the seven bridges exactly once. Everyone who tried it ended up either skipping or recrossing at least one bridge. Most people came to the conclusion that a path crossing every bridge exactly once did not exist, but they did not know why.

Eventually, the problem of the Königsberg bridges came to the attention of the Swiss mathematician Leonhard Euler. Euler found the puzzle interesting and, in an article published in 1736, proved that it could not be solved. He began by representing each area of land with a capital letter and each bridge with a small letter. The map can be simplified even further by representing the four areas of land by four points and the seven bridges by seven lines connecting the points. This is shown in the diagram at the right, which we will call a *network*.

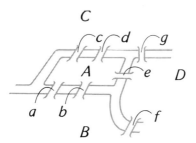

►A **network** is a figure consisting of points, called *vertices,** connected by lines, called *arcs*.

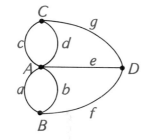

The problem of the Königsberg bridges is equivalent to the problem of drawing this network without retracing any arc or taking the pencil off the paper. Euler showed that whether or not this problem has a solution depends on the *degree* of each vertex of the network.

►The **degree** of a vertex of a network is the *number* of arcs that meet at it.

In the network for the bridge problem, five arcs meet at vertex A and so the degree of vertex A is 5. Three arcs meet at each of the vertices B, C, and D, and so the degree of each of these vertices is 3. Euler's results showed that this network, which has four vertices of an odd degree, cannot be drawn with one continuous stroke of a pencil.

Exercises

Set I

A network can be traveled if it is possible to draw it without retracing any arc or taking your pencil off the paper. It is easy to see that the left-hand network at the right can be traveled by starting at any vertex. The right-hand network can be traveled only if you start on one of the vertices marked with an arrow.

Networks

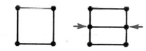

Example paths

---

* Vertices is the plural of vertex.

Each of the networks below is topologically equivalent to a simple closed curve.

1. Do you think that every network topologically equivalent to a simple closed curve can be traveled?

2. In what way are all of the vertices of such networks alike?

Each of these networks is topologically equivalent to a line.

3. Do you think that every network topologically equivalent to a line can be traveled?

4. How many vertices of odd degree do such networks have?

5. In what way are the rest of the vertices of such networks alike?

Each of the networks at the left has four vertices.

6. Copy and complete the following table for these networks.

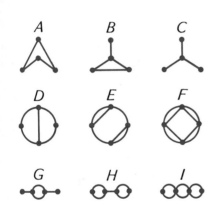

Network	Number of even vertices	Number of odd vertices	Can the network be traveled?
A	4	0	Yes
B			
C			

(Continue the table to show all nine networks.)

7. What do all the vertices of the networks that cannot be traveled have in common?

Each of the following networks also has four vertices.

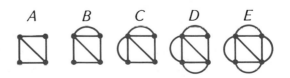

Chapter 10: TOPICS IN TOPOLOGY

8. Which networks have two even vertices and two odd vertices?

9. Which have four even vertices?

10. Which have four odd vertices?

11. Which network(s) cannot be traveled?

Set II

In 1809, the French mathematician Louis Poinsot noted that the figure shown at the right cannot be drawn in one continuous stroke of a pencil.

1. What do the numbers at the vertices of this network represent?

2. How many vertices of even degree does it have?

3. How many vertices of odd degree does it have?

Place your paper over the networks below and try to travel each one. (Some of the networks can be traveled only if you start from certain vertices.) Keep a record of those that can be traveled and those that cannot.

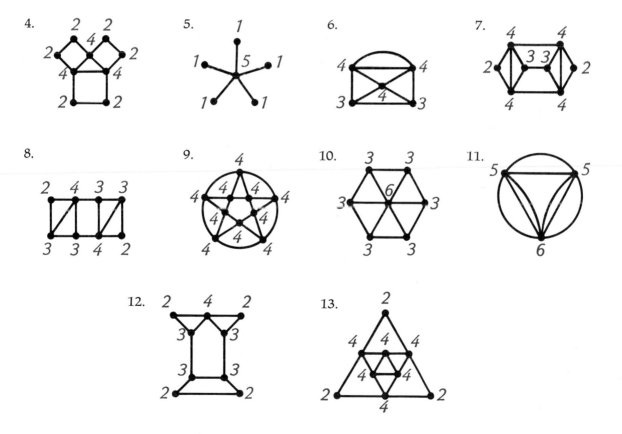

14. Copy and complete the following table for these networks.

Network	Number of even vertices	Number of odd vertices	Can the network be traveled?
4	9	0	Yes
5			

(Continue the table to show all ten networks.)

Of the ten networks, six can be traveled. If your table shows less than six, try drawing each of the networks that you have marked "no" again.

On the basis of the information in your table, does it seem that a network can be traveled if

15. all of its vertices are even?

16. it has more even vertices than odd vertices?

17. it has two odd vertices?

18. it has more than two odd vertices?

19. Copy and complete the following table for the networks in exercises 4 through 6.

Network	Number of arcs	Sum of degrees of vertices
4	12	24
5		
6		

20. What do you notice?

Set III

An amusement park has a fun house in the form of a maze with the floor plan shown at the top of the next page. The maze contains a room with trick mirrors, labeled M in the figure, together with seven smaller rooms.

1. To show how the eight rooms of the maze are connected, draw a network in which the entrance, exit, and the rooms are represented as vertices and the possible paths between rooms are

Chapter 10: TOPICS IN TOPOLOGY

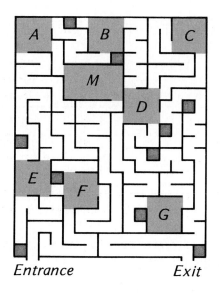

Entrance   *Exit*

represented as arcs. For example, there is one path connecting
rooms A and B, and so an arc would be drawn between vertices
A and B of your network as shown here.

Use your network to answer the following questions.

2. Is it possible to enter the maze, go to the room of mirrors, and
   exit without traveling any of the corridors more than once and
   without entering any room more than once? If it is, through
   what rooms would the path go?

3. Is it possible to travel the maze as described in exercise 2 in
   more than one way?

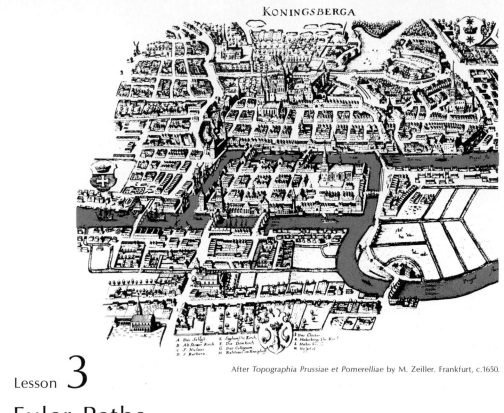

Lesson **3**

Euler Paths

It was in 1736 that Leonhard Euler proved that it was impossible to walk through the city of Königsberg and cross each of its seven bridges exactly once. In 1875, an eighth bridge was built. It is not difficult to show that the addition of this new bridge made it possible to solve the problem.

Networks representing the city before and after the new bridge are shown below. The first network contains four odd vertices.

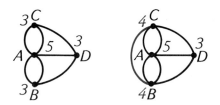

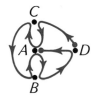

Adding an arc between vertices B and C to form the second network changes two of these vertices to even. One way to travel the new network is to start at vertex D and follow the path shown in the adjoining figure, ending at vertex A. Such a path is called an *Euler path*.

▶ An **Euler path** is a continuous path that passes through every arc of a network exactly once.

Notice that the Euler path for the network with eight bridges begins at one of the odd vertices and ends at the other one. To see why, look at what would happen if we did not start at vertex D. In traveling the network we must go over each arc, including the three that end at D, exactly once. The first time we come to D, it will be along one of these three arcs and we can leave along either of the other two. This leaves one arc untraveled and when we cover it we will be "stuck" at D because there are no arcs remaining along which we can leave. In other words, if our trip does not *begin* at vertex D, it must *end* there.

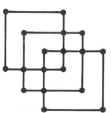

Reasoning in the same way, we can show that the path must also either begin or end at the other odd vertex, A. Because a path has exactly one beginning and one ending, it follows that it must begin at one of the odd vertices and end at the other. It also follows that a network with more than two odd vertices cannot be traveled in a single trip. Euler proved that every network having no more than two odd vertices has an Euler path.

Exercises

Set I

The first figure at the right is an old Chinese symbol for good luck.

1. Does it have an Euler path?

2. How can this be determined without trying to draw the path?

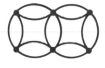

This figure of three squares was a favorite puzzle of Lewis Carroll.

3. Try drawing it several times, starting from a different corner each time.

4. The network has 18 vertices. What do they have in common?

5. Do you think that the network can always be traveled, regardless of the vertex from which you start?

6. Do you think that every Euler path on this network ends up at the same vertex from which it begins?

The following questions refer to the four networks below.

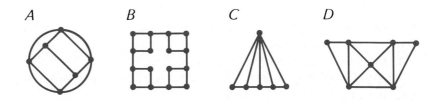

7. Which networks have Euler paths?

8. How can this be determined without trying to draw the paths?

9. Which networks have Euler paths that end up at the same vertex from which they begin?

To draw network A at the left, two trips are necessary. In other words, the figure can be drawn without retracing any arc if the pencil is removed from the paper once. One way of doing this is shown in the second figure. The leap of the pencil is equivalent to adding an arc. By adding the arc in color as in the third figure, we can connect two odd vertices and thus make both of them even.

10. Find the least number of trips necessary to travel each of the networks below by drawing them.

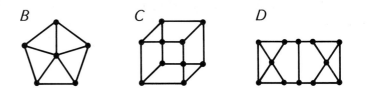

11. Copy and complete the following table for these networks.

Network	Number of odd vertices	Least number of trips
A	4	2
B	▓▓▓	▓▓▓
C	▓▓▓	▓▓▓
D	▓▓▓	▓▓▓

12. How is the number of trips necessary to travel a network related to the number of odd vertices that it has?

13. How many trips would it take to travel a giant network having a million odd vertices?

Network A at the right separates the paper into three regions: two are inside the network and the third is the region outside. Network B has just one region: the one outside.

The number of regions of a network is related to the numbers of vertices and arcs that it contains.

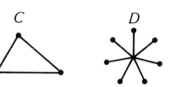

14. Copy and complete the following table for the nine networks at the right.

Network	Number of regions	Number of vertices	Number of arcs
A	3	2	3
B	1	6	5

(Continue the table to show all nine networks.)

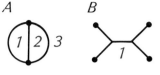

15. Write a formula relating the number of regions, R, in a network, the number of vertices, V, and the number of arcs, A.

Set II

In his article on the Königsberg bridges, Euler considered another bridge problem, which is illustrated below.*

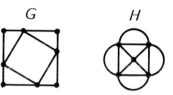

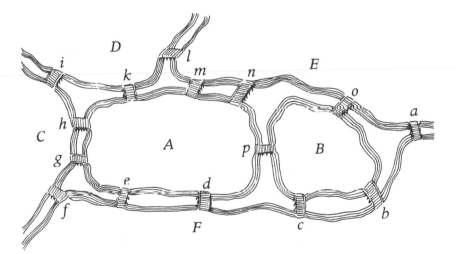

---

*Euler's article is included in *Mathematics: An Introduction to Its Spirit and Use*, with introductions by Morris Kline (W. H. Freeman and Company, 1979).

Two islands, A and B, are surrounded by water that leads to four rivers. Fifteen bridges cross the rivers and the water surrounding the islands. Is it possible to make a trip that crosses each bridge exactly once?

1. Draw a network in which 6 vertices represent the regions of land and 15 arcs represent the bridges.

2. What do you notice about the vertices of the network?

3. Can a trip be made that crosses each bridge exactly once? If so, can it begin and end on any of the regions of land?

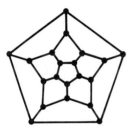

In 1857, the Irish mathematician William Rowan Hamilton invented a game based on the network at the left. One object of the game was to find a path that goes through every *vertex* of the network exactly once and ends in the vertex in which it started. Such a path is called a *Hamilton path,* in contrast with an *Euler path,* which goes along every arc of a network. The figures below show a network that has both an Euler path and a Hamilton path.

Network   *An Euler path*   *A Hamilton path*

4. Which of the networks below have Euler paths?

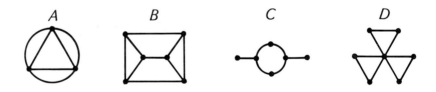

A   *B*   *C*   *D*

5. Which networks have Hamilton paths?

6. Can a network have an Euler path without having a Hamilton path?

7. Can a network have a Hamilton path without having an Euler path?

Look again at the network for Hamilton's game.

8. Does it have an Euler path? Explain why or why not.

9. Place your paper over the figure and see if you can discover a Hamilton path for it.

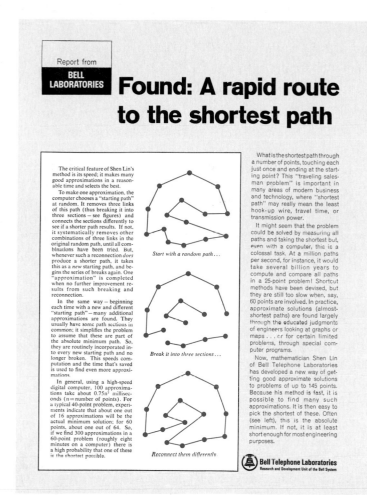

By permission of Bell Telephone Laboratories, Inc.

This advertisement of the Bell Telephone Laboratories concerns finding the shortest Hamilton path for a given network. According to the advertisement, this problem is "important in many areas of modern business and technology, where 'shortest path' may really mean the least hook-up wire, travel time, or transmission power."

The numbers on the arcs of the network at the right represent the distances between the vertices.

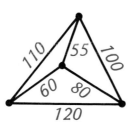

1. How many Hamilton paths does the network have?

2. Place your paper over the figure and make a separate drawing of each path.

3. What is the length of the shortest Hamilton path?

Lesson 4

Trees

The word *tree* is used in mathematics to mean a certain kind of network.* Most of the networks considered in the preceding lessons contain simple closed curves, which make it possible to begin at a vertex, travel over part (or all) of the network, and return to the original vertex without passing over any arc twice.

▶ A **tree** is a network that does not contain any simple closed curves.

A real tree, the kind Peter and B. C. are looking at, is like a topological tree in that it consists of a trunk, branches, and twigs that ordinarily do not grow back together. A cat on one branch of a tree cannot climb around the rest of the tree and come back to where it started without returning along the same branches.

The examples below include some networks that are trees and some that are not.

Trees   *Not trees*

---

* It was introduced in 1857 by the English mathematician Arthur Cayley in a paper titled "On the Theory of the Analytical Forms Called Trees."

596

There is just one tree that contains two vertices and one tree that contains three. They are shown as figures A and B at the right. There are two different trees that contain four vertices and they are shown as figures C and D.

The longest path on any of these trees is clearly on the one labeled C: it consists of three arcs.

► The number of arcs in the longest possible path on a tree is called its **diameter.**

The diameters of the four trees above are 1, 2, 3, and 2, respectively.

Tree networks appear in a wide variety of subjects. You have already seen trees used in mathematics in solving problems in counting and in probability.* Map makers draw complicated trees to represent rivers and their tributaries. The map below shows the Mississippi River and some of the other rivers that flow into it.

---

* See, for example, the trees on pages 378 and 432.

Genealogists use trees to show the relationships between members of different generations of a family. The process used by the postal service to sort mail can be represented by a tree: the zip code identifies its main branches. Tree diagrams are drawn by chemists to show the arrangements of atoms in molecules.

Exercises

Set I

Three generations in a person's family tree are shown in the figure at the left.

1. How many vertices does the tree have?

2. How many arcs does it have?

The adjoining tree, drawn from a photograph, shows part of a stroke of lightning.

3. How many vertices does this tree have?

4. How many arcs does it have?

5. In what way are your answers to exercises 1 and 2 like your answers to exercises 3 and 4?

The pattern of points shown at the left can be connected to form a tree in many different ways. Some of them are shown below.

6. Make two copies of the same pattern of points on your paper and connect them to form trees different from those shown above.

7. How many vertices does each of the three trees above and each of your trees have?

8. How many arcs does each of the three trees above and each of your trees have?

9. On the basis of your answers to exercises 1 through 8, write a formula relating the number of vertices, V, in a tree to the number of arcs, A.

The figure at the right shows some of the veins in a leaf from a fig tree.

10. Does your formula hold true for this network? Explain why or why not.

This figure shows the arrangement of the atoms in a benzene molecule.

11. Does your formula hold true for this network? Explain why or why not.

The diameter of a tree is the number of arcs in the longest possible path on it. For example, the diameter of the tree showing part of a stroke of lightning is 6.

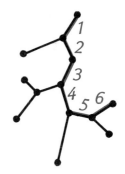

12. Do any paths on this tree other than the one shown in color also contain 6 arcs?

The trees below are adapted from figures in *Patterns in Nature* by Peter S. Stevens.*

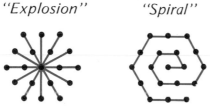

13. What is the diameter of "Explosion"?

14. What is the diameter of "Spiral"?

Many snowflakes have the shapes of trees. At the right is a photograph of one taken through a microscope, together with a tree diagram of its structure.

15. What is the diameter of this tree?

* Little, Brown, 1974.

Set II

Each of the following trees contains four vertices. Although trees A,

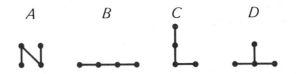

A   B   C   D

B, and C look different, they are topologically equivalent because they can be twisted and stretched into the same shape. For this reason, they are considered to be the same tree. Tree D, however, *is* different because it cannot be made from the others without disconnecting and reconnecting one arc.

Each of the trees below also has four vertices.

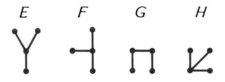

E   F   G   H

1. Which one is different from the other three?

2. Which ones are topologically equivalent to tree D above?

3. Which of the eight trees above have diameters of 3?

4. Which have diameters of 2?

There are three different trees that contain five vertices, as shown below.

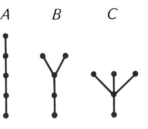

A   B   C

5. Which of the following trees are topologically equivalent to tree A?

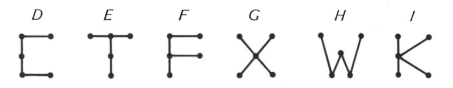

D   E   F   G   H   I

Which are topologically equivalent to

6. tree B?

7. tree C?

There are six different trees that contain six vertices. Five of them are shown at the right.

A   B   C   D   E

8. What is the diameter of each of these trees?

To which of the trees above is each of the following trees topologically equivalent?

9.   10.   11.   12.

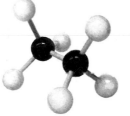

Set III

Chemists use networks called *structural formulas* to show the ways in which atoms are linked together in molecules. Photographs of models of several hydrocarbon molecules, along with trees showing their structures, are shown below. The vertices of the trees labeled C represent carbon atoms and the vertices labeled H represent hydrogen atoms.

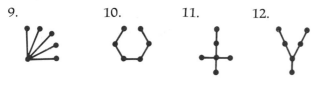

Methane   *Ethane*   *Propane*

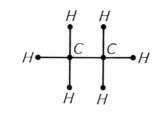 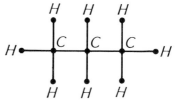

1. What is the degree of each C vertex?

2. What is the degree of each H vertex?

The "skeleton" of each of these molecules consists of its carbon atoms and the bonds between them. Here is the skeleton of the propane molecule:

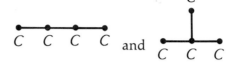

C   C   C

There are two different butane molecules containing four carbon atoms each. Their skeletons are:

C

|

C   C   C   C   and   C   C   C

3. Copy these figures and then add enough arcs and vertices representing hydrogen atoms to them so that the degrees of the vertices are the numbers you gave as the answers to exercises 1 and 2.

There are three different pentane molecules containing five carbon atoms each.

4. Draw tree diagrams of their skeletons and then complete each tree to show the rest of the molecule.

5. Does each of your three drawings for exercise 4 contain the same number of hydrogen atoms?

6. Refer to the figures above and your drawings to copy and complete the following table.

Hydrocarbon	Number of carbon atoms	Number of hydrogen atoms
Methane	1	4
Ethane	▦	▦
Propane	▦	▦
Butane	▦	▦
Pentane	▦	▦

Notice that a *pentane molecule* contains five carbon atoms and that a *pentagon* contains five vertices.

7. How many carbon atoms do you think an octane molecule has?

8. How many hydrogen atoms do you think an octane molecule has?

Chapter 10: TOPICS IN TOPOLOGY

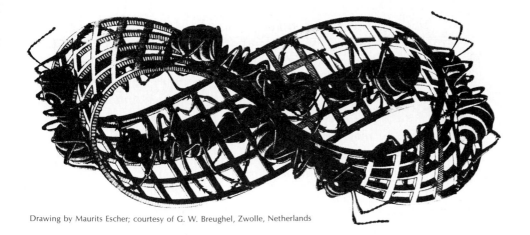

Drawing by Maurits Escher; courtesy of G. W. Breughel, Zwolle, Netherlands

Lesson 5

The Moebius Strip and Other Surfaces

There is something rather remarkable about the band pictured in the woodcut by Maurits Escher shown above. Look carefully at the procession of ants crawling around it. Escher wrote:

> *An endless ring-shaped band usually has two distinct surfaces, one inside and one outside. Yet on this strip nine red ants crawl after each other and travel the front side as well as the reverse side. Therefore the strip has only one surface.* *

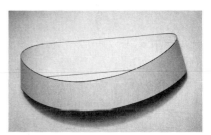

This band, called a *Moebius strip,* is named after one of the pioneers in topology, the German mathematician and astronomer August Ferdinand Moebius. The idea of the Moebius strip is so simple that it is rather surprising that it did not become widely known until 1865 when Moebius wrote a paper about its properties.

The two adjoining figures show an ordinary "belt shaped" loop and a Moebius strip. The belt-shaped loop has two sides and two edges; it is impossible to travel on this loop from one side to the other without crossing over an edge.

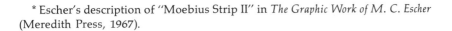

* Escher's description of "Moebius Strip II" in *The Graphic Work of M. C. Escher* (Meredith Press, 1967).

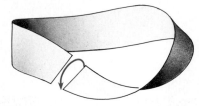

The Moebius strip contains a "half twist," so-called because it can be made by turning one end of a rectangular strip through an angle of 180° (half of 360°) before taping it to the other end. As a result of this half-twist, the Moebius strip has only one side and one edge. The strip's onesidedness gives it a number of strange properties that are easily discovered by experiment.

Some of these properties have been put to practical use. The B. F. Goodrich Company has patented a conveyor belt in the shape of a Moebius strip—the belt lasts longer because both "sides" are actually one and receive equal wear. A continuous-loop recording tape sealed in a cartridge will play twice as long if it has a twist in it. The advertisement below reveals that the Moebius strip has been put to use in the design of electronic resistors. An acrobatic trick performed by free-style skiers who make a twist while doing a somersault is now commonly known as the Moebius flip.

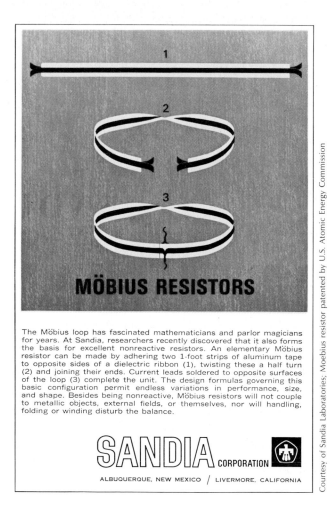

Chapter 10: TOPICS IN TOPOLOGY

Exercises

Set I

Experiment: THE MOEBIUS STRIP

The best way to understand the properties of the Moebius strip is to make one and discover them yourself.

Cut two strips, each as long as possible and 1 inch or 2 centimeters wide, from graph paper. Take one of the strips, make it into a loop and turn one end over before taping the two ends together. The result, a loop with a half-twist, is a Moebius strip. Do the same thing with the other strip so that you have two Moebius strips.

Put your pencil down midway between the edges of one of the Moebius strips and draw a line down its center, continuing the line until you return to the point at which you started.

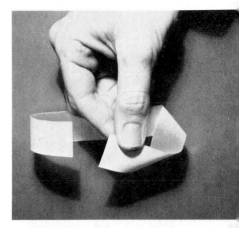

1. How many times does the line go around the strip before returning to the original point?

2. How many sides of the strip does the line seem to be on?

3. In drawing the line, you never crossed over the edge to get to the other side; so how many sides does the strip have?

4. If you tried to paint just one side of the strip, what would happen?

Cut the band along the line you have drawn.

5. What is strange about the result?

Put your pencil down midway between the edges of the resulting strip and draw a line down its center, continuing the line until you return to the point at which you started.

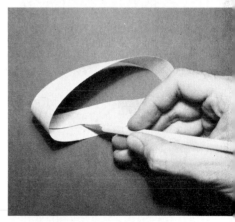

6. How many sides of the strip is the line on?

7. How many sides does the strip have?

8. Is it a Moebius strip? Explain why or why not.

Cut the new band along the line that you drew.

9. What is the result?

Cut the other Moebius strip that you made parallel to and about one-third of the way from the edge. When you have cut all the way around the loop you will find that you are across from the point at which you started. Continue cutting, staying the same distance from the edge as before, until you come back to where you began.

10. What is the result?

11. How do they compare in width?

12. How do they compare in length?

13. Which one is a Moebius strip?

This woodcut of a Moebius strip was created by Escher several years before the one shown at the beginning of this lesson.

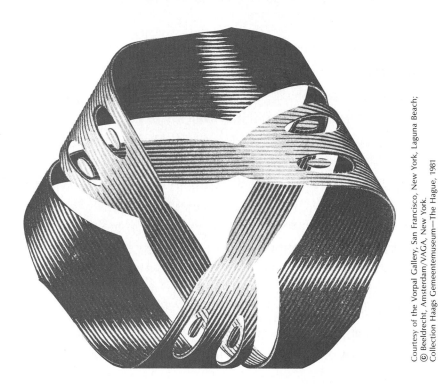

Courtesy of the Vorpal Gallery, San Francisco, New York, Laguna Beach; © Beeldrecht, Amsterdam/VAGA, New York. Collection Haags Gemeentemuseum—The Hague, 1981

14. What property of a Moebius strip that you discovered in the preceding exercises does it illustrate?

Set II

Experiment: OTHER SURFACES

By doing the Set I exercises, you discovered some of the properties of the Moebius strip, a band that contains a half-twist. The experiment that follows will reveal some of the properties of surfaces with more than one half-twist.

Cut six more strips of the same size as those made for Set I from

Chapter 10: TOPICS IN TOPOLOGY

graph paper. Make two of the strips into bands having two half-twists. Draw a line down the center of one of the bands.

1. Would it be possible to paint just one side of this band?

2. How many sides does a band with two half-twists in it have?

Cut the band along the line.

3. What is the result?

4. How do they compare in width and length?

Cut the other band one-third of the way from an edge.

5. What is the result?

6. How do they compare in width and length?

Make two additional strips into bands having three half-twists. Draw a line down the center of one of the bands.

7. How many sides does a band with three half-twists in it have?

Cut the band along the line.

8. What is the result? (Look carefully. It is unlike any of the previous results.)

Cut the other band one-third of the way from an edge.

9. What is the result? (The result will be so tangled up that you will have to pull at it a bit and study it closely in order to figure out what it is.)

Make another strip into a band having four half-twists.

10. How many sides does it have?

Cut the band down the center.

11. What is the result?

Make the last strip into a band having five half-twists. Cut the band down the center.

12. What is the result?

Answer the following questions on the basis of the results of both experiments.

13. How is the number of sides that a band has related to the number of half-twists in it?

14. What do you think would be the result if a long band with 100 half-twists in it were cut down the center?

15. What do you think would be the result if a long band with 101 half-twists in it were cut down the center?

Set III

Experiment: AN ANTITWISTER PRINCIPLE

Tape one end of a strip of paper of about the same dimensions as those used in Sets I and II of this lesson to your desk. Then, beginning with the free end level, give the free end four half-twists counterclockwise as shown in this diagram. Now hold the free end

Free end   *Taped-down end*

down on the desk with your finger as shown in the first figure below.

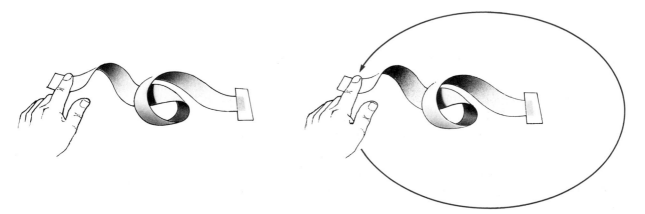

1. Do you think that it is possible to untangle the strip as long as both ends remain fixed in the positions shown?

   Keeping the free end pointing to the left, slide it with your finger in a circle counterclockwise around the taped end as shown in the second figure above.

2. What happens to the strip?

   This principle has been applied to an invention that makes it possible to transmit electricity from a fixed source to something that is turning.*

---

*The device, patented by D. A. Adams, is described and illustrated in "The Amateur Scientist" by C. L. Stong, *Scientific American*, December 1975.

*The Moebius strip on a 1967
Brazilian postage stamp*

Chapter 10 / Summary and Review

In this chapter we have studied some ideas in topology.

The mathematics of distortion (*Lesson 1*)   A simple closed curve
is a curve in which it is possible to start at any point and move
continuously around the curve, passing through every other point
exactly once before returning to the starting point.

Figures are topologically equivalent if they can be twisted and
stretched into the same shape without connecting or disconnecting
any points.

The Jordan Curve Theorem says that a simple closed curve in a
plane divides the plane into two regions: an inside and an outside.

Networks (*Lessons 2, 3, and 4*)   A network is a figure consisting of
points, called vertices, connected by lines, called arcs.

The degree of a vertex of a network is the number of arcs that
meet at it.

An Euler path is a continuous path that passes through every arc
of a network exactly once. A network has an Euler path only if it
has no more than two vertices of odd degree.

A tree is a network that does not have any simple closed curves.

The diameter of a tree is the number of arcs in the longest
possible path on it.

The Moebius strip and other surfaces (*Lesson 5*) A Moebius strip is a band that contains a half-twist. It has only one side and one edge.

Exercises

Set I

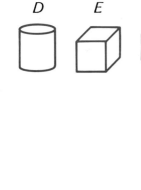

In a test of a child's intelligence, the child is asked to copy a set of figures, some of which are shown at the left.*

1. Which of these figures are simple closed curves?

2. Which figures are trees?

3. Which figures are topologically equivalent?

The maps below show the streets in two hilly regions in which a truck delivers newspapers. The papers are thrown from both sides of the truck so that it does not retravel any street unless it is necessary.

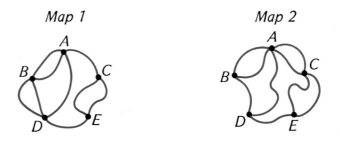

4. Is it possible to plan a route on map 1 that covers each street exactly once?

5. Is it possible to plan a route on map 2 that covers each street exactly once?

6. Explain how it is possible to tell, without trying to discover it, that one of the maps does not have such a route.

7. On the map that has such a route, can the route begin at any intersection? If not, where should it begin?

---

* The Gesell Figure Copying Test, devised by Dr. Arnold Gesell of Yale University.

Chapter 10: TOPICS IN TOPOLOGY

These figures from an article in *Scientific American* illustrate a double-stranded molecule with cross-links that might be joined as shown in part A to form the arrangement shown in part B.*

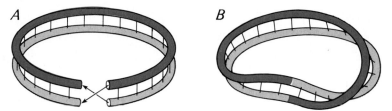

8. What does the molecule shown in part B look like?

9. Thought of as a surface, how many edges does it have?

10. What would be the result if all of the cross-links were broken?

The points in the figure at the right represent islands and the lines represent routes of a ferryboat company. The company is planning to make some changes in the routes. Copy the five points and make some drawings to determine the answers to the following questions.

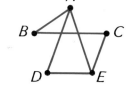

11. How could someone get from island A to island C by traveling two of the routes?

12. What is the *least* number of routes needed to make it possible to travel between any two islands?

13. What kind of network would such a system of routes be?

14. What is the smallest diameter that such a network could have?

15. If the company puts in a route linking every pair of islands directly, how many routes would there be?

Set II

If we think of the peculiar figure on the cover of *Mad* magazine as a network instead of solid, it will look like the figure below.

* "Chemical Topology" by Edel Wasserman, *Scientific American*, November 1962.

1. How many vertices does it have?

2. How many arcs does it have?

3. How many regions does it have? (Remember that one of the regions of a network is the region outside it.)

In doing the Set I exercises in Lesson 3, you discovered a formula relating the number of regions, R, of a network, to its number of vertices, V, and the number of arcs, A. One way to write the formula is

$$R + V = A + 2.$$

4. Does this formula apply to the peculiar figure pictured on *Mad* magazine?

An old puzzle requires drawing a continuous line that crosses each arc of the network at the left exactly once, without going through any of its vertices. A couple of unsuccessful attempts to solve the puzzle are shown below. In the first figure, one of the arcs has not

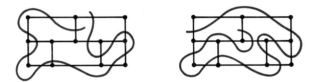

been crossed and, in the second one, an arc has been crossed more than once.

To discover whether the puzzle can be solved, place your paper over the networks below and try to draw such a line on each. The line should cross each arc exactly once and should not pass through any vertex. Keep a record of which networks have a line.

5.　　　6.　　　7.　　　8.　　　9.

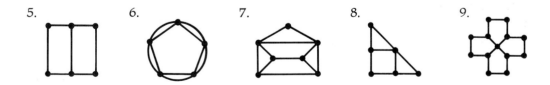

Each network above contains two or more internal regions. We will consider each region to be of an even or odd degree according to whether it is surrounded by an even or odd number of arcs. For example, the network in exercise 5 contains two even regions because each is surrounded by four arcs.

10. Copy and complete the following table for the five networks.

Network	Number of even regions	Number of odd regions	Does the network have a path?
5	2	0	Yes
6	▨	▨	▨

(Continue the table to show all five networks.)

11. What seems to determine whether or not a continuous line can be drawn that crosses each arc of a network exactly once?

12. Do you think the original puzzle has a solution? Explain why or why not.

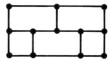

Set III

In 1976, it was proved that every map on a plane can be colored in four colors or less so that no two regions sharing a border have the same color. The adjoining figure, from a paper written in 1910, shows that more than four colors may be needed to color a map on a Moebius strip.*

First, imagine that the figure is flat as shown.

5		6	
3	1	2	
	4		

1. Are there any pairs of numbered regions that do not share part of their borders? If so, which pairs are they? (For your convenience, every pair of numbered regions in the map is given below.)

   1-2, 1-3, 1-4, 1-5, 1-6, 2-3, 2-4, 2-5, 2-6,
   3-4, 3-5, 3-6, 4-5, 4-6, 5-6

Imagine that the rectangle was cut out and made into a Moebius strip by giving its left edge a half-twist and bringing it around and taping it to the right edge. Because of the half-twist, the unnumbered region at the lower right would be connected to region 5, becoming part of that region. The unnumbered region originally at the lower left would be connected to region 6, becoming part of that region. The adjoining figure has been numbered to indicate this.

5		6	
3	1	2	
6	4		5

---

*The map-coloring problem on a plane was discussed in Chapter 1, Lesson 3. The above figure is from "Some Remarks on the Problem of Map-Coloring on One-Sided Surfaces" by Heinrich Tietze, a paper in the Annual Report of the German Mathematics Association, 1910.

Also think of the Moebius strip as being made of transparent material so that the regions show through it.

2. Are there any pairs of regions on the Moebius strip map that do not share parts of their borders? If so, which pairs are they?

Because the Moebius strip is transparent, color in any region will also show on the other side.

3. How many colors are needed to color this map so that no two regions sharing a border have the same color?

Chapter 10 / Problems for Further Exploration

Lesson 1

1. Experiment: *The Borromean Rings*

The first figure below shows two rings linked together; they cannot be separated without cutting one of the rings. The second figure shows two rings that are not linked at all.

Strange as it may seem, *three* rings can be linked together *without any two of the rings being linked with each other.* In other words, the three rings cannot be separated without cutting one of them, yet, if any one of the rings were to disappear, the other two would immediately come apart.

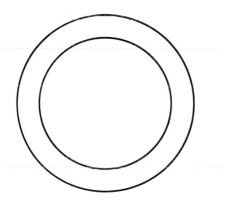

Use a compass to draw three identical rings of about the size shown in the first figure at the right on a large file card. Cut the rings out and then cut one of them as shown in the second figure

In the diagram below, the three rings are linked together in such a way that the brown ring is not linked to either white ring. The white rings, however, are linked to each other so that, if the brown ring were cut and removed, they would not come apart.

a) Can you link the three rings that you have made together so that *no pair* of them is linked? If so, tape the result to your paper.

Three rings linked in this way are called Borromean rings after an Italian family of the Renaissance named Borromeo, in whose coat of arms they appeared.

The adjoining photograph shows three lizards carved from wood.

b) What is interesting about the way in which they are linked together?

c) How do you suppose that these "linked" lizards were made?

2. The fingerprint classification system used by the F.B.I. uses letters and numbers to identify the topological patterns in fingerprints.* Each fingerprint is assigned a letter according to its basic pattern.

Pattern	Letter
Arch	A
Tented arch	T
Radial loop	R
Ulnar loop	U
Whorl	W

The figure below illustrates the steps in classifying a person's fingerprints: first with letters and then numbers to determine the file classification.

Example 1

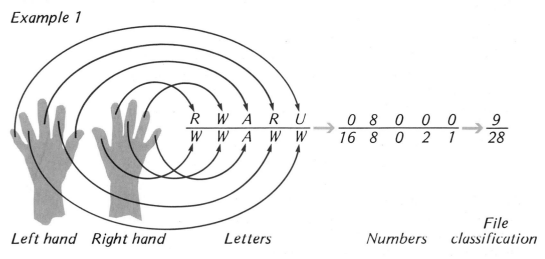

Left hand   Right hand   *Letters*   *Numbers*   *File classification*

---

* The examples in this exercise are taken from *Fingerprint Techniques* by Andre A. Moenssens (Chilton, 1971).

Chapter 10: TOPICS IN TOPOLOGY

Additional examples are shown below.

Example 2

$$\frac{T\ U\ U\ T\ W}{W\ W\ W\ W\ U} \rightarrow \frac{0\ 0\ 0\ 0\ 1}{16\ 8\ 4\ 2\ 0} \rightarrow \frac{2}{31}$$

Example 3

$$\frac{W\ W\ T\ W\ W}{R\ U\ U\ W\ W} \rightarrow \frac{16\ 8\ 0\ 2\ 1}{0\ 0\ 0\ 2\ 1} \rightarrow \frac{28}{4}$$

a) How do the numbers seem to be assigned?
b) How do the file classifications seem to be determined?
   What can you conclude about a person's fingerprints if their file classification is

c) $\frac{1}{1}$?

d) $\frac{17}{9}$?

Lesson 2

1. A letter carrier is supposed to deliver mail to all of the buildings on both sides of the streets shown on the map at the right. The carrier starts from the post office at the upper-left corner and would like to travel each side of each street exactly once before returning to the post office. Find out whether this is possible by doing each of the following.

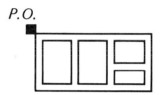

P.O.

   a) Draw a network to represent the letter carrier's territory. Label the corner representing the location of the post office P. (Note that each street must be represented by two arcs because the carrier must deliver mail on both sides.)
   b) Can the network be traveled in a path that begins and ends at P? If it can, trace the map and draw such a path.
   c) What do you notice about the vertices of the network?
   d) Do you think every letter carrier's route can be traveled in a path that ends where it begins? Explain.

2. In 1967, John Horton Conway and Michael Stewart Patterson, two mathematicians at Cambridge University in England, invented a topological game called Sprouts.* The game is played by two people who take turns drawing a network.

_____
* *Mathematical Carnival* by Martin Gardner (Knopf, 1975).

First, several points, which serve as the original vertices of the network, are marked on a piece of paper. The players then take turns drawing arcs, following these rules.

1. Each arc either must connect two vertices or must connect one vertex to itself.

2. When an arc is drawn, a new vertex must be chosen somewhere on it.

3. No arc may cross itself, cross another arc, or pass through any vertex.

4. No vertex may have a degree of more than 3.

The last person able to play wins the game. The figures below show one way in which a game that starts with two points might be played.

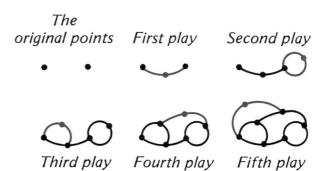

The original points   *First play*   *Second play*

Third play   *Fourth play*   *Fifth play*

This game ended on the fifth play because no more arcs can be added without breaking the rules.

Is there a limit to the number of moves that a Sprouts game can last if the players want to continue it as long as possible? To find out, try playing it with yourself or with someone else. Play several games starting with just two points and see if you can draw any conclusions. Then try playing games that start with three points and games that start with four points.

a) Is it possible for a game that starts with two points to go on indefinitely? If not, what is the greatest number of moves it can last?

What conclusions can you draw about a game that starts with

b) three points?

c) four points?

d) ten points?

Lesson 3

1. In his book *On the Trail of the Bushongo,* Emil Torday wrote of encountering a circle of African children playing with sand:

The children were drawing, and I was at once asked to perform certain impossible tasks; great was their joy when the white man failed to accomplish them.*

One task was to trace the figure at the right in the sand with one continuous sweep of the finger.
a) Draw an outline of the figure on graph paper and then see if you can figure out how to draw it.
b) Why is it unlikely that someone without any knowledge of networks would be able to do this on the first attempt?
c) What was the children's secret for drawing the network?

2. *The Air-Conditioning Inspector*†

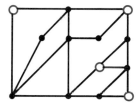

To locate the source of an obnoxious odor in an office building, an air-conditioning inspector is called in and asked to inspect the air-conditioning ducts thoroughly. Because the ducts are cramped, the inspector prefers not to go through any section more than once if it can be avoided. A scale map of the system is shown at the right with the access points shown in color. An access point can be used either as an entrance to the system or as an exit from it.
a) Can the inspector check all of the ducts without crawling through any of them more than once? Explain.
b) Place your paper over the figure and, by drawing several possible paths, see if you can find one that you think is best.

Lesson 4

1. A family gathering consists of father, mother, son, daughter, brother, sister, cousin, nephew, niece, uncle and aunt. But only two men and two women are present. They have a common ancestor and there has been no marriage between relatives.‡

   Explain by drawing a tree diagram how the four people are related.

2. Seven secret agents need to be able to communicate with each other either directly or indirectly.§ These communications can

* Related by Claudia Zaslavsky in *Africa Counts* (Prindle, Weber, and Schmidt, 1973).

† From *A Sourcebook of Applications of School Mathematics* by Donald Bushaw, Max Bell, Henry O. Pollak, Maynard Thompson, and Zalman Usiskin (N. C. T. M., 1980).

‡ This puzzle, by Pierre Berloquin, is from his book *One Hundred Games of Logic* (Scribner's, 1977).

§ Adapted from a problem by Gary Chartrand in *Graphs as Mathematical Models* (Prindle, Weber, and Schmidt, 1977).

incur a certain amount of risk. The following table lists the "risk factors" of direct communication between certain pairs of agents.

Agent pair	Risk factor	Agent pair	Risk factor
A-C	7	B-E	5
A-D	4	B-G	6
A-E	2	C-E	8
A-F	4	D-E	3
A-G	5	D-F	5
B-D	6	E-G	4

All other direct communications are either impossible or too dangerous.

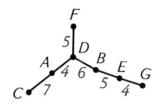

The tree at the left, in which each agent is represented as a vertex and each direct communication as an arc, shows one possible communications network for the agents. It has a total risk of 31.

a) Try to discover a communications network with as small a total risk as possible.

b) Does it have to be a tree? Explain.

c) What is the total risk of the network that you discovered?

d) What method did you use to find it?

Lesson 5

1. Experiment: *A Pair of Topological Surprises*

Part 1. Cut out a strip of paper about 1.5 inches (or 3 centimeters) wide and as long as possible. Fold the strip in half and cut two slits into both ends as shown in the first figure below.

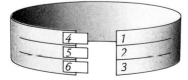

Unfold the strip and number the ends as shown in the second figure. Put three short pieces of tape on ends 4, 5, and 6, and then make a loop as shown in the figure at the left.

Chapter 10: TOPICS IN TOPOLOGY

Now tape end 4 to end 1.

Pass end 5 *over* end 4, and end 2 *under* end 1 and tape together.

Pass end 6 *between* ends 4 and 5; pass end 3 *over* end 1 and tape together.

Finish both cuts so that they go all the way around the band.

a) What is the result?

Part 2. Prepare another strip in exactly the same way as described in the first paragraph of Part 1.

Then turn end 6 over (give it a half-twist) and tape it to end 1.

Turn end 4 over and tape it to end 2.

Turn end 5 over and tape it to end 3.

Finish both cuts so that they go all the way around the band.

b) What is the result?

2. Experiment: *More Topological Surprises**

Cut out four large crosses like the one at the right from graph paper. Each arm should be about 1 inch (or 2 centimeters) wide and extend about 3 inches (or 6 centimeters) from the center of the cross.

Part 1. Tape the ends of one cross together to make two connected bands as shown in the figure below. Cut along the center of each band.

a) What is the result?

Part 2. Repeat the directions of Part 1 with another cross, but put a half-twist in one of the bands as shown in the figure below.

---

*Most of this experiment is derived from material in *Mathematical Magic Show* by Martin Gardner (Scribner's, 1977).

b) What is the result?

Part 3. Repeat the directions of Part 1 with another cross, but put a half-twist in *each* band as shown in the figure below.

c) What is the result?

Part 4. Tape the ends of the last cross together to make two connected bands, one with a half-twist, as shown at the bottom of page 621. Then cut the twisted band *one-third of the way from the edge* and finally cut the other band along its *center*.

d) What is the result?

APPENDIX

Basic Ideas and Operations

1 ANGLES AND THEIR MEASUREMENT

Imagine a large pie that has been divided by a very sharp knife into 360 equal slices. Viewed from above, one of the slices would look like the figure below. The slice's two straight edges form an *angle* and the point at which they meet is called the *vertex*. The straight edges lie along the *sides* of the angle.

The *measure* of an angle gives the size of the "opening" between its sides. The measure of the angle of the slice shown above is one degree, which is written as 1°. If all 360 slices were left together, they would completely surround the center of the pie, and so the number of degrees about a point is 360.

The degree, then, is the basic unit used to express the measures of all angles. Here are other examples. Consider the angle formed by the sides of one-half of a pie. Because one-half of a pie is equal to 180 of the small slices, its angle has a measure of 180°. Such an angle is called a *straight angle* because its sides lie along a straight line.

Next consider a pie from which a quarter slice has been cut. The angle of the quarter slice has a measure of 90° and is called a *right angle*. The rest of the pie can be thought of as another slice; the measure of its angle is 360° − 90° = 270°.

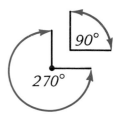

Angles are usually measured with a protractor. Although some protractors are circular in shape, most are semicircular like the one pictured here. Protractors usually have two scales so that they can

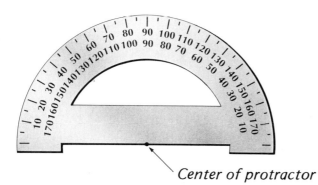

Center of protractor

be used to measure angles in either a clockwise or a counterclockwise direction.

To measure an angle with a protractor, first place the center of the protractor on the vertex of the angle. Next, line up the edge of the protractor with one side of the angle (which side does not

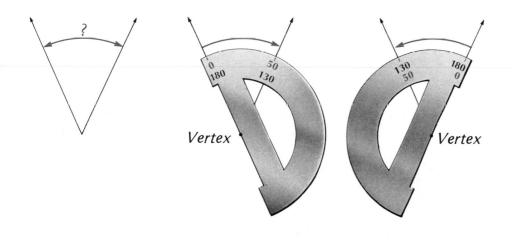

matter; both possibilities are shown above). Finally, read the measure of the angle by looking at where its other side falls under the scale; in the example, it is 50°.

For practice, measure the angles below. The measures of the angles are given at the bottom of this page so that you will know whether or not you are measuring them correctly.

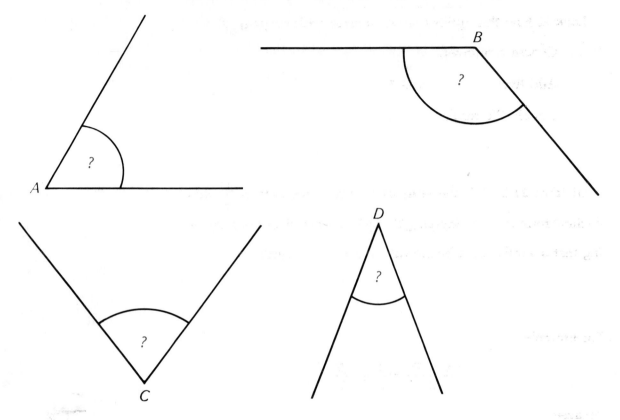

2 THE DISTRIBUTIVE RULE

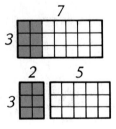

The distributive rule relates multiplication and addition. It is illustrated by the adjoining figure. The number of small squares in the figure can be expressed as either

$$3 \times (2 + 5) \quad \text{or} \quad (3 \times 2) + (3 \times 5).$$

Notice that

$$3 \times (2 + 5) = 3 \times 7 = 21$$

and

$$(3 \times 2) + (3 \times 5) = 6 + 15 = 21.$$

Angle A = 60°, Angle B = 130°, Angle C = 75°, Angle D = 42°.

In general,

$$a \times (b + c) = a \times b + a \times c.$$

Look at how this applies to the number trick on page 40.

Choose a number. n

Add five. $n + 5$

Double the result. $2 \times (n + 5) = (2 \times n) + (2 \times 5)$

$$= 2n + 10$$

Dividing 21 by 3 is the same as changing the fraction $\dfrac{21}{3}$ into decimal form or as multiplying 21 by $\dfrac{1}{3}$. It is therefore not surprising that a similar distributive rule applies to fractions:

$$\frac{b + c}{a} = \frac{b}{a} + \frac{c}{a}$$

For example,

$$\frac{6 + 15}{3} = \frac{6}{3} + \frac{15}{3}$$

because

$$\frac{6 + 15}{3} = \frac{21}{3} = 7 \quad \text{and} \quad \frac{6}{3} + \frac{15}{3} = 2 + 5 = 7.$$

Continuing with the number trick on page 40,

$$2n + 10$$

Subtract four. $2n + 6$

Divide by two. $\dfrac{2n + 6}{2} = \dfrac{2n}{2} + \dfrac{6}{2}$

$$= n + 3$$

Subtract the number
first thought of. 3

The result is three.

Check your understanding of the distributive rule by trying these problems. The answers are given on page 632.

1. Multiply by 2: $n + 3$.
2. Multiply by 3: $n + 5$.
3. Divide by 2: $2n + 10$.
4. Divide by 4: $4n + 12$.

3   SIGNED NUMBERS

Everyone knows that the whole numbers have a definite order. After you learned to count, you were able to say, for example, what number comes after 12 or what number comes before 8. The order of the whole numbers can be shown by representing them as evenly spaced points along a line.

$$0 \quad 1 \quad 2 \quad 3 \quad 4 \quad 5 \quad 6 \quad 7$$

If the line is extended beyond 0 in the opposite direction, how should the points on the other side be numbered? The customary way of doing it looks like this:

$$-4 \quad -3 \quad -2 \quad -1 \quad 0 \quad 1 \quad 2 \quad 3 \quad 4$$

The numbers of the points on the line left of the zero are *negative* and the numbers on the right of the zero are *positive*. Because they are identified by the signs – and +, these numbers are often called *signed numbers*. (A number without a sign is assumed to be positive.)

If you have taken an algebra course, you have learned but may have forgotten the rules for calculating with signed numbers. Only those that are needed for the exercises in this book are explained here.

Addition   An easy way to add two signed numbers is to think of gains and losses. You might think of a football field to help in picturing the situation. For example, what is the sum of 2 and –5?

Appendix: BASIC IDEAS AND OPERATIONS

A gain of 2 yards followed by a loss of 5 yards is the same as a loss of 3 yards:

$$2 + -5 = -3$$

What is the sum of –3 and 7? A loss of 3 yards followed by a gain of 7 yards is equivalent to a net gain of 4 yards:

$$-3 + 7 = 4$$

What is the sum of –8 and –4? A loss of 8 yards followed by a loss of 4 yards is the same as a net loss of 12 yards:

$$-8 + -4 = -12$$

Try the following exercises. The answers are on page 632.

5. $-6 + 5$
6. $9 + -2$
7. $-3 + -15$
8. $12 + -27$
9. $-64 + 10$

Subtraction   To subtract a positive number, think of it as a loss. For example, what is 10 subtracted from 3? A gain of 3 yards followed by a loss of 10 yards is the same as a net loss of 7 yards:

$$3 - 10 = -7$$

What is 6 subtracted from 0? A gain of 0 yards followed by a loss of 6 yards is the same as a loss of 6 yards:

$$0 - 6 = -6$$

Try these exercises. The answers are on page 632.

10. $8 - 9$
11. $1 - 12$
12. $5 - 20$

LUCY, HOW MUCH IS SIX FROM FOUR?

SIX FROM FOUR?! YOU CAN'T SUBTRACT SIX FROM FOUR..

YOU CAN'T SUBTRACT A BIGGER NUMBER FROM A SMALLER NUMBER

YOU CAN IF YOU'RE STUPID!

SCHULZ

Multiplication The multiplication of signed numbers is easily illustrated by a multiplication table.

×	3	2	1	0	−1	−2	−3
3	9	6	3	0	−3	−6	−9
2	6	4	2	0	−2	−4	−6
1	3	2	1	0	−1	−2	−3
0	0	0	0	0	0	0	0
−1	−3	−2	−1	0	1	2	3
−2	−6	−4	−2	0	2	4	6
−3	−9	−6	−3	0	3	6	9

This table shows, for example, that

$$3 \times -2 = -6$$

and that

$$-1 \times -3 = 3.$$

In general, the product of two numbers having *opposite* signs is *negative*, and the product of two numbers having the *same* sign is *positive*.

Try these exercises. The answers are on page 632.

13. -4×3

14. -1×-8

15. 7×-2

16. $(-5)^2$

Division The rules for dividing signed numbers, or changing signed fractions into decimal form, are the same as those for multiplying them. The quotient of two numbers having *opposite* signs is *negative*; the quotient of two numbers having the *same* sign is *positive*.

For example,

$$\frac{12}{-3} = -4 \quad \text{and} \quad \frac{-20}{-2} = 10.$$

Try these exercises. The answers are on page 632.

17. $\dfrac{-35}{5}$

18. $\dfrac{-24}{-8}$

19. $\dfrac{10}{-4}$

4   PERCENT

The word *percent* literally means *per hundred*. For example, to say that there is a 50% chance of a coin coming up heads when it is tossed means that it will turn up heads about 50 times per 100 tosses. If we change 50% to a fraction, we get

$$\frac{50}{100} = 0.50.$$

To change a fraction into a percentage, *simply multiply it by 100*:

$$0.50 \times 100 = 50\%$$

What is $\dfrac{1}{38}$ expressed as a percentage? Here are the steps in finding out:

Step 1. $\dfrac{1}{38} \times 100 = \dfrac{100}{38}$

Step 2.

$$
\begin{array}{r}
2.6 \\
38\overline{)100.0} \\
\underline{76} \\
24\ 0 \\
\underline{22\ 8} \\
1\ 2
\end{array}
$$

Step 3. 2.6 rounded to the nearest whole number is 3.

$$\text{So } \frac{1}{38} \approx 3\%.$$

Check your understanding of percent by changing each of the following numbers into percentages. (Where necessary, round to the nearest whole number.) The answers are given below.

20. $\dfrac{3}{4}$

21. 1

22. $\dfrac{2}{25}$

23. $\dfrac{1}{36}$

24. $\dfrac{15}{64}$

ANSWERS TO EXERCISES IN THE APPENDIX

1. $2n + 6$.	9. −54.	17. −7.
2. $3n + 15$.	10. −1.	18. 3.
3. $n + 5$.	11. −11.	19. −2.5.
4. $n + 3$.	12. −15.	20. 75%.
5. −1.	13. −12.	21. 100%.
6. 7.	14. 8.	22. 8%.
7. −18.	15. −14.	23. 3%.
8. −15.	16. 25.	24. 23%.

Appendix: BASIC IDEAS AND OPERATIONS

ANSWERS TO
SELECTED EXERCISES

Answers to Selected Exercises

CHAPTER 1, LESSON 1

Set I
3.

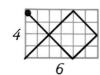

Set II
4. Because 5 does not divide evenly into 12.

CHAPTER 1, LESSON 2

Set I
7. The ball ends up in the upper-left corner if the length is even; it ends up in the upper-right corner if the length is odd.

Set II
1. Yes, the ball ends up in the upper-left corner because the length is even and the width is odd.   **6.** Reduce the length and width to 6 and 1; the ball ends up in the upper-left corner because the reduced length is even and the reduced width is odd.

CHAPTER 1, LESSON 3

Set I
3. $1 + 3 + 5 + 7 + 9 + 11 + 13 = 7 \times 7$.   **9.** $12 \times 8 + 2 = 98$.

Set II
4. No.   **7.** It is divided by 4.

CHAPTER 1, LESSON 4

Set I
14. They seem to be equal.

Set II
11. No.

CHAPTER 1, LESSON 5

Set I
1. Seven of the squares must be brown and seven of the squares must be white.
5. Six are white and eight are brown.
9. Yes. (It would have to be white.)
10. No.

Set II
4. 125.   **5.** At the corners of the large cube.   **13.** 1,728.

CHAPTER 1, LESSON 6

Set I

3. $\square$°

10. $5n$.

13. (Proof with boxes and circles only.)

Choose a number. □

Add the next larger number. □□∘

Add seven. □□∷∷

Divide by two. □∘∘∘∘

Subtract the number first thought of. ∘∘∘∘

The result is four.

Set II

6. The result is a six-digit number consisting of the original three-digit number written twice.

CHAPTER 2, LESSON 1

Set I

17. 10, 25, 40, 55, 70. **23.** Yes; the common difference is 0. **27.** Yes.

Set II

2. $3 + 9 \cdot 7 = 66.$
5. $5 + 10 \cdot 6 = 65.$

CHAPTER 2, LESSON 2

Set I

13. 0, 0, 0, 0. **15.** 1, 1,000, 1,000,000.
26. Yes; the common ratio is 1.5. **27.** No.

Set II

2. Six. **6.** 5, 10, 20.
14. $5 \cdot 2^0, 5 \cdot 2^1, 5 \cdot 2^2, 5 \cdot 2^3, 5 \cdot 2^4.$
16. $7 \cdot 4^{10}.$

CHAPTER 2, LESSON 3

Set I

4. $8 + 4 + 2 + 1 = 15.$ **9.** $16 + 4 = 20.$

Set II

2. $1 + 2 + 4 + 8 + 16 = 32 - 1 = 31.$
9. $1 + 2 = 3.$ **13.** $4 - 1 = 3.$

CHAPTER 2, LESSON 4

Set I

1. 1, 4, 9, 16, 25. **5.** 2.
9. It is multiplied by four.

Set II

10. 49. **13.** 17.

CHAPTER 2, LESSON 5

Set I

4. It is multiplied by 8.

Set II

2. The odd numbers (or an arithmetic sequence). **14.** $3^4 + 4^4 + 5^4 + 6^4 = 7^4.$

CHAPTER 2, LESSON 6

Set I

3. They are equal.

Set II

2. Every fourth term.
13. $1^2 + 1^2 + 2^2 + 3^2 + 5^2 + 8^2 = 104 = 8 \cdot 13.$ **18.** Pattern C.

CHAPTER 3, LESSON 1

Set I

5.

x	0	1	2	3	4
y	1	12	23	34	45

Set II

7. $y = 10x + 1.$ **11.** $y = 100 - x.$
14. 300,000 kilometers per second.

17.

d	1	2	3	4
s	1	4	9	16

CHAPTER 3, LESSON 2

Set I

1. A (5, 6); B (1, 8); C (−2, 7); D (−3, 6); E (−4, 3); F (−3, 0); and so forth.

Set II
2. E.   8. (3, 4).   13. They have the same size and shape.

CHAPTER 3, LESSON 3

Set I
2. Five.
6. Function A:

x	0	1	2	3	4	5
y	0	1	2	3	4	5

Function B:

x	0	1	2	3	4	5
y	0	0.5	1	1.5	2	2.5

Function C:

x	0	1	2	3	4	5
y	0	2	4	6	8	10

8. All three lines go through the origin.
14. Line $y = x + 1$ meets the y-axis at 1, and so forth.

Set II
3. If the lightning strikes 0 kilometers away, the time between the flash and the thunder is 0 seconds.
7.

t	0	1	2	3	4	5
h	10	8	6	4	2	0

9. The height of the candle before it is lit.
15. The woman.

CHAPTER 3, LESSON 4

Set I
5. $y = x^2 + 1$.
11.

x	−3	−2	−1	0	1	2	3
y	3	8	11	12	11	8	3

14.

x	0	1	2	3	4	5	6
y	9	4	1	0	1	4	9

Set II
1. About 3.2 square feet.
4.

s	0	1	2	3	4
l	0	2	8	18	32

8.

t	0	1	2	3	4	5	6	7	8
h	0	7	12	15	16	15	12	7	0

CHAPTER 3, LESSON 5

Set I
4. It would get very close to the x-axis.
5.

x	−6	−5	−4	−3	−2	−1
y	−1	−1.2	−1.5	−2	−3	−6

11. The curves have the same shape.

Set II
1. It gets smaller.   3. They are very small.   4. $w = e^3$.
8.

s	2	4	6	8	10
t	50	25	16.7	12.5	10

12.

t	0	8	16	24	32	40	48	56
s	1	2	4	8	16	32	64	128

14. It rapidly increases.

CHAPTER 3, LESSON 6

Set I
4. Interpolate. The value estimated is between values that are known.   6. They lie along a line.

Set II
2. 13 centimeters.   7. It gets steeper and steeper.

CHAPTER 4, LESSON 1

Set I
8. 10^4.   16. 10^{23}.   19. 10^{26}.
22. 2×10^{46}.   24. $(10^3)^2 = 10^3 \times 10^3 = 10^6$.
30.

Seconds ago	Years ago
0	0
10^8	3.2
10^9	32
etc.	

Set II
2. Ten thousand times ten thousand.
3. One hundred million.
9. $10^{800,000,000}$.   12. The eighth.

CHAPTER 4, LESSON 2

Set I
3. 12,000,000.   **17.** 4.2×10^{13}.

Set II
5. 6×10^{15}.   **7.** 4.5×10^9.
10. 80,000,000,000 grams.   **14.** 6×10^6.
16. 3.8×10^{25}.   **19.** 1.5×10^{11}.

CHAPTER 4, LESSON 3

Set I
8. 16,384.   **12.** 1.

Set II
4. $4 \times 64 = 256$.   **5.** $2 + 6 = 8$.
6. $32^2 = 1,024$.

CHAPTER 4, LESSON 4

Set I
1. $0.301 + 0.778 = 1.079$.   **7.** 15.

Set II
3. 12.301.   **5.** 7×10^9.   **9.** 3.903.

CHAPTER 4, LESSON 5

Set I
2. 4.061.   **3.** 6.061.   **7.** 3.65.
8. 3.65×10^1 (or 36.5).

Set II
7. $8.352 + 9.833 = 18.185$; 1.53×10^{18} kilo-meters.

CHAPTER 4, LESSON 6

Set I
2. It becomes less steep.   **3.** They are spaced closer and closer together.

Set II
5. It is ten times as large.   **6.** 100.

CHAPTER 5, LESSON 1

Set I
8. Yes. It looks the same if it is rotated 120° or 240°.   **13.** None.   **21.** Line symmetry (6 lines) and rotational symmetry.

Set II
5. A tilted square.   **12.** 360°.   **14.** Their product is equal to 360.

CHAPTER 5, LESSON 2

Set I
7.
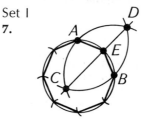

CHAPTER 5, LESSON 3

Set I
2. Three.   **3.** It increases.   **7.** 360°.
9. 150°.   **10.** Three triangles and two squares. .

Set II
1. 3-12-12.
5.
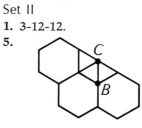

6. There are three triangles around C and there are supposed to be only two.
9. It cannot exist because point C is not sur-rounded by the right polygons.

CHAPTER 5, LESSON 4

Set I
1. Four.   **4.** Three.   **10.** Three.

Set II
1. 3-3-3.　　**7.** Three.　　**8.** 180°.
14. $6 \times 4 = 24$ corners and $6 \times 4 = 24$ sides; $\frac{24}{3} = 8$ corners and $\frac{24}{2} = 12$ edges.
20. An octahedron each of whose six corners touches one of the six faces of a cube.

CHAPTER 5, LESSON 5

Set I
6. 3-4-3-4.　　**16.** A regular octagon.
20. All of their faces are equilateral triangles.　　**26.** Eight

Set II
2. Squares, hexagons, and octagons.
5. $(12 \times 4) + (8 \times 6) + (6 \times 8) = 144$ corners and 144 sides; $\frac{144}{3} = 48$ corners and $\frac{144}{2} = 72$ edges.

CHAPTER 5, LESSON 6

Set I
2. Rectangles.　　**3.** A right equilateral triangular prism.　　**18.** Yes.　　**19.** 4-4-7.
22. No.

Set II
2. Yes.　　**3.** 13.　　**10.** 12.

CHAPTER 6, LESSON 1

Set I
4. Major.　　**5.** 12.　　**12.** They become more elongated.

Set II
1.

Curve	a	b
A	6	6
B	4	6
C	2	6

CHAPTER 6, LESSON 2

Set I
8. 4,000 feet.

Set II
4. The farther the ray, the smaller the angle.　　**7.** In front of the focus.

CHAPTER 6, LESSON 3

Set I
2. Two.

Set II
3. It increases.　　**4.** They become more pointed.

CHAPTER 6, LESSON 4

Set I
4. 90° and 270°.　　**5.** It is roughly horizontal.

Set II
2. $y = 4$ sine x.　　**6.** It becomes smaller.
9. The frequency (or wavelength).

CHAPTER 6, LESSON 5

Set I
2. 450°.
9.

x	0	1	2
y	1	3	9

Set II
1. 5, 10, 15, 20, 25.
2. 1, 2, 4, 8, 16, 32.　　**9.** 234.

CHAPTER 6, LESSON 6

Set I
1. At the top of the wheel.　　**3.** No.

Set II
2. Three.

CHAPTER 7, LESSON 1

Set I
5. Independent.　**7.** $4 \times 3 \times 2 \times 1$.
11. 4×4.　**13.** 9.　**18.** 25.

Set II
2. 2.　**7.** 64.　**10.** 1,352.　**13.** 18.

CHAPTER 7, LESSON 2

Set I
6. True.　**7.** 120.

Set II
2. 4×3.　**5.** 870.　**7.** 56.
12. 1,005,006.

CHAPTER 7, LESSON 3

Set I
1. $4 \times 3 \times 2 \times 1 = 24$.　**8.** 2,520.

Set II
3. 6.　**6.** 35.　**10.** 24.

CHAPTER 7, LESSON 4

Set I
2. 4×3.　**4.** $\dfrac{4 \times 3}{2 \times 1}$.　**5.** 210.
7. 720.　**10.** 15.

Set II
1. 28.　**3.** 1,820.　**6.** 21.

CHAPTER 8, LESSON 1

Set I
2. 5 to 1.　**5.** $\dfrac{1}{3}$.　**13.** One out of
three.　**15.** $\dfrac{3}{20}$.　**16.** 17 to 3.

Set II
2. 5%　**6.** $\dfrac{1}{8}$.　**9.** $\dfrac{1}{38}$, 3%.

CHAPTER 8, LESSON 2

Set I

1. Number of boys	2	1	0
Number of possibilities	1	2	1
Probability	$\frac{1}{4}$	$\frac{1}{2}$	$\frac{1}{4}$
Percent probability	25%	50%	25%

2. 50%.　**3.** 75%.　**5.** 50%.　**9.** They
are equally likely.

Set II
1. 1, or 100%.　**2.** 0, or 0%.

CHAPTER 8, LESSON 3

Set I
5. A family with four children has three boys
and a girl.　**9.** 5 and 0.　**15.** $\dfrac{5}{16}$.
17. 27%.

Set II
5. It has a similar shape.

CHAPTER 8, LESSON 4

Set I
2. 8.　**3.** $\dfrac{2}{9}$, or 22%.　**8.** Rolling a 7 is
more likely.　**12.** $\dfrac{1}{9}$.

Set II
6. 1,296.　**8.** 14.

CHAPTER 8, LESSON 5

Set I
3. $\dfrac{1}{9}$.　**8.** There are 36 cards in the deck
that are not tens, jacks, queens, or kings.

There are 52 cards altogether. 275 bridge hands. **15.** $\frac{3}{20}$.

10. Once in in the "liquor jugs" sentence than it is in ordinary English.

CHAPTER 9, LESSON 3

Set I
2. About 5.3. **8.** 7. **17.** Each one is 100 more.

Set II
1. The mode. **5.** Each of them is 2. **9.** 22.

CHAPTER 8, LESSON 6

Set I
2. $\frac{49}{100}$, 49%. **4.** $\frac{42}{100}$ or $\frac{21}{50}$, 42%.

6. $\frac{9}{10}$. **9.** No; one is not the opposite of the other.

CHAPTER 9, LESSON 4

Set I
2. 120. **5.** 4. **6.** 70%. **18.** 33.

Set II
3. 94%. **10.** A normal curve.
12. 100. **20.** 546. **26.** 17.

Set II
3. $\left(\frac{7}{10}\right)^8$. **8.** It would reach 100%.

12. Telephone numbers can end in 100 different pairs of digits; the Cracker Jack boxes contain 100 different prizes.

CHAPTER 9, LESSON 5

Set I
11. Almost four times as tall.

Set II
2. The area of the larger bag is four times the area of the smaller one. **11.** Graph A.
12. Yes. All of them show this.

15. It was about $3\frac{1}{2}$ times as much.

CHAPTER 9, LESSON 1

Set I
2.

11. About 57%. **18.** 15–19.

CHAPTER 9, LESSON 6

Set I
1. $83. **5.** No. Every voter did not have an equal chance of being included in the sample.

Set II
8. 0–1 kilometer. **10.** About 30%.

Set II
6. They would make the figure reported too large. **11.** Yes.

CHAPTER 9, LESSON 2

Set I
5. Q and Z. **6.** 24. **16.** It is much less

CHAPTER 10, LESSON 1

Set I
1. A and C.   7. K and X.

Set II
2. The cardioid and the octagon.
10. A simple closed curve.

CHAPTER 10, LESSON 2

Set I
2. Each of them has a degree of 2.
8. A, B, D, and E.   11. C.

Set II
14. (The networks that can be traveled are 4, 6, 7, 9, 11, and 13.)   16. No.

CHAPTER 10, LESSON 3

Set I
2. By noticing that all of its vertices are even.   9. Network D.

Set II
2. Vertices D and E are odd; the other vertices are even.   5. A and B.

CHAPTER 10, LESSON 4

Set I
12. Yes.   13. 4.

Set II
2. E, F, and H.   3. A, B, C, and G.
6. E and F.   12. B.

CHAPTER 10, LESSON 5

Set I
1. Twice.   2. Both.   3. Just one.
4. You would paint the entire strip.
10. Two interlocking loops.

Set II
1. Yes.   2. Two.

INDEX

Index